CISM COURSES AND LECTURES

The series presents lecture notes, monographs, edited works and proceedings in the field of Mechanics, Engineering, Computer Science and Applied Mathematics.
Purpose of the series is to make known in the international scientific and technical community results obtained in some of the activities organized by CISM, the International Centre for Mechanical Sciences.

INTERNATIONAL CENTRE FOR MECHANICAL SCIENCES

COURSES AND LECTURES - No. 427

CONFIGURATIONAL MECHANICS OF MATERIALS

EDITED BY

REINHOLD KIENZLER
UNIVERSITY OF BREMEN

GÉRARD A. MAUGIN
PIERRE ET MARIE CURIE UNIVERSITY

Springer-Verlag Wien GmbH

This volume contains 123 illustrations

Originally published by Springer-Verlag Wien New York in 2001

SPIN 10839118

In order to make this volume available as economically and as rapidly as possible the authors' typescripts have been reproduced in their original forms. This method unfortunately has its typographical limitations but it is hoped that they in no way distract the reader.

ISBN 978-3-211-83338-4 ISBN 978-3-7091-2576-2 (eBook)
DOI 10.1007/978-3-7091-2576-2

PREFACE

A rational analysis and design of contemporary materials and mechanical systems requires an ever finer scale of mathematical modeling and experimental observation and measurement than has been the case in the past. It has to be recognized that it is, in many circumstances, no longer sufficient to consider the material at hand to be a perfect continuum on some scale of examination. One has, rather, to descend deeper into the conditions of matter in order to make reliable and realistic predictions regarding the possible behaviour of the material or structural element under a variety of environmental constitution encountered during the lifetime of the piece. In particular, the so-called phenomenological theories of material response, with their simplistically postulated criteria of strength and failure, have to be replaced today by consideration of a variety of defects and imperfections always existing, such as micro-cracks, inclusions, vacancies, voids, dislocations etc.

One feature of these defects is that, under the influence of external sources such as, e. g., loadings and temperature changes, they can move (by mass diffusion) or change their shape within the material in which they find themselves. In the case of void nucleation and growth or micro-crack advance, new surfaces are created, i. e., the configurational of the material changes, giving rise to the encompassing term "Configurational Mechanics". An alternative designation of this quite novel branch of engineering science could be "Mechanics in Material Space", emphasizing that "objects" such as voids or interfaces between some two phases of a material may move within the material, rather then with respect to "Physical Space", (as is the case in usual (Newtonian) mechanics), in which deformation and motion of bodies with mass is considered. Here file term space is to be understood in a purely descriptive rather than in some strict mathematical sense. Still a third designation which is used for the same body of knowledge and inquiry is the term "Eshelbian Mechanics", which honours the late J. D. Eshelby who was most successful in originating (in a seminal paper of 1951) and developing the concept of a force on a material defect as the change in total energy of a given mechanical system with respect to a possible displacement of such a defect within the material.

In view of an ever broadening and widening of this "Configurational Mechanics of Materials" it was deemed appropriate and even desirable to organize a one-week course held at the International Centre for Mechanical Sciences (CISM) in Udine, Italy, September 11 - 15, 2000, in order to present to an interested audience a summary of some latest developments and advances in the field.

The Lecture Notes before you embody a distillate of presentations and discussions held at CISM during the week indicated. It is obvious that the 6 lecturers could not

cover in a comprehensive fashion all the analytical, experimental and numerical aspects of the field, yet it is believed that most of the important recent advances were included.
In a first introductory lecture on "Conservation laws and their application in configurational mechanics", delivered by the first author, G. Herrmann and R. Kienzler attempted to provide, in an as simple fashion as possible, the mathematical framework required to carry out a theoretical investigation of defects moving in materials.
A more general and more encompassing approach was taken in the lecture on "Elements of field theory in inhomogeneous and defective materials" by G. A. Maugin and C. Trimarco, delivered by the first author, in which numerous topics were discussed such as driving forces on field singularities, with application to the propagation of shock waves, phase-transition fronts and localized nonlinear waves, as well as computational schemes.
"Material mechanics of electromagnetic solids" by C. Trimarco and G. A. Maugin focussed on the topic of the title, and laid the foundation for an analytical treatment of defective electromagnetic materials, and was presented by the first-named author.
It is a surprising fact that the determination of some energy-release rates and stressintensity factors can be carried out not only within the framework of mechanics of extended continua, but even within theories of strength-of-materials concerned with bars, shafts and beams. Such considerations were presented in a paper "Configurational mechanics applied to strength-of-materials" by R. Kienzler and G. Herrmann, and given by the first-named author.
As was shown in a paper by D. Gross entitled "Morphological equilibrium and kinetics of 2-phase materials", configurational forces play a decisive role also in determining the final equilibrium shape of precipitates in a variety of two-phase materials. Analytical findings were corroborated by experimental evidence.
In a final paper by A. Chudnovsky and S. Preston on "Variational formulation of a material ageing model", delivered by the first author, a classification of the many fracture mechanisms was undertaken, followed by fatigue crack-growth equations and life-time prediction and supported by a body of experimental results.
For a smooth and pleasant organization of the Course as an advanced school, the coordinators should express their gratitude, also in the name of all participants, to the Secretary General of CISM, Professor G. Bianchi and his efficient staff, as well as to Professor C. Tasso, editor of the CISM series of publications, for his help in setting the standards of production of this volume.

Reinhold Kienzler
Gerard A. Maugin

Participants in the CISM Advanced School "Configurational Mechanics of Materials"

CONTENTS

Page

Preface

Photo of the participants to the school

Conservation Laws and their Application in Configurational Mechanics
by G. Herrmann and R. Kienzler 1

Elements of Field Theory in Inhomogeneous and Defective Materials
by G.A. Maugin and C. Trimarco 55

Material Mechanics of Electromagnetic Solids
by C. Trimarco and G.A. Maugin 129

Configurational Mechanics Applied to Strength - of - Materials
by R. Kienzler and G. Herrmann 173

Morphological Equilibrium and Kinetics of Two-Phase Materials
by D. Gross 221

Variational Formulation of a Material Ageing Model
by A. Chudnovsky and S. Preston 273

Conservation Laws and Their Application in Configurational Mechanics

G. Herrmann[1], and R. Kienzler[2]

[1]Stanford University, Stanford, CA, USA
[2]University of Bremen, Bremen, Germany

Abstract: Conservation laws play a leading role in establishing the necessary mathematical apparatus for the analysis of various problems in configurational mechanics. Two essentially different, yet related methodologies are offered for the establishment of such laws and those methologies are then applied to several simple one-dimensional problems, as well as to plane elastostatics and elastodynamics.

Introduction

Conservation (or balance) laws constitute the mathematical foundation for most basic principles in various branches of physics and mechanics. The one important exception is perhaps the 2nd law of thermodynamics which is expressed as an inequality. Thus it is essential to be concerned with such laws also in the special area of Configurational Mechanics (CM). As discussed more fully in this text, CM is concerned with that class of mechanical continua, which undergo processes in which not only the deformation of a body takes place, but also its configuration changes. As prime examples of such processes one might mention crack advance, void nucleation and growth, change in geometry of a free boundary, as well as motion of an interface surface between two phases of a material.

In the present contribution, an attempt has been made to discuss not only the possible mathematical methodologies for establishment of conservation laws, but also to include some applications, particularly to one-dimensional elastic bars and to plane elastostatics and elastodynamics. Applications to coupled fields, such as piezoelectricity and thermoelasticity should be discussed in other parts of this text and are also considered in the monograph by Kienzler and Herrmann (2000).

In order to begin on a firm foundation, the first Section here is concerned with the definition of conservation (or balance) laws, both in differential (strong) and integral (weak) form.

The possible methodologies of establishing conservation laws are presented in the following two sections. If a Lagrangian function can somehow be postulated for a system under consideration, then Noether's method, together with the important extension by Bessel-Hagen, represents a suitable and powerful tool for the establishment of conservation laws.

The method is based on Noether's theorem on invariant variational problems. This theorem associates every conservation law of a system with an underlying symmetry property that results from an invariance of (Hamilton's) action integral under a continuous group (in Lie's sense) of transformations. Bessel-Hagen showed that the strict invariance of the action integral might be weakend by inclusion of the divergence of a function defined on the space of independent and dependent variables and their derivatives, (jet bundle space). For a detailed treatment of this material reference should be made to the books by Olver (1993), as well as Bluman and Kumei (1989).

If a Langrangian function for the system under consideration is not available (e.g., because dissipation is present) and the system is described merely by a set of differential equations, then a recently developed systematic procedure, called the Neutral Action method, has become available, (cf. T. Honein et al., 1991, Chien, 1992, Chien et al., 1993 a). This method amounts essentially to replacing the so-called characteristic of a conservation law, completely determined in Noether's method in terms of the transformation selected, by a function to be determined in such a way that its (inner) product with the equations of motion represents formally a null Lagrangian, and whose action therefore, as is known, does not change variationally, i. e., the action behaves neutrally under its variation. Thus the term "Neutral Action" has been adopted for this methodology.

It can be shown that if the Neutral Action method is applied to the Euler-Lagrange equation of a variational problem, i. e., a Lagrangian function exists, then the same conservation laws ensue as if the Noether method had been applied, but together with the Bessel-Hagen extension.

Applications of these two different methodologies to establish conservation laws, partly already given as illustrative examples in the respective Sections 2 and 3, are given in the following 2 sections. Section 4 is concerned with conservation laws for homogeneous and inhomogeneous plane elastostatics. Here the by now classical J, L and M integrals are rederived for a homogeneous linearly elastic body and then Rice's J integral is extended to a class of inhomogeneous elastic materials. Relations to stress-intensity factors of fracture mechanics are also investigated for this class and several specific examples are worked out. Section 5 is devoted to elastodynamics. Here only one possible methodology of inclusion of the additional independent variable time is investigated, and illustrated with a simple example of an elastic bar. The section concludes with a brief discussion of the complete energy-momentum tensor appropriate for linear elastodynamics. The final Section 6 concludes with some general remarks.

1 Definition of a Conservation Law

Before discussing the establishment and application of conservation laws, let us make clear what kind of a mathematical expression should or can be called a conservation law.

We consider a system involving m independent variables $x_i(i = 1, 2, ..., m)$ and μ dependent variables $v_\alpha(\alpha = 1, 2, ..., \mu)$ and let it be governed by a set of q differential equations represented by

$$\Delta_\beta(x_i, v_\alpha, v_{\alpha,i}) = 0 \qquad (\beta = 1, 2, ..., q), \tag{1.1}$$

where Δ_β is a set of differential operators acting on x_i, v_α and derivatives of v_α with respect to x_i.

If a set of m associated function $P_j(x_i, v_\alpha, v_{\alpha,i}, ...)(j = 1, 2, ..., m)$, satisfies the equation

$$\frac{dP_1}{dx_1} + \frac{dP_2}{dx_2} + ... + \frac{dP_m}{dx_m} = \frac{dP_i}{dx_i} = P_{i,i} = 0 \qquad (i = 1, 2, ..., m) \tag{1.2}$$

for all solutions of (1.1), then such a differential form is called a conservation law (or a divergence expression).

Here d/dx_i is the total differential operator, for which the comma symbol is also used.

For example, if $$P_2 = P_2\left(x_1, x_2, \frac{\partial v_3(x_2, x_4)}{\partial x_4}, v_4(x_1, x_2, x_3)\right)$$

then $$P_{2,2} = \frac{dP_2}{dx_2} = \frac{\partial P_2}{\partial x_2} + \frac{\partial P_2}{\partial\left(\frac{\partial v_3}{\partial x_4}\right)}\frac{\partial^2 v_3}{\partial x_2 \partial x_4} + \frac{\partial P_2}{\partial v_4}\frac{\partial v_4}{\partial x_2}.$$

By contrast $\partial/\partial x_i$ is here the symbol for partial differentiation with respect to the variable indicated, implying that in this operation all other forms of variables, whether independent, dependent, or partial derivatives of dependent with respect to the independent ones, remain fixed.

Taking the function P_2 indicated above,

$$\frac{\partial P_2}{\partial x_3} = 0, \text{ even though } v_4 = v_4(x_1, x_2, x_3),$$

but
$$\frac{dP_2}{dx_3} = \frac{\partial P_2}{\partial v_4}\frac{\partial v_4}{\partial x_3}.$$

In equation (1.2), the summation convention over repeated indices has been introduced. The range of summation of latin indices is 1 though m, that of latin indices is 1 though μ.

The general conservation law (1.2) may also be written in symbolic notation

$$\textbf{div}\, \boldsymbol{P} = 0\,, \tag{1.3}$$

where $\boldsymbol{P}$ is the m -component vector $P_j(j = 1, 2, ..., m)$.

If $m = 1$, we are dealing with a set of μ ordinary differential equations involving a single independent variable x and the conservation law is

$$\frac{dP}{dx} = 0 \qquad \text{or} \qquad P = const. \tag{1.4}$$

and provides a first integral. The number μ of dependent variables is not restricted. Occasionally, instead of d/dx we use the abbreviation

$$\frac{d(\,)}{dx} = (\,)'.$$

If we are concerned with a dynamical system in three-dimensional space referred to Cartesian coordinates $x_i(i = 1, 2, 3)$ and time t, such that $m = 4$, the conservation law (1.3) reads

$$\textbf{div}\, \boldsymbol{P} = \frac{dP_t}{dt} + \frac{dP_1}{dx_1} + \frac{dP_2}{dx_2} + \frac{dP_3}{dx_3} = 0\,. \tag{1.5}$$

It indicates that the time rate of change of a density P_t is balanced by the divergence of the associated fluxes (or currents) P_1, P_2, and P_3.

For example, if P_t is the mass density ρ and there are no mass fluxes across the unit volume occupied by the mass ρ, then $d\rho / dt = 0$ and $\rho =$ const. If P_t is the energy density of a unit volume (or a unit mass), then the conservation law (1.5) states that the time rate of change of the energy density dP_t / dt is balanced by the divergence of the fluxes of energy out of the unit volume $dP_1 / dx_1 + dP_2 / dx_2 + dP_3 / dx_3$, i. e., energy is conserved, which is a statement of the first law of thermodynamics.

A conservation law in differential form (1.5) may also be written in integral form. We first consider the divergence theorem in three-dimensional space

$$\int_B \frac{df_i}{dx_i} dV = \int_S f_i \, n_i \, dA \qquad (i = 1, 2, 3), \tag{1.6}$$

where f_i are "well behaved" functions defined in a body B with the infinitesimal volume element $dV = dx_1 \, dx_2 \, dx_3$, which is enclosed by a surface S with area element dA and unit outward normal vector $\boldsymbol{n}$ (see Figure 1.1)

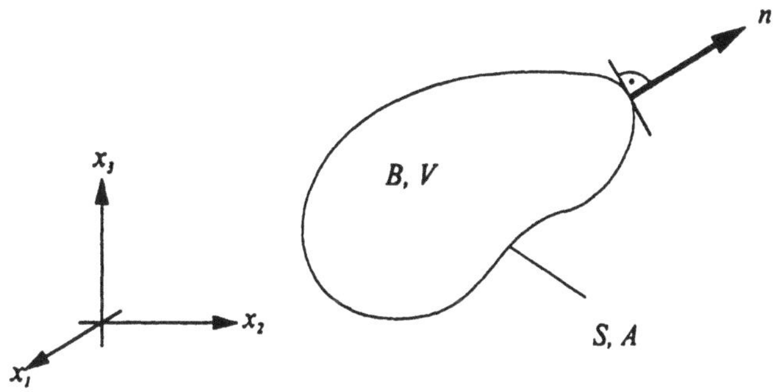

Fig. 1.1 Body B with volume V, surrounding surface S with area A and unit outward normal vector $\boldsymbol{n}$

If only the components P_1, P_2 and P_3 are transformed in (1.5), the conservation law reads in integral form

$$\int_B \frac{dP_t}{dt} dV + \int_S (P_1 \, n_1 + P_2 \, n_2 + P_3 \, n_3) \, dA = 0 . \tag{1.7}$$

Relation (1.5) is now expressed with respect to a finite volume V, rather than a infinitesimal volume dV, i. e., conservation laws may be formulated either in the local (strong) or global (weak) form. In two spatial dimensions the divergence theorem is the integral over a plane surface S in terms of an integral along a (closed) line (or path), Γ (arc length s), which encloses the area (see Figure 1.2), i. e.,

$$\int_S f_{i,i} \, dA = \oint_\Gamma f_i \, n_i \, ds \qquad (i = 1.2).$$

Again, it is presumed that the vector point function f_i is sufficiently well-behaved (no singularities, jumps, etc.)

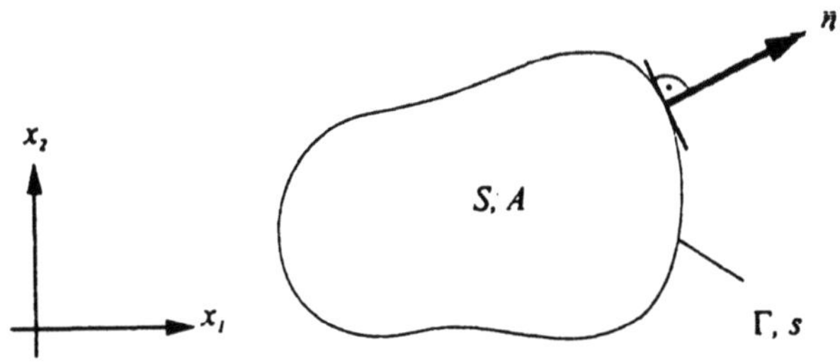

Fig. 1.2 Plane surface S with area A, surrounding curve Γ with arc length s and unit outward normal vector $\boldsymbol{n}$

2 Noether's Method of Establishing Conservation Laws

This method can be applied only if a Lagrangian function is available for the system under consideration. Such a Lagrangian function may be looked upon as a potential and, for example in elastostatics (cf. Section 4), may be identified as the negative of the strain (or internal) energy density W (per unit of volume) of the system.

$$L = -W(u_{i,j}) \tag{2.1}$$

for a body which is homogeneous and

$$L = -W(x_i, u_{i,j}), \tag{2.2}$$

if the properties of the body are not constant, but rather functions of the independent variables (coordinates) x_i. In (2.1) and (2.2) $u_{i,j}$ is the displacement gradient.

In certain dynamic problems, the Lagrangian L might be the so-called kinetic potential

$$L = T - W \tag{2.3}$$

where T is the kinetic energy.

Systems with no dissipation usually possess a Lagrangian and, in a quasi-static process, the work of external forces (body forces and surface tractions) is balanced by changes in the internal energy.

2.1 The Euler-Lagrange Equation; One-Dimensional Case

If a Lagrangian for a general system is postulated, the equations of equilibrium (or motion) are already predetermined and they are derived as follws:

One considers the action integral A which is the integral of L over an arbitrary domain Ω in the space of independent variables x_i

$$A = \int_{\Omega} L(x_i, v_\alpha, v_{\alpha,i})\, d\Omega\,. \tag{2.4}$$

Next one applies Hamilton's principle by performing "the variation" δA of A, i. e., by considering different values of the functional A for different sets of functions $v_\alpha(x_i)$, with the subsidiary conditions that the domain Ω remains fixed during variations. Thus, among all possible functions $v_\alpha(x_i)$, each of which supplies a number A, the solution function is distinguished by the feature, that the associated A is stationary, i. e., $\delta A = 0$. To perform the variation δA of A, the rules of the calculus of variations are to be applied, originally developed by Euler (1744), and presented in numerous texts, e. g., Gelfand & Fomin (1963). One needs to know these rules to perform the variation, just as one needs to know the rules of differential calculus to perform differentiation.

Let us consider first the special case $m = 1$, $\mu = 1$, then

$$L = L(x, v, v_{,x}) \tag{2.5}$$

and

$$A = \int_a^b L(x, v, v_{,x})\, dx \tag{2.6}$$

where a and b designate some arbitrary points on the x-axis. Without knowledge of the specific form of L, i. e., its specific dependence on $x, v, v_{,x}$, we can calculate the variation δA of A by varying under the integral sign v and $v_{,x}$, according to

$$v \to v + \delta v \tag{2.7a}$$

$$v_{,x} \to v_{,x} + \delta v_{,x} \tag{2.7b}$$

but doing nothing to x. The function δv is the so-called variation of v and is arbitrary, except that it vanishes at the end points a and b of the interval under consideration (admissible functions), and that it is small in comparison to v (see Figure 2.1). Analogous features are valid for $\delta v_{,x}$.

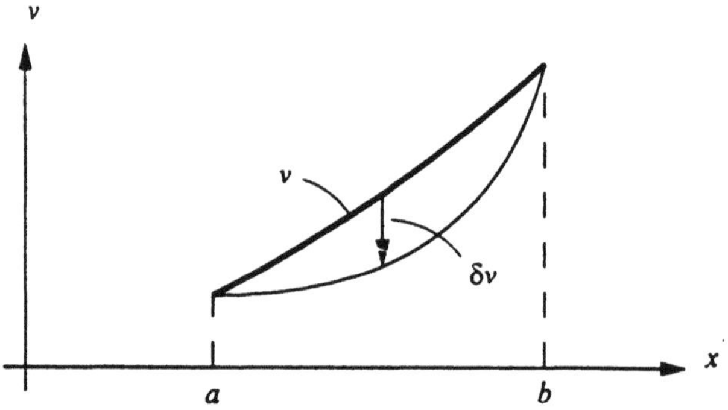

Fig. 2.1 Function v and its variation δv

The variation of the action integral results in

$$\delta A = \int_a^b L\left(x, v+\delta v, v_{,x}+\delta v_{,x}\right)dx - \int_a^b L\left(x, v, v_{,x}\right)dx . \tag{2.8}$$

The integrand L at $x, v+\delta v, v_{,x}+\delta v_{,x}$ will now be expanded in a Taylor series about the values at $x, v, v_{,x}$, just as the value of a sufficiently smooth function of one variable $f(x)$ at $x+a$ can be expressed in terms of its value at x and its derivatives at x, namely

$$f(x+a) = f(x) + \frac{a}{1!}\frac{df(x)}{dx} + \frac{a^2}{2!}\frac{d^2 f(x)}{dx^2} + \cdots .$$

Since δv and $\delta v_{,x}$ are assumed to be small, the Taylor series is truncated after the linear term in δv and $\delta v_{,x}$, i. e., terms involving $(\delta v)^2$, $\delta v\, \delta v_{,x}$, $(\delta v_{,x})^2$ and higher order products, abbreviated by $O(\delta^2)$, are neglected.

Equation (2.8), then changes to

$$\delta A = \int_a^b \left(L\left(x, v, v_{,x}\right) + \frac{\partial L}{\partial v}\delta v + \frac{\partial L}{\partial v_{,x}}\delta v_{,x} - L\left(x, v, v_{,x}\right)\right) dx + O\left(\delta^2\right) = \int_a^b \left(\frac{\partial L}{\partial v}\delta v + \frac{\partial L}{\partial v_{,x}}\delta v_{,x}\right)\delta x. \tag{2.9}$$

Next we wish that the second term be also a variation δv and not the variation δv_{x}. This wish can be fulfilled by knowing that the operation of variation δ and differention $d()/dx = ()_{,x}$ may be interchanged and by considering the differentiation of a product

$$\left(\frac{\partial L}{\partial v_{,x}}\delta v\right)_{,x} = \frac{d}{dx}\left(\frac{\partial L}{\partial v_{,x}}\delta v\right) = \frac{d}{dx}\left(\frac{\partial L}{\partial v_{,x}}\right)\delta v + \frac{\partial L}{\partial v_{,x}}\delta v_{,x}$$

or

$$\int_a^b \left(\frac{\partial L}{\partial v_{,x}}\delta v\right)_{,x} dx = \frac{\partial L}{\partial v_{,x}}\delta v\Bigg|_a^b = \int_a^b \frac{d}{dx}\left(\frac{\partial L}{\partial v_{,x}}\right)\delta v\, dx + \int_a^b \frac{\partial L}{\partial v_{,x}}\delta v_{,x}\, dx .$$

Since $\delta v = 0$, by requirement, at a and b, we have

$$\int_a^b \frac{\partial L}{\partial v_{,x}}\delta v_{,x}\, dx = -\int_a^b \frac{d}{dx}\left(\frac{\partial L}{\partial v_{,x}}\right)\delta v\, dx$$

and therefore

$$\delta A = \int_a^b \left(\frac{\partial L}{\partial v}\delta v - \frac{d}{dx}\left(\frac{\partial L}{\partial v_{,x}}\right)\delta v\right) dx . \tag{2.10}$$

For $\delta A = 0$ and δv arbitrary the consequence is

$$\frac{\partial L}{\partial v} - \frac{d}{dx}\left(\frac{\partial L}{\partial v_{,x}}\right) = 0 . \tag{2.11}$$

This is called the Euler-Lagrange equation of the variational problem $\delta A = 0$, which is identical to the equilibrium equation of the system. Note that d/dx is the total differential operator

$$\frac{d}{dx} = \frac{\partial}{\partial x} + v_{,x}\frac{\partial}{\partial v} + v_{,xx}\frac{\partial}{\partial v_{,x}} . \tag{2.12}$$

The operator

$$\mathbf{E}(\;) = \left[\frac{\partial}{\partial v} - \frac{d}{dx}\frac{\partial}{\partial v_{,x}}\right](\;) \tag{2.13}$$

is referred to as the Euler operator (which is always acting on the Lagrangian L). Thus the Euler-Lagrange equation above may be written for short

$$\mathbf{E}(L)=0. \tag{2.14}$$

2.2 Transformation of Dependent and Independent Variables; One-Dimensional Case

Next we consider again equation (2.5)

$$L = L\left(x, v, v_{,x}\right)$$

and (2.6)

$$A = \int_a^b L\, dx .$$

We now do something which is usually not done within a standard course of study in engineering science or applied and engineering mechanics, namely we subject the Lagrangian L to an infinitesimal transformation of *both* the independent *and* dependent variables, i. e., we pass from the usual, unstarred quantities x, v to starred quantities x^*, v^* according to the prescription

$$x \to x^* = x + \varepsilon\,\xi\left(x, v\right), \tag{2.15a}$$

$$v \to v^* = v + \varepsilon\,\varphi\left(x, v\right). \tag{2.15b}$$

Here the single constant parameter ε is supposed to be small in the sense, that the ensuing terms without ε and linear in ε will be retained while terms with ε^2 and with higher powers will be omitted. The functions ξ and φ, with the arguments indicated, are completely arbitrary. the justification for carrying out this transformation is for the moment obscure and will become clear only when the result is considered. But note the difference between the transformation (2.7) which took us from L to the associated differential equations, and the transformation (2.15).

The transformed A shall be called A^* and reads

$$A^* = \int_{a^*}^{b^*} L\left(x^*, v^*, \frac{dv^*}{dx^*}\right) dx^* . \tag{2.16}$$

We next wish to express all the starred quantities in terms of the original, unstarred ones. This means, in particular, that the transformed domain $a^* - b^*$ will be expressed in terms of the original domain $a-b$ and the differential dx^* can be expressed as $\left(dx^*/dx\right)dx$. Using the transformation prescription (2.15 a)

$$x^* = x + \varepsilon\,\xi\left(x, v\right),$$

it is readily seen that the derivative dx^*/dx is

$$\frac{dx^*}{dx} = 1 + \varepsilon\frac{d\xi}{dx}. \qquad (2.17)$$

Thus

$$A^* = \int_a^b L\left(x + \varepsilon\,\xi\left(x, v\right), v + \varepsilon\,\varphi\left(x, v\right), ?\,\right)\frac{dx^*}{dx}\,dx.$$

The question mark ? indicates the dependence of L on dv^*/dx^*, but expressed in terms of v, ξ, φ and ε, which we do not know yet. We proceed to find it as follows:

$$\frac{dv^*}{dx^*} = \frac{dv^*}{dx}\frac{dx}{dx^*} = \frac{dv^*}{dx}\frac{1}{\dfrac{dx^*}{dx}}.$$

dv^*/dx is again readily found from the relation

$$v^* = v + \varepsilon\,\varphi\left(x, v\right)$$

as

$$\frac{dv^*}{dx} = \frac{dv}{dx} + \varepsilon\frac{d\varphi}{dx} = v_{,x} + \varepsilon\frac{d\varphi}{dx}.$$

We also need

$$\frac{1}{\dfrac{dx^*}{dx}} = \frac{1}{1 + \varepsilon\dfrac{d\xi}{dx}}.$$

We recall that, by binominal expansion

$$\left(1 + \gamma\right)^{-1} = 1 - \gamma + O\left(\gamma^2\right).$$

Thus finally,

$$\frac{dv^*}{dx^*}=\left[v_{,x}+\varepsilon\frac{d\varphi}{dx}\right]\left[1-\varepsilon\frac{d\xi}{dx}\right]+0\left(\varepsilon^2\right)=v_{,x}+\varepsilon\left(\frac{d\varphi}{dx}-v_{,x}\frac{d\xi}{dx}\right)+O\left(\varepsilon^2\right) \tag{2.18}$$

and A^* becomes

$$A^*=\int_a^b L\left(x+\varepsilon\,\xi\,(x,v),v+\varepsilon\varphi\,(x,v),v_{,x}+\varepsilon\left(\frac{d\varphi}{dx}-v_{,x}\frac{d\xi}{dx}\right)\right)\frac{dx^*}{dx}\,dx\,. \tag{2.19}$$

The integrand L at x^*, v^* and dv^*/dx^* will now be, again, expanded in a Taylor series about the value at $x, v, v_{,x}$.

The result is first, with (2.17)

$$A^*=\int_a^b\left\{L\left(x,v,v_{,x}\right)+\varepsilon\left[\frac{\partial L}{\partial x}\xi+\frac{\partial L}{\partial v}\varphi+\frac{\partial L}{\partial v_{,x}}\left(\frac{d\varphi}{dx}-v_{,x}\frac{d\xi}{dx}\right)\right]+O\left(\varepsilon^2\right)\right\}\left[1+\varepsilon\frac{d\xi}{dx}\right]dx\,.$$

and finally, omitting terms in ε^2 and higher,

$$A^*=A+\varepsilon\int_a^b\left\{\left(\xi\frac{\partial}{\partial x}+\varphi\frac{\partial}{\partial v}\right)L+\frac{\partial L}{\partial v_{,x}}\left(\frac{d\varphi}{dx}-v_{,x}\frac{d\xi}{dx}\right)+\frac{d\xi}{dx}L\right\}dx$$

$$=A+\varepsilon\int_a^b\left\{\left[\xi\frac{\partial}{\partial x}+\varphi\frac{\partial}{\partial v}+\left(\frac{d\varphi}{dx}-v_{,x}\frac{d\xi}{dx}\right)\frac{\partial}{\partial v_{,x}}\right]L+L\frac{d\xi}{dx}\right\}dx\,. \tag{2.20}$$

As a side remark, the operator $\xi\,\partial/\partial x+\varphi\,\partial/\partial v$ is referred to as the infinitesimal generator $\mathbf{w}$ (of a Lie group in the space x, v)

$$\mathbf{w}=\xi\frac{\partial}{\partial x}+\varphi\frac{\partial}{\partial v} \tag{2.21}$$

and the operator

$$\mathbf{pr}^{(1)}\mathbf{w}=\mathbf{w}+\left(\frac{d\varphi}{dx}-v_{,x}\frac{d\xi}{dx}\right)\frac{\partial}{\partial v_{,x}} \tag{2.22}$$

is referred to as the first prolongation $\mathbf{pr}^{(1)}\mathbf{w}$ of the group into the jet bundle space $x, v, v_{,x}$.

These designations belong to the theory of continuous Lie groups, whose knowledge is not essential for us and which we simply use here. But to explore this background, the reader is referred to, e. g., Olver (1993), where for $\mathbf{w}$, the letter $\mathbf{v}$ is used, and vice versa.

Thus we can write

$$A^* = A + \varepsilon \int_a^b \left[\mathbf{pr}^{(1)}\mathbf{w} + \frac{d\xi}{dx} \right] L\, dx . \tag{2.23}$$

The first term in the integrand above describes the change of L to L^* in the domain $a-b$ and is a differential operator, while the second term is a factor which describes the change of domain $a-b$ to a^*-b^*.

In the integrand of equation (2.20), we rewrite the 3 terms involving d/dx as follows:

$$\left(\frac{d\varphi}{dx} - v_{,x} \frac{d\xi}{dx} \right) \frac{\partial L}{\partial v_{,x}} + L \frac{d\xi}{dx} = \frac{d}{dx} \left[\left(\varphi - v_{,x}\, \xi \right) \frac{\partial L}{\partial v_{,x}} + L\xi \right] -$$

$$- \varphi \frac{d}{dx} \left(\frac{\partial L}{\partial v_{,x}} \right) + v_{,xx}\, \xi \frac{\partial L}{\partial v_{,x}} + v_{,x}\, \xi \frac{d}{dx} \left(\frac{\partial L}{\partial v_{,x}} \right) - \xi \left(\frac{\partial L}{\partial x} + v_{,x} \frac{\partial L}{\partial v} + v_{,xx} \frac{\partial L}{\partial v_{,x}} \right), \tag{2.24}$$

where (2.12) and the product rule $fg_{,x} = (fg)_{,x} - f_{,x} g$ was used. It is noted that the terms $\xi v_{,xx} \partial L / \partial v_{,x}$ cancel. Finally we can rewrite (2.20) with (2.24) as

$$A^* = A + \varepsilon \int_a^b \left\{ \frac{d}{dx} \left[\varphi \frac{\partial L}{\partial v_{,x}} + \xi \left(L - v_{,x} \frac{\partial L}{\partial v_{,x}} \right) \right] + \left(\varphi - \xi v_{,x} \right) \left(\frac{\partial L}{\partial v} - \frac{d}{dx} \frac{\partial L}{\partial v_{,x}} \right) \right\} dx . \tag{2.25}$$

It is noted that $\partial L / \partial v - d/dx \left(\partial L / \partial v_{,x} \right)$ is the Euler-Lagrange expression $\mathbf{E}(L)$ (2.11, 2.14)

$$\mathbf{E}(L) = \frac{\partial L}{\partial v} - \frac{d}{dx} \left(\frac{\partial L}{\partial v_{,x}} \right). \tag{2.26}$$

The factor $\left(\varphi - \xi v_{,x} \right)$ is referred to as the characteristic Q

$$Q = \varphi - \xi v_{,x} . \tag{2.27}$$

The bracket to which the total differential operator d/dx is applied shall be referred to as the current (or flux) P

$$P = \varphi \frac{\partial L}{\partial v_{,x}} + \xi \left(L - \frac{\partial L}{\partial v_{,x}} v_{,x} \right) \tag{2.28a}$$

$$= \xi L + Q \frac{\partial L}{\partial v_{,x}} . \tag{2.28b}$$

Thus, in "short hand" notation we can write

$$A^* = A + \varepsilon \int_a^b \left[\frac{dP}{dx} + Q\mathbf{E}(L) \right] dx . \tag{2.29}$$

Along solutions $v(x)$, we must have $\mathbf{E}(L) = 0$. Then only dP/dx is left in the integrand which can be readily integrated; (in two or more dimensions we would apply the divergence theorem). We then have

$$A^* = A + \varepsilon [P^b - P^a] . \tag{2.30}$$

Here P^b is the value of P at b and P^a is the value of P at a. We note that if

$$\frac{dP}{dx} = 0 \rightarrow P = const. \qquad \text{or} \qquad P^b = P^a , \tag{2.31}$$

i.e., if a conservation law exists, then $A^* = A$. This expresses the converse part of Noether's first theorem (1918), which, in essence, states that if the (action) integral A is invariant with respect to the action of an infinitesimal continuous one-parameter Lie group, then there exists a conservation law.

2.3 The General Case

We consider now the general case with an arbitrary number of independent variables m and an arbitrary number of dependent variables μ, with L still depending on the first order derivatives only

$$L = L\left(x_{i,}v_{\alpha}, v_{\alpha,i}\right) \qquad \begin{array}{l}(i = 1, 2, ..., m),\\ (\alpha = 1, 2, ..., \mu).\end{array} \tag{2.32}$$

The Lagrangian is defined over an arbitrary domain Ω. The one-parameter group of (point) transformations is defined as

$$x_i^* = x_i + \varepsilon\xi_i\left(x_j, v_\beta\right) \tag{2.33a}$$

$$v_\alpha^* = v_\alpha + \varepsilon\varphi_\alpha\left(x_{j,} v_\beta\right). \tag{2.33b}$$

We omit intermediate manipulations analogous to the simpler cases discussed in detail above. The change of the action integral A due to this transformation then follows to be

$$A^* = A + \varepsilon\int\limits_\Omega\left(\mathbf{pr}^{(1)}\,\mathbf{w} + \xi_{i,i}\right)L\,d\Omega \tag{2.34a}$$

$$= A + \varepsilon\int\limits_\Omega\left(P_{i,i} + Q_\alpha\,\mathbf{E}_\alpha\,(L)\right)d\Omega\,. \tag{2.34b}$$

The infinitesimal generator $\mathbf{w}$ and its first prolongation are, respectively, given by

$$\mathbf{w} = \xi_i\frac{\partial}{\partial x_i} + \varphi_\alpha\frac{\partial}{\partial v_\alpha}, \tag{2.35a}$$

$$\mathbf{pr}^{(1)}\,\mathbf{w} = \mathbf{w} + \left(\frac{d\varphi_\alpha}{dx_i} - v_{\alpha,j}\frac{d\xi_j}{dx_i}\right)\frac{\partial}{\partial v_{\alpha,i}}. \tag{2.35b}$$

The transformation functions ξ_i and φ_α are determined from the invariance conditions (2.34) with $A^* = A$ as

$$\left(\mathbf{pr}^{(1)}\,\mathbf{w} + \xi_{i,i}\right)L = 0, \tag{2.36}$$

which leads to an overdetermined system of partial differential equations for these functions. Once ξ_i and φ_α are known, the characteristics Q_α and the conserved current P_i are given as

$$Q_\alpha = \varphi_\alpha - \xi_i v_{\alpha,i}\,, \tag{2.37}$$

$$P_i = \varphi_\alpha \frac{\partial L}{\partial v_{\alpha,i}} + \xi_j \left(L\delta_{ij} - v_{\alpha,j} \frac{\partial L}{\partial v_{\alpha,i}} \right). \tag{2.38}$$

(δ_{ij} is the Kronecker symbol with $\delta_{ij} = 1$ if $i = j$ and $\delta_{ij} = 0$ for $i \neq j$.)

Along solutions, all the μ Euler-Lagrange expressions should vanish

$$\mathbf{E}_\alpha(L) = \frac{\partial L}{\partial v_\alpha} - \frac{d}{dx_i}\left(\frac{\partial L}{\partial v_{\alpha,i}} \right) = 0 . \tag{2.39}$$

From $A^* = A$, (2.34 b) and (2.39) the desired conservation law follows to be

$$P_{i,i} = \frac{dP_i}{dx_i} = \mathbf{div}\, \boldsymbol{P} = 0 , \tag{2.40}$$

with P_i given by (2.38). Application of the divergence theorem leads to

$$\int_\Omega P_{i,i}\, d\Omega = \int_{\partial\Omega} P_i\, n_i\, d\, \partial\Omega + \sum_{r=1}^{r=p} R_r = 0 , \tag{2.41}$$

where $\partial\Omega$ is the surface bounding Ω and R_r is the *r-th* of the p singularities within Ω, i. e., enclosed by $\partial\Omega$.

Already at this stage it is possible, and probably even desirable, to identify some quantities introduced above in physical terms. In elastostatics, where v_α are displacements and x_i, e. g., Cartesian coordinates, the expression

$$\sigma_{i\alpha} = -\frac{\partial L}{\partial v_{\alpha,i}} \tag{2.42}$$

shall be called the physical momentum tensor or Cauchy stress tensor and

$$b_{ij} = v_{\alpha,j} \frac{\partial L}{\partial v_{\alpha,i}} - L\delta_{ij} \tag{2.43}$$

the material momentum tensor or Eshelby tensor. Thus (2.38) reads

$$-P = \varphi_\alpha\, \sigma_{i\alpha} + \xi_j\, b_{ij} . \tag{2.44}$$

If we recall

$$v_\alpha^* = v_\alpha + \xi\varphi_\alpha ,$$

$$x_j^* = x_j + \varepsilon\xi_j$$

and take as transformation constant (physical) translations, i. e., $\varphi_\alpha = c_\alpha = const$. and $\xi_j = 0$, the physical momentum is conserved

$$P_{i,i} = 0 \quad \Rightarrow \quad \sigma_{i\alpha,i} = 0 . \tag{2.45}$$

On the other hand, if we take constant coordinate transformations, or material translations, i. e., $\varphi_\alpha = 0$ and $\xi_j = c_j = const.$, the material momentum is conserved

$$P_{i,i} = 0 \quad \Rightarrow \quad b_{ij,i} = 0 . \tag{2.46}$$

In general φ_α and ξ_j have to be determined from the condition (2.36).

2.4 Bessel-Hagen Extension

An important extension of Noether's first theorem has been proposed by Bessel-Hagen (1921) who weakened the requirement

$$A^* = A$$

and showed that it is sufficient to demand

$$A^* = A + \int_\Omega \overline{K}_{i,i}\, d\Omega \qquad \text{or} \qquad A^* = A + \varepsilon\int_\Omega K_{i,i}\, d\Omega \tag{2.47}$$

in order to obtain conservation laws. Here K_i is an arbitrary vector point function, whose divergence enters the integrand and the underlying symmetries are called divergence symmetries, cf. Olver (1993).

Indeed, instead of

$$\mathbf{div}\, \boldsymbol{P} = P_{i,i} = 0$$

we now have

$$\mathbf{div}\left(\boldsymbol{P} + \boldsymbol{K}\right) = \left(P_i + K_i\right)_i = 0 . \tag{2.48}$$

The Bessel-Hagen extension will be employed in the example below.

2.5 Example

The question arises as to what the most general functions $\xi = \xi(x, v)$ and $\varphi = \varphi(x, v)$ might look like, but it does turn out that to answer this question a specific form of the Lagrangian must be introduced, e. g., examples of specific problems have to be considered.

We pick as a simple example, the elastic, homogeneous bar of constant compressional stiffness EA with the Lagrangian L

$$L = \frac{1}{2} = EA\,u'^2 . \tag{2.49}$$

In the product EA, E is Young's modulus and A the cross-sectional area. The displacement is given by u and $u' = u_{,x} = du/dx$ is its first derivative, i. e., the strain.

Obviously, the Lagrangian is given as

$$L = L\left(u_{,x}\right) = L(u') \tag{2.50}$$

and the associated Euler-Lagrange equation (cf. 2.13 and 2.14) consists of a single term

$$\frac{d}{dx}\left(\frac{\partial L}{\partial u'}\right) = \frac{d}{dx}(EAu') = EAu'' = 0 ,$$

where $u'' = d^2u/dx^2$. Hence

$$u'' = 0 . \tag{2.51}$$

We are interested in nontrivial conservation laws only, i. e., laws which are valid only along the solution $u'' = 0$.

The conserved current P is given by (cf. e. g., 2.28 b)

$$P = \xi L + Q\frac{\partial L}{\partial u'} , \tag{2.52}$$

where Q is the characteristic given by (2.27)

$$Q = \varphi - \xi u' \tag{2.53}$$

or, upon substitution

$$P = -\frac{1}{2}\xi EAu'^2 + \varphi EAu' . \qquad (2.54)$$

We now wish to determine the most general ξ and φ, such that the conservation law $dP/dx = 0$ still prevails. Thus ξ and φ have to be determined from the invariance condition (2.20) which leads with (2.49) to

$$EA\left(\frac{d\varphi}{dx}u' - \frac{1}{2}\frac{d\xi}{dx}u'^2\right) = 0 . \qquad (2.55)$$

The same condition may be obtained directly from the conservation law $dP/dx = 0$, together with the equation of equilibrium (2.51).

We consider first the transformation (symmetry)

$$\xi = \xi(x, u), \qquad (2.56a)$$

$$\varphi = \varphi(x, u). \qquad (2.56b)$$

Substitution into the above equation leads to

$$-\frac{1}{2}u'^2\left(\frac{\partial\xi}{\partial x} + \frac{\partial\xi}{\partial u}u'\right) + u'\left(\frac{\partial\varphi}{\partial x} + \frac{\partial\varphi}{\partial u}u'\right) = u'\frac{\partial\varphi}{\partial x} + u'^2\left(\frac{\partial\varphi}{\partial u} - \frac{1}{2}\frac{\partial\xi}{\partial x}\right) + u'^3\left(-\frac{1}{2}\frac{\partial\xi}{\partial u}\right) = 0 .$$

Since ξ and φ, by definition, do not depend on u', the above equation is satisfied only if the factors of each "form" of u' vanish separately.

Thus

$$\frac{\partial\varphi}{\partial x} = 0 \rightarrow \varphi = \varphi(u),$$

$$\frac{\partial\xi}{\partial u} = 0 \rightarrow \xi = \xi(x),$$

$$\frac{\partial\varphi}{\partial u} - \frac{1}{2}\frac{\partial\xi}{\partial x} = 0 \rightarrow \left\{\begin{aligned} \frac{\partial\varphi}{\partial u} &= \frac{1}{2}C_2 \\ \frac{\partial\xi}{\partial x} &= C_2 \end{aligned}\right\} \rightarrow \quad \begin{aligned} \varphi &= \frac{1}{2}C_2 u + C_4 \\ \xi &= C_2 x + C_1 , \end{aligned}$$

where the arrow $\rightarrow$ indicates "leads to" and the C_i's are arbitrary constants.

Now ξ and φ are completely determined and the current P is

$$P=\left(C_2 x+C_1\right)\left(-\frac{1}{2}EAu'^2\right)-C_2\left(-\frac{1}{2}EAuu'\right)+C_4 EAu' . \tag{2.57}$$

We next introduce the designations already employed when discussing elastic bars in Kienzler (1993)

$-\frac{1}{2}EAu'^2=B$ is the material force. (2.58)

$-\frac{1}{2}EAuu'=H$ is the material scalar moment (virial), (2.59)

$EAu'=N$ is the physical force. (2.60)

The current P is then

$$P=C_1 B+C_2\left(Bx-H\right)+C_4 N=const . \tag{2.61}$$

and there follow three conservation laws

$$B' \quad =0 , \tag{2.62a}$$

$$(Bx-H)'=0 , \tag{2.62b}$$

$$N' \quad =0 . \tag{2.62c}$$

Let us now consider the extension of Bessel-Hagen mentioned in Section 2.4. We introduce a function $K=K\left(x,u,u'\right)$ and the conserved current is now augmented by K

$$P+K=const . \tag{2.63}$$

which, by differentiation, (cf. 2.48 and 2.55) becomes

$$EA\left(-\frac{1}{2}\frac{d\xi}{dx}u'^2+\frac{d\varphi}{dx}u'\right)+\frac{\partial K}{\partial x}+\frac{\partial K}{\partial u}u'+\frac{\partial K}{\partial u'}u''=0. \tag{2.64}$$

We notice immediately that K cannot depend on u', because $\partial K/\partial u'$ will introduce u'' as a factor and the product has to vanish. Ordered by the "forms" of u', equation (2.64) can be written as

$$\frac{1}{EA}(P+K)' = u'\left(\frac{\partial\varphi}{\partial x}+\frac{1}{EA}\frac{\partial K}{\partial u}\right)+u'^2\left(\frac{\partial\varphi}{\partial u}-\frac{1}{2}\frac{\partial\xi}{\partial u}\right)+u'^3\left(-\frac{1}{2}\frac{\partial\xi}{\partial u}\right)+\frac{1}{EA}\frac{\partial K}{\partial x}=0.$$

From the last term above we conclude $K = K(u)$ and from the next to the last $\xi = \xi(x)$. The second term implies

$$\frac{\partial\varphi}{\partial u}-\frac{1}{2}\frac{\partial\xi}{\partial x}=0$$

and we conclude

$$\varphi = uf_1(x)+f_2(x),$$

$$\xi = 2\int f_1(x)\,dx + C_1 .$$

The first term in the conservation law implies

$$\frac{\partial\varphi}{\partial x}=-\frac{1}{EA}\frac{\partial K}{\partial u}$$

or, by substitution

$$uf_1'+f_2'=-\frac{1}{EA}\frac{\partial K}{\partial u}.$$

It follows that

$$f_1' = const. = C_3$$

and

$$f_2' = const. = C_5$$

or, by substitution

$$u\,C_3 + C_5 = -\frac{1}{EA}\frac{\partial K}{\partial u}.$$

By integration

$$\frac{K}{EA}=-\frac{C_3}{2}u^2 - C_5 u - C_6$$

and

$$f_1(x)=C_3 x+\frac{C_2}{2},$$

$$f_2(x)=C_5 x + C_4 .$$

Thus

$$\xi = C_3 x^2 + C_2 x + C_1, \tag{2.65a}$$

$$\varphi = C_3 ux + \frac{1}{2} C_2 u + C_5 x + C_4 , \tag{2.65b}$$

$$K = -\frac{1}{2} C_3 \, EAu^2 - C_5 \, EAu + C_6 \, EA . \tag{2.65c}$$

The constant C_6 does not contribute anything of interest.

The conserved current is

$$P + K = \left(C_1 + C_2 x + C_3 x^2\right)\left(-\frac{1}{2} EAu'^2\right) + \left(\frac{1}{2} C_2 u + C_3 xu + C_4 + C_5 x\right) EAu' +$$

$$+ C_3 \left(-\frac{1}{2} EA^2\right) - C_5 EAu. \tag{2.66}$$

We introduce the additional notation

$$R = -\frac{1}{2} EAu^2 , \tag{2.67}$$

$$V = EAu . \tag{2.68}$$

Thus

$$P + K = C_1 B + C_2 (Bx - H) + C_3 \left(Bx^2 - 2Hx + R\right) + C_4 N + C_5 (Nx - V) , \tag{2.69}$$

which leads to the five independent conservation laws

$$B' = 0 , \tag{2.70a}$$

$$(Bx - H)' = 0 , \tag{2.70b}$$

$$(Bx^2 - 2Hx + R)' = 0 , \tag{2.70c}$$

$$N' = 0 , \tag{2.70d}$$

$$(Nx - V)' = 0 . \tag{2.70e}$$

We observe that the employment of the Bessel-Hagen extension was indeed fruitful, since the number of conservation laws has increased from three to five.

3 Neutral Action Method for Systems Without a Lagrangian

For systems without a Lagrangian (e. g., for some systems with dissipation) the governing equations are no longer the Euler-Lagrange equations of a variational problem, but rather a set of partial differential equations which has been obtained on the basis of some balance laws (e. g., balance of linear momentum) and some postulated constitutive relations (e. g., standard linear solid in viscoelasticity).

Until recently, no systematic procedure existed to derive conservation laws for such systems, but a few years ago, a novel methodology was advanced, cf. T. Honein et al. (1991).

3.1 The Procedure

In order to develop this new procedure, a result of the calculus of variations is required as discussed below. The calculus itself will, however, not be needed for the application of the new method. This result states that if a Lagrangian is itself a divergence (then called a "null Lagrangian"), then it satisfies identically the Euler-Lagrange equation (2.11). This means that if some Lagrangian $\tilde{L}$ can expressed as $\tilde{L} = q_{i,i}$, then it follows $E_\alpha(\tilde{L}) = 0$

The basic idea of the new procedure is as follows: As we have seen in Section 2.3, a conservation law $P_{i,i} = 0$ is expressible in terms of the Euler-Lagrange equation $\mathbf{E}_\alpha(L)$ as (cf. 2.34 b)

$$P_{i,i} = -Q_\alpha \mathbf{E}_\alpha(L) = 0, \tag{3.1}$$

where the elements of the set Q_α are the characteristics of the conservation law which are fixed by the transformation functions ξ_i and φ_α. as

$$Q_\alpha = \varphi_\alpha - \xi_i \, v_{\alpha,i} \, .$$

Now, instead of the Euler-Lagrange equation $\mathbf{E}_\alpha(L) = 0$, we have a set of μ equations on μ dependent variables $v_\alpha(\alpha = 1, 2, \ldots, \mu)$ which we can abbreviate as

$$\Delta_\beta(x_i, v_\alpha, v_{\alpha,i}) = 0 \, . \tag{3.2}$$

The possibility is not excluded that they may be Euler-Lagrange equations.

Now, instead of the characteristics Q being specified by the transformation functions ξ_i and ϕ_α, we determine Q_α, employing the symbol $-f_\alpha$ instead (in order to avoid confusion), such that

$$f_\alpha \mathbf{\Delta}_\alpha = P_{i,i}, \tag{3.3}$$

where the functions f_α are not considered to be pre-determined as Q_α were.

Thus it follows from (3.3) that a requirement for the existence of conservation laws is that

$$\mathbf{E}_\beta(f_\alpha \mathbf{\Delta}) = 0, \tag{3.4}$$

since $\mathbf{E}_\beta(P_{i,i}) = 0$ as mentioned above. The sum $f_\alpha \mathbf{\Delta}_\alpha$ implies that it is formally a null Lagrangian whose action integral

$$A = \int_B \tilde{L}\, dV = \int_B f_\alpha\, \mathbf{\Delta}_\alpha\, dV = \int_S P_j\, n_j\, dA$$

has vanishing variation for any dependent variable v_α, i. e., $\delta A = 0$. In other words, in order to construct conservation laws for any system (whether or not it possesses a Lagrangian), governed by a set of differential equations $\mathbf{\Delta}_\alpha = 0$, we try to construct a product $f_\alpha \mathbf{\Delta}_\alpha$ whose action A does not change variationally, i. e., the action A behaves neutrally under its variation. Thus the name "Neutral Action" (NA) method was given to this procedure (Chien, 1992).

3.2 Example 1: Elastic Bar

As an application of the NA method let us consider again the elastic bar, but start not with the Lagrangian L, but rather with the differential equation $\mathbf{\Delta} = u'' = 0$. In following the NA method we multiply the differential equation by $f = f(x, u, u')$, apply the Euler operator to the product fu'', and set it equal to zero, i. e.,

$$\mathbf{E}(fu'') = 0 \tag{3.5}$$

or, explicitely (cf. Kienzler and Herrmann, 2000)

$$\mathbf{E}(fu'') = \frac{\partial(fu'')}{\partial u} - \frac{d}{dx}\left(\frac{\partial(fu'')}{\partial u'}\right) + \frac{d^2}{dx^2}\left(\frac{\partial(fu'')}{\partial u''}\right) = \frac{\partial f}{\partial u}u'' - \frac{d}{dx}\left(\frac{\partial f}{\partial u'}u''\right) + \frac{d^2}{dx^2}f = 0\,.$$

The first two terms above vanish and the last may be evaluated in steps as

$$\frac{d^2 f}{dx^2} = \frac{d}{dx}\frac{df}{dx} = \frac{d}{dx}\left(\frac{\partial f}{\partial x} + \frac{\partial f}{\partial u}u' + \frac{\partial f}{\partial u'}u''\right)$$

$$=\frac{\partial^2 f}{\partial x^2}+\frac{\partial^2 f}{\partial x\,\partial u}u'+\frac{\partial^2 f}{\partial x\,\partial u'}u''+$$

$$+\frac{\partial^2 f}{\partial x\,\partial u}u'+\frac{\partial^2 f}{\partial u^2}u'^2+\frac{\partial^2 f}{\partial u\,\partial u'}u''u'+\frac{\partial f}{\partial u}u''+$$

$$+\frac{\partial^2 f}{\partial x\,\partial u'}u''+\frac{\partial^2 f}{\partial u\,\partial u'}u'u''+\frac{\partial^2 f}{\partial u'^2}u''^2+\frac{\partial f}{\partial u'}u'''.$$

The only term which may be set equal to zero at this stage is the last, because it contains a derivative higher than the highest appearing in the differential equation and should be eliminated by substitution into this equation. Thus

$$\frac{d^2 f}{dx^2}=\frac{\partial^2 f}{\partial u'^2}u''^2+u''\left(\frac{\partial f}{\partial u}+2\frac{\partial^2 f}{\partial x\,\partial u'}+2\frac{\partial^2 f}{\partial u\,\partial u'}u'\right)+\frac{\partial^2 f}{\partial x^2}+2u'\frac{\partial^2 f}{\partial x\,\partial u}+\frac{\partial^2 f}{\partial u^2}u'^2=0.$$

Applying the same arguments as before, the factor of u''^2 and u'' must vanish separately, i. e.,

$$\frac{\partial^2 f}{\partial u'^2}=0;\quad \frac{\partial f}{\partial u}+2\frac{\partial^2 f}{\partial x\,\partial u'}+2\frac{\partial^2 f}{\partial u\,\partial u'}u'=0.$$

From the fist above, by integration, we find

$$f=f_1(x,u)\,u'+g(x,u).$$

Substitution into the second yields

$$3\frac{\partial f_1}{\partial u}u'+2\frac{\partial g}{\partial u}+2\frac{\partial f_1}{\partial x}=0..$$

Since f_1 and g do not depend on u', we conclude

$$\frac{\partial f_1}{\partial u}=0\rightarrow f_1=f_1(x),$$

$$\frac{\partial g}{\partial u}+2\frac{\partial f_1}{\partial x}=0\rightarrow\frac{\partial g}{\partial u}=f_2(x)\rightarrow g=f_2(x)\,u+f_3(x).$$

and finally

$$f(x,u,u') = f_1(x)u' + f_2(x)u + f_3(x). \tag{3.6}$$

Substitution back into $d^2f/dx^2 = 0$ and full use, now, of the governing differential equation $u'' = 0$ results in

$$f_1''u' + f_2''u + f_3'' + 2u'f_2' = 0 .$$

The factors of u, u' and the term without these must vanish separately, because f_1, f_2 and f_3 are all functions of x only

$$f_2'' = 0 \qquad \rightarrow f_2 = \frac{1}{2}C_3x + C_0 ,$$

$$f_3'' = 0 \qquad \rightarrow f_3 = C_5x + C_4 ,$$

$$f_1'' + 2f_2' = 0 \qquad \rightarrow f_1'' = -C_3 \qquad \rightarrow f_1 = -\frac{1}{2}C_3x^2 - \frac{1}{2}C_2x - \frac{1}{2}C_1 .$$

The labelling of constants has been chosen to coincide with those of the same example, but treated via Noether's theorem in the previous Section.

Before proceeding we modify the product $f\Delta$ as follows

$$f\Delta = fu'' = \frac{d}{dx}(fu') - u'\frac{df}{dx} = \frac{d}{dx}(fu') - \frac{d}{dx}\left(\frac{df}{dx}u\right) + u\frac{d^2f}{dx^2}$$

and, since $d^2f/dx^2 = 0$, the current P is found to be

$$P = fu' - \frac{'df}{dx}u .$$

The complete f is

$$f = u'\left(-\frac{1}{2}C_3x^2 - \frac{1}{2}C_2x - \frac{1}{2}C_1\right) + u\left(\frac{1}{2}C_3x + C_0\right) + C_5x + C_4 \tag{3.7}$$

and its derivative is, because $u'' = 0$, given by

$$\frac{df}{dx} = u'\left(-C_3x - \frac{1}{2}C_2 + \frac{1}{2}C_3x + C_0\right) + \frac{1}{2}uC_3 + C_5.$$

It follows that P can be expressed as

$$P = u'^2\left(-\frac{1}{2}C_3x^2 - \frac{1}{2}C_2x - \frac{1}{2}C_1\right) + uu'\left(\frac{1}{2}C_3x + C_0\right) + u'(C_5x + C_4) -$$

$$- uu'\left(-\frac{1}{2}C_3x - \frac{1}{2}C_2 + C_0\right) - u^2\frac{1}{2}C_3 - uC_5$$

$$= -\frac{1}{2}u'^2\left(C_1 + C_2x + C_3x^2\right) + u'\left(\frac{1}{2}C_2u + C_3xu + C_4 + C_5x\right) - \frac{1}{2}u^2C_3 - uC_5 . \tag{3.8}$$

Comparison of the above expression for the current P shows that it is up to the constant factor EA identical to the corresponding expression obtained by the application of Noether's theorem together with the Bessel-Hagen extension (2.46). The comparison of the two procedures which lead to the same result indicates that the NA method appears to be more straightforward since it deals only with one function f, rather than with three, namely ξ, φ and K.

3.3 Example 2: One-Dimensional Linear Viscoelasticity

We consider one-dimensional linear viscoelasticity on the basis of the Kelvin-Voigt model (Chien et al., 1993). In the absence of body forces, the stress equation of equilibrium is $\sigma' = 0$ and the relation between stress σ and derivatives of displacement u is

$$\sigma = Eu' + \eta u'^{\bullet} , \tag{3.9}$$

where a dot indicates differentiation with respect to the time t. Here E is Young's modulus and η the viscosity coefficient of a Kelvin-Voigt element depicted in Figure 3.1

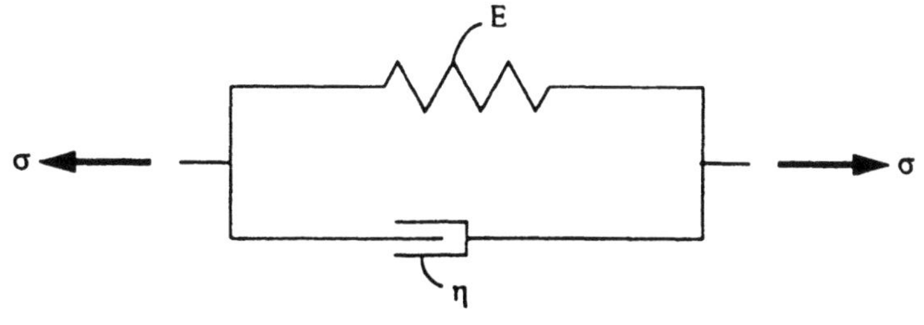

Fig. 3.1 Kelvin-Voigt model

It is convenient to introduce a "potential" φ as

$$\varphi = Eu + \eta u^{\bullet} \tag{3.10}$$

and the governing equation $\mathbf{\Delta}$ is then

$$\mathbf{\Delta} = \varphi'' = 0\,. \tag{3.11}$$

The characteristic f of the conservation law is taken to be

$$f = f(\varphi, \varphi', \varphi^{\bullet}) \tag{3.12}$$

and the condition for existence of conservation laws becomes

$$\mathbf{E}(f\mathbf{\Delta}) = \frac{\partial f}{\partial \varphi}\varphi'' - \left(\frac{\partial f}{\partial \varphi'}\varphi''\right)' - \left(\frac{\partial f}{\partial \varphi^{\bullet}}\varphi''\right)^{\bullet} + f'' = 0\,. \tag{3.13}$$

Since the only unknown in the above equation is the characteristic f which depends on φ, φ' and $\varphi^{\bullet}$, it follows that all coefficients of second and higher order derivatives of φ in this equation must be set equal to zero independently. The resulting set of equations is as follows:

coefficient	equation
$\varphi''\varphi^{\bullet\bullet}$	$\frac{\partial^2 f}{\partial \varphi^{\bullet 2}} = 0,$
$\varphi'^{\bullet 2}$	$\frac{\partial^2 f}{\partial \varphi^{\bullet 2}} = 0,$
$\varphi'^{\bullet}$	$2\frac{\partial^2 f}{\partial \varphi\,\partial \varphi^{\bullet}}\varphi' = 0\,,$ (3.14)
φ''	$2\frac{\partial f}{\partial \varphi} - \frac{\partial^2 f}{\partial \varphi\,\partial \varphi^{\bullet}}\varphi^{\bullet} + \frac{\partial^2 f}{\partial \varphi\,\partial \varphi'}\varphi' = 0\,,$

leaving

$$\frac{\partial^2 f}{\partial \varphi^2}\varphi'^2 = 0\,.$$

After some mathematical manipulations, the solution of the above set of equations is found to be

$$f = f^1(\varphi')\varphi^{\bullet} + f^2(\varphi') + C\frac{\varphi}{\varphi'^2}. \tag{3.15}$$

Since f^1 and f^2 are arbitrary functions of φ', it is permissible to rename them as

$$f^1(\varphi') = -\frac{\partial^2 g(\varphi')}{\partial \varphi'^2},$$
$$f^2(\varphi') = \frac{\partial h(\varphi')}{\partial \varphi'}, \tag{3.16}$$

Defining f^1 and f^2 in this way will allow the conservation laws derived below to have a simpler appearance.

With

$$f = \frac{\partial^2 g\,(\varphi')}{\partial \varphi'^2}\varphi^{\bullet} + \frac{\partial h(\varphi')}{\partial \varphi'} + C\frac{\varphi}{\varphi'^2} \tag{3.17}$$

and $\Delta = \varphi''$ it is now possible to construct divergence-free expressions, $P_x' + P_t^{\bullet} = 0$, out of the product $f\Delta$. The resulting currents P_x and P_t are

$$P_x = \frac{\partial g(\varphi')}{\partial \varphi'}\varphi^{\bullet} + h(\varphi') - C\left(\frac{\varphi}{\varphi'} - x\right), \tag{3.18a}$$

$$P_t = g(\varphi'). \tag{3.18b}$$

Since the stress, $\sigma = \varphi'$, of this system is constant with respect to x, the conservation law pertaining to only $h\,(\varphi') \neq 0$,

$$P_x = h(\varphi') = A_1(t), \tag{3.19}$$

implies that any function of the constant stress is a constant with respect to x, where $A_1(t)$ is an arbitrary function of time.

The conservation law corresponding to only $C \neq 0$ can be rewritten as

$$P_x = \varphi - x \ \varphi' = A_2(t), \tag{3.20}$$

where $A_2(t)$ is an arbitrary function in t. In terms of stress and displacement,

$$P_x = Eu + \eta u^{\bullet} - x\sigma = A_2(t), \tag{3.21}$$

This relation expresses the dependence between stress, displacement and velocity which can be obtained by integrating the constitutive equation of the system and taking into account the equilibrium equation.

In terms of stress, the currents that pertain to only $g(\varphi') \neq 0$ can be written as

$$P_x = -\frac{\partial g(\sigma)}{\partial \sigma} \int \sigma^{\bullet} dx, \tag{3.22a}$$

$$P_t = g(\sigma), \tag{3.22b}$$

where g is any function of stress.

If one chooses $g(\sigma) = \sigma^2 /(2E)$ and splits the stress σ into an elastic component $\sigma_e = Eu'$ and a viscous component $\sigma_\eta = \eta \, u'^{\bullet}$, i. e.,

$$\sigma = \sigma_e + \sigma_\eta \tag{3.23}$$

and, furthermore, introduces the strain rate as

$$\varepsilon^{\bullet} = u'^{\bullet}, \tag{3.24}$$

then the conservation law $P_x' + P_t^{\bullet} = 0$ can be written as

$$\frac{d}{dt}\left(\frac{\sigma_e^2}{2E}\right) = \sigma\varepsilon^{\bullet} - \eta\varepsilon^{\bullet 2}, \tag{3.25}$$

which states that the rate of elastic energy $\sigma_e^2 / 2E$ is equal to the rate of work $\sigma\varepsilon^{\bullet}$ done by the tractions, minus the energy dissipation $\sigma_\eta \varepsilon^{\bullet} = \eta\varepsilon^{\bullet 2}$. This is a form of the statement of the principle of conservation (or rather balance) of energy (first law of thermodynamics) for the mechanical system at hand. By choosing a different function $g(\sigma)$, other conservation laws might be constructed.

3.4 Discussion

The establishment of conservation laws may be schematically illustrated by means of a flow diagram, Fig. 3.2

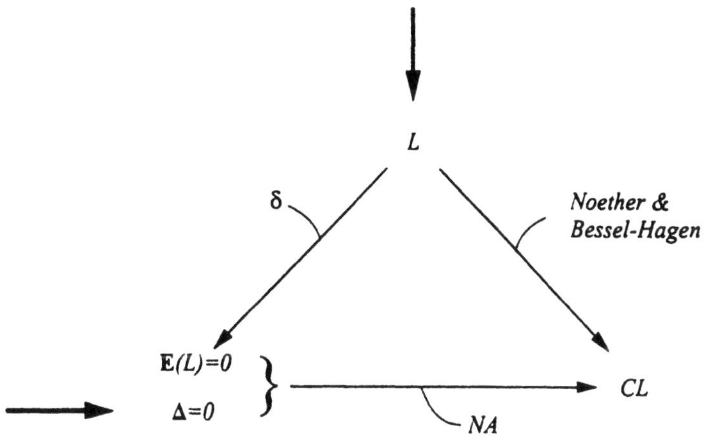

Fig. 3.2 Flow diagram pertaining to conservation laws

If a Lagrangian function L is available (the arrow to L from above is supposed to indicate that L has been constructed based on some external, most likely physical, considerations) then conservation laws CL can be established by the application of Noether's theorem. A broader class of such laws is obtainable if Bessel-Hagen's extension is used, as indicated by the arrow from L to CL.

On the other hand, based on L and applying the rules of the calculus of variations, symbolized by $\boldsymbol{\delta}$, the associated Euler-Lagrange differential equation indicated by $\mathbf{E}(L)=0$ can also be derived, made graphic by the arrow from L to $\mathbf{E}(L)=0$.

The governing differential equations of a problem may, however, be not the result of a variational procedure (in the absence of L), but could have been formulated on the basis of some balance laws and some constitutive relations describing material behaviour. This will occur for most dissipative systems and the corresponding equations are symbolized by $\boldsymbol{\Delta}=0$, the arrow *to* this equation indicating their establishment based on outside considerations.

The arrow leading from the corner of differential equations (either $\mathbf{E}(L)=0$ or $\boldsymbol{\Delta}=0$) to the CL corner and labelled NA symbolizes the fact, that the Neutral Action method can be applied equally well to $\mathbf{E}(L)=0$ or $\boldsymbol{\Delta}=0$ and leads to conservation laws. If it is applied to $\mathbf{E}(L)=0$, then one obtains the same conservation laws as if Noether's method, but together with the Bessel-Hagen extension, were applied to the corresponding Lagrangian L.

It should be reemphasized in this context that for equations of the type $\mathbf{\Delta} = 0$, no procedure existed for a systematic construction of conservation laws, until the NA method was advanced. What was done was to concoct, by hook or by crook, an artificial L (cf., e. g., the "mirror image" system used in Morse & Feshbach, 1953) and then apply Noether's theorem, with or without the Bessel-Hagen extension.

But, as it turns out, even if a "natural" L exists and thus $\mathbf{E}(L) = 0$ is available, the NA method applied to $\mathbf{E}(L) = 0$ leads to conservation laws in a more straight-forward manner than application of Noether's theorem with the extension of Bessel-Hagen. In the former (NA) method, as already mentioned, one deals with only one set of functions to be determined, while in the latter, three sets of functions are to be found, leading to identical results.

4 Conservation Laws for Homogeneous and Inhomogeneous Plane Elastostatics

4.1 General Remarks

The material presented in this Section should be useful whenever the scale of a mechanics problem dictates the modeling of materials by smoothly varying parameters. Some soil mechanics and geophysics applications offer an example of such a situation. It is also recognized that, due to material diffusion, the elastic moduli near the interface of two bonded materials are sometimes better modelled by smooth, yet rapidly changing functions of the spatial coordinates. Furthermore, "functionally graded materials" have recently acquired prominence.

In what follows, we shall not aim at presenting many possible (and probably relevant) aspects of conservation laws for inhomogeneous materials, but rather concentrate on establishing a path-independent J integral for several classes of inhomogeneity for plane elastostatics. This might be useful in evaluating energy-release rates for plane cracks in materials whose properties change in the direction of possible crack advance.

We begin by recalling that in terms of a Lagrangian function (2.32) the conserved currents (or fluxes) P_i leading to the conservation law $P_{i,i} = 0$ are expressed quite generally in equation (2.38) where the functions ξ_j and φ_a, which describe the infinitesimal transformation, were introduced in equation (2.33).

In plane elastostatics the dependent variables $v_\alpha = u_\alpha$ are the components of displacement in Cartesian coordinates and $\alpha = 1, 2$. The independent variables x_j are the material coordinates and $j = 1, 2$ also. Therefore, it is not necessary to distinguish between greek and latin indices in the following.

The explicit expression for L (i. e., the negative of the strain energy density W) in plane elastostatics, for plane strain is

$$W = \frac{\lambda}{2} \varepsilon_{ii} \, \varepsilon_{jj} + \mu \, \varepsilon_{ij} \, \varepsilon_{ij} \tag{4.1}$$

with the strain $\varepsilon_{ij} = \left(u_{j,i} + u_{i,j}\right)/2$, or, in terms of derivatives of u_i

$$W = \frac{1}{2}(\lambda + 2\mu)\left(u_{1,1}^2 + u_{2,2}^2\right) + \frac{1}{2}\mu \left(u_{1,2}^2 + u_{2,1}^2\right) + \lambda u_{1,1} \, u_{2,2} + \mu u_{1,2} \, u_{2,1}, \tag{4.2}$$

with λ and μ are Lamé's constants. For plane stress, λ may be replaced by

$$\lambda^* = \lambda(1 - 2v)/(1 - v)$$

with Poisson's ratio v.

We wish to be concerned with a smoothly inhomogeneous body whose properties can be described by

$$\lambda = \lambda(x_1, x_2), \tag{4.3a}$$

$$\mu = \mu(x_1, x_2), \tag{4.3b}$$

and consider the homogeneous body as a special case $\lambda = const$, $\mu = const$.

The form of P_i given by (2.38), together with the requirement that $P_{i,i} = 0$, leads to an overdetermined system of linear partial differential equations for ξ_i and φ_i.

The general solution of this system is easily obtained and is given by (Honein and Herrmann, 1997)

$$\begin{aligned} \varphi_1 &= Ax_2 + Bu_2 + Cu_1 + A_0, \\ \varphi_2 &= -Ax_1 - Bu_1 + Cu_2 + B_0, \\ \xi_1 &= Bx_2 + Dx_1 + C_0, \\ \xi_2 &= Bx_1 + Dx_2 + D_0, \end{aligned} \tag{4.4}$$

where the arbitrary constants $A, A_0, B, B_0, C, C_0, D$ and D_0 are related to Lamé's moduli λ and μ through the equations

$$
\begin{aligned}
-2C &= \left(Bx_2 + Dx_1 + C_0\right)\mu_{,1}/\mu + \left(-Bx_1 + Dx_2 + D_0\right)\mu_{,2}/\mu \\
&= \left(Bx_2 + Dx_1 + C_0\right)\lambda_{,1}/\lambda + \left(-Bx_1 + Dx_2 + D_0\right)\lambda_{,2}/\lambda .
\end{aligned} \tag{4.5}
$$

Having solved for ξ_i and φ_i, we can immediately obtain conservation laws by substituting equations (4.4) and (4.2) into (2.38), provided (4.5) are satisfied. Thus, it is seen that equations (4.5) may be regarded as a condition that restricts the class of inhomogeneous materials admitting conservation laws under the special version of Noether's theorem considered there. For every material in this class, the number of conservation laws corresponds to the number of independent parameters $A, A_0, \ldots, D_0$. We note that the constants A, A_0 and B_0 do not appear in equations (4.5) and that they correspond to transformations (variations) of the dependent variables only. Conservation laws arising from such variations $(\xi_i = 0)$ refer to conservation laws in physical space, in contradistinction to those obtained when $\xi_i \neq 0$ and which are termed here conservation laws in material space.

Indeed, one may verify that, in this case, the conservation law corresponding to A_0 (translations along the u_1-axis) is nothing but the first equilibrium equation (conservation of linear momentum in the x_1-direction). Similarly, the conservation law associated with B_0 (translation along the u_2-axis) is the second equilibrium equation (conservation of linear momentum in the x_2-direction). The conservation of angular momentum is associated with the parameter A.

In the sequel, (by putting $A = A_0 = B_0 = 0$) we will concentrate exclusively on conservation laws in material space. We shall start by recovering known results for a homogeneous body.

4.2 The Homogeneous Case

By equations (4.5), C must vanish, while B, D, C_0 and D_0 may be chosen arbitrarily. Upon setting in succession these last constants, except one of them, equal to zero, one can show, by substituting (4.4) and (4.2) into (2.38) and applying the divergence theorem, that the resulting conservation laws lead to

(1) The L integral for $B \neq 0$, $\quad L = \oint_\Gamma \varepsilon_{3kj}\left(x_k b_{ij} + u_k \sigma_{ij}\right) n_i ds$,

(2) The M integral for $D \neq 0$, $\quad M = \oint_\Gamma x_j b_{ij} n_i ds$,

(3) The J_1 integral for $C_0 \neq 0$, $\quad J_1 = \oint_\Gamma b_{i1} n_i ds$,

(4) The J_2 integral for $D_0 \neq 0$, $J_2 = \oint_\Gamma b_{i2} n_i ds$,

with the Eshelby tensor b_{ij} introduced in (2.43) and the permutation tensor ε_{ikj}.

These results agree with those obtained previously by Günther (1962) and Knowles and Sternberg (1972) and are derived in detail, fore example, in Kienzler and Herrmann (2000).

4.3 The Inhomogeneous Case

The general solution of equations (4.5) can easily be effected by means of the method of characteristics. This, however, will not be discussed in detail, for we are not looking for completeness at this point. Rather, our aim is to examine some situations that allow an extension of the important J-integral to inhomogeneous materials; therefore, symmetries related to translation, rather than rotation, will be emphasized.

Thus we let $B = 0$, and also $D_0 = 0$ since it is associated with the translation in the x_2-direction and not in the x_1-direction. Subsequently, we examine two cases:

a: $D = 0$

In this instance, equations (4.5) are equivalent to

$$\frac{\lambda_{,1}}{\lambda} = \frac{\mu_{,1}}{\mu} \tag{4.6}$$

and

$$-2C = C_0 \, \mu_{,1} / \mu , \tag{4.7}$$

which requires, as can be verified, that Poisson's ratio v be independent of x_1 and the shear modulus μ be an exponential function in x_1, i. e.,

$$\mu(x_1, x_2) = \mu_0(x_2) e^{\alpha x_1} , \tag{4.8}$$

where μ_0 is an arbitrary function of x_2 and α is an arbitrary constant. To arrive at physically meaningful results, however, these must be chosen in a manner which insures that $\mu > 0$ in the region of interest.

Now, the components of $\boldsymbol{P}$ can readily be calculated from (2.38). And it is not difficult to show, by applying the divergence theorem, that the ensuing path-independent integral is given by

$$J_e = \oint_\Gamma \left[\left(Wn_1 - u_{k,1}\, t_k \right) - \frac{1}{2} \alpha u_k\, t_k \right] ds\,, \tag{4.9}$$

where $t_k = \sigma_{jk}\, n_j$ are the components of tractions along the boundary Γ of a domain S and n_1 is the first component of the unit outward normal to Γ.

This result remains valid for plane stress (with a proper adjustment of the strain energy density, as discussed above), and it reduces to Rice's J integral for a homogeneous body, Rice (1968), as can easily be seen upon setting $\alpha = 0$. On the other hand, one may verify, by direct calculation, that the obvious three-dimensional version of the integrand of (4.9) remains divergence-free, provided that Poisson's ratio is independent of x_1 and the shear modulus satisfies

$$\mu\left(x_1, x_2, x_3\right) = \mu_0\left(x_2, x_3\right) e^{\alpha x_1}\,, \tag{4.10}$$

where μ_0 is an arbitrary function of the coordinates x_2, x_3 and α is an arbitrary constant.

b: $D \neq 0$, *but* μ *and* λ *are indpenent of* x_2

In this case, equations (4.5) can be readily integrated and the result requires that Poisson's ratio be constant and the shear modulus have the form

$$\mu = c\left(x_1 + C_0 / D\right)^{-2C/D}\,, \tag{4.11}$$

which represents a power-law dependence, c being an arbitrary constant.

For later convenience, we let

$$\begin{aligned} -2C/D &= \alpha\,, \\ C_0 &= 1\,, \\ D &= m\,. \end{aligned} \tag{4.12}$$

Now, it is readily shown that the resulting path-independent integral emanating from the associated conservation law is given by

$$J_e = \oint_\Gamma \left[Wn_1 - t_k u_{k,1} + m\left(Wx_k n_k - t_k x_i u_{,i} - \frac{1}{2} \alpha u_k t_k \right) \right] ds\,, \tag{4.13}$$

which, in terms of the standard J and M integrals, and

$$U = \frac{1}{2} \oint_{\Gamma} u_k \, t_k \, ds \,, \tag{4.14}$$

can be written as

$$J_e = J + m\left(M - \alpha U\right), \tag{4.15}$$

where we recall that M is defined in two dimensions by

$$M = \oint_{\Gamma} \left(W x_k n_k - t_k x_i u_{k,i}\right) ds \,. \tag{4.16}$$

The shear modulus μ may, in this case, be expressed as

$$\mu = \mu_0 \left(m x_1 + 1\right)^{\alpha} \tag{4.17}$$

and must be subjected to the condition $\mu > 0$ in the region of interest, μ_0 being a constant.

It is interesting to note that the homogeneous case is now recovered not by setting $\alpha = 0$, which leads to a linear combination of the J and M integrals, but rather by putting $m = 0$.

Under the same restrictions on the material parameters, one may verify, on one hand, that the result (4.15) holds true under plane stress conditions and, on the other hand, that the obvious three-dimensional version of the integrand of (4.15) remains divergence-free, provided M is replaced by its three-dimensional analogue given by

$$M = \oint_{\Gamma} \left(W x_k n_k - t_k x_i u_{k,i} - \frac{1}{2} u_k t_k \right) ds \,. \tag{4.18}$$

4.4 Relation to Stress-Intensity Factors

One of the major concerns pertaining to linear elastic fracture mechanics analysis is the accurate prediction of stress-intensity factors at crack tips. In this endeavour, the standard J integral has proved to be an extremely useful tool. Under combined opening (mode I) and sliding (mode II) this integral, around a right crack tip, is related to the stress-intensity factors K_I for mode I and K_{II} for mode II, by the following relation, cf. Rice (1968)

$$J = \frac{1}{E^*}\left(K_I^2 + K_{II}^2\right), \tag{4.19}$$

For plane stress E^* is equal to Young's modulus E, for plane strain, E^* equals $E/(1-\upsilon^2)$.

In this Sub-Section, we will prove that the same relation holds true for the J_e integrals derived in the previous Sub-Section under the conditions of their respective validity.

The literature concerned with analytical determination of stress-intensity factors for inhomogeneous materials with smoothly varying elastic moduli is quite sparse. The work by Delale and Erdogan (1983) appears to be the first to consider the problem of mode I deformation in the presence of a modulus variation in the direction parallel to the crack line. Young's modulus was assumed to vary exponentially while Poisson's ratio was constant. In that paper, the authors stated: "Even though no systematic study of the problem appears to have been made, it is reasonable to expect that in nonhomogeneous materials with continuous and generally differentiable elastic moduli the nature of the stress singularity at a crack tip would be identical to that of a homogeneous solid." A similar statement was also made by Erdogan (1983).

The Delale-Erdogan conjecture was proved to be true by Eischen (1987). Using an eigenfunction-expansion technique similar to that of Williams (1957), Eischen considered a constant Poisson's ratio and a general functional form of the Young's modulus variation, and was able to show that an $r^{-1/2}$ stress and strain singularity exists at the crack tip, r being the radial distance measured from that tip. Moreover, the angular variation of the singular stress field and the associate displacements round a crack tip in an inhomogeneous material were shown to be exactly the same as the angular variation in a homogeneous material. Eischen noted also that these conclusions are not altered if Poisson's ratio is allowed to vary smoothly.

Armed with these results, it is now a simple matter to relate the J_e integrals to stress-intensity factors. Taking the path of integration to be a small circular loop of radius ε centered at the crack tip and noting that the stress and displacement fields behave there, respectively, as $0(r^{-1/2})$ and $0(r^{+1/2})$, it is easily seen that the contribution of the last term of the integrand of both (4.9) and (4.13) vanishes as ε tends to zero.

Also, by taking the origin to coincide with the crack tip in Case b, the contribution due to the M integral under the same limiting procedure is also seen to vanish. Thus, in the limit, we are left only with the terms that form the standard J integral and we obtain the following result

$$J_e = \frac{1}{E^*}\left(K_I^2 + K_{II}^2\right), \tag{4.20}$$

which follows by recalling that the functional form of the local elastic field at the crack tip is unaffected by the inhomogeneity.

Note that, relations (4.20), E^* is evaluated taking the crack-tip values of Young's modulus and Poisson's ratio.

Being truly path-independent integrals in their respective region of validity, the extended J_e integral derived here can be used directly in numerical calculations, such as the finite element method. Stress-intensity factors for graded materials can be calculated accurately based on remote fields, rather than on less accurate numerical crack-tip fields. The details are not different, in principle, from those for calculating the J integral for homogeneous materials.

4.5 Examples

In this Sub-Section, we will illustrate the use of the J_e integrals derived in Sub-Section 4.3 by applying them to two simple examples. Both of these involve the double cantilever beam shown in Figure 4.1.

Each of the arms will be modelled as a built-in beam of unit width bent either by a force F_0 or a couple M_0 applied at the end. We recall that the elementary beam equations may be written for the coordinate system depicted in Fig. 4.1 as

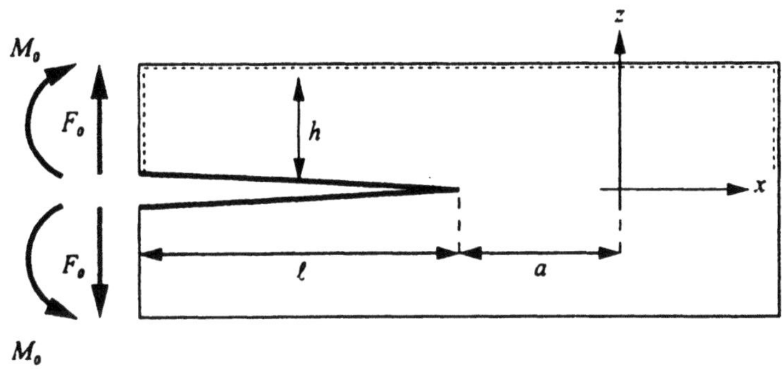

Fig. 4.1 Double cantilever beam

$$\frac{dQ}{dx} + q = 0,$$

$$\frac{dM}{dx} - Q = 0,$$

$$\frac{d^2 w}{dx^2} = -\frac{M}{E(x)\,I},$$

$$\frac{d^2}{dx^2}\left(E(x)\,I\frac{d^2 w}{dx^2}\right) = q\,, \tag{4.21}$$

where, with appropriate sign convention, q is the lateral load intensity, Q the shear force, M the bending moment, w the deflection and I the area moment of intertia. The dependence of Young's modulus E on the coordinate x has been noted explicitly, for we will be concerned with inhomogeneous materials.

The governing equations must be supplemented by boundary conditions. At the crack tip, $x = -a$, we have

$$w\Big|_{x=-a} = 0 \tag{4.22}$$

$$\frac{dw}{dx}\Big|_{x=-a} = w'\Big|_{x=-a} = 0\,. \tag{4.23}$$

The boundary conditions at the left end, $x = -a - \ell$, will be specified shortly.

Example 1

As our first example, we will assume that the specimen is subjected to the forces F_0 only. In this case the boundary conditions at the end of the upper arm read

$$\frac{d^2 w}{dx^2}\Big|_{end} = 0\,, \tag{4.24}$$

$$\frac{d}{dx}\left(E(x)\,I\frac{d^2 w}{dx^2}\right)\Big|_{end} = F_0\,. \tag{4.25}$$

We assume further that Young's modulus is an exponential function in x and we write

$$E(x) = E_0\, e^{\beta x}\,, \tag{4.26}$$

where β is a real constant and E_0 is a positive number.

Under these conditions, the integration of the field equations for the upper arm is easily carried out and leads to the following expressions for the deflection and its derivative at the end $x = -a - \ell$

$$w\Big|_{end} = -\frac{F_0\, e^{\beta a}}{\beta^3\, E_0 I}\left(2 - 2e^{\beta \ell} + 2\beta\ell + \beta^2\ell^2\right), \tag{4.27}$$

$$w'\Big|_{end} = -\frac{F_0\, e^{\beta a}}{\beta^2\, E_0 I}\left(1 - e^{\beta \ell} + \beta\ell\right). \tag{4.28}$$

Now, we can evaluate the J_e integral given by equation (4.9) along the contour surrounding the whole specimen. Due to symmetry, the calculation needs to be performed only for half of this contour, which is depicted by the dotted curve in Figure 4.1. In view of the kinematic assumptions of the elementary beam theory, the strain energy density vanishes at the left end and the following result is obtained

$$\frac{1}{2} J_e = -\left(w'\Big|_{end} + \frac{1}{2}\beta w\Big|_{end}\right) F_0, \tag{4.29}$$

which, on substituting equations (4.27) and (4.28), results in

$$J_e = \frac{\ell^2 F_0^2}{E(-a) I}. \tag{4.30}$$

The stress-intensity factor K_I is now calculated, by applying formula (4.20), to be

$$K_I = \frac{\ell F_0}{\sqrt{I}}. \tag{4.31}$$

We note that K_I has the same dependence on the load as in the homogeneous case. However, for a given end deflection, K_I depends on β, a, etc. As mentioned before, a beam of unit thickness is considered. For an arbitrary thickness, b say, J_e in (4.30) is to be divided by b, and K_I (4.31) must be multiplied by $1/\sqrt{b}$ (cf. Kienzler, 1993).

Example 2

For our second example, we assume that the beam is loaded by couples only. Thus, the boundary conditions at the end of the upper arm read

$$E(x)\,I\left.\frac{d^2w}{dx^2}\right|_{end}=M_0\,, \tag{4.32}$$

$$\left.\frac{d}{dx}\left(E(x)\,I\frac{d^2w}{dx^2}\right)\right|_{end}=0\,. \tag{4.33}$$

In anticipation of applying equation (4.13), we choose, by letting $a=0$, the origin to coincide with the crack tip, and we assume that Young's modulus varies according to

$$E(x)=\frac{E_0}{mx+1}\,, \tag{4.34}$$

where E_0 and m are arbitrary constants subjected to the requirement that E be positive in the material body.

The field equations are now easily integrated and yield the following results for the upper arm:

$$w=\frac{M_0}{E_0\,I}\left(m\frac{x^3}{6}+\frac{x^2}{2}\right), \tag{4.35}$$

$$\frac{dw}{dx}=\frac{M_0}{E_0\,I}\left(m\frac{x^2}{2}+x\right), \tag{4.36}$$

$$\frac{d^2w}{dx^2}=\frac{M_0}{E_0\,I}(mx+1)\,. \tag{4.37}$$

The longitudinal displacement u, which vanishes along the central line $z=h/2$, may be calculated by requiring that the shear strain ε_{xz} vanishes and may be written as

$$u=-(z-h/2)\frac{dw}{dx}\,. \tag{4.38}$$

The only nonvanishing component of the stress field is σ_{xx}, which is given by

$$\sigma_{xx}=-\frac{M_0}{I}(z-h/2)\,. \tag{4.39}$$

Having these formulae, an evaluation of the J_e integral given by equation (4.13) may be carried out to yield the following result:

i) along a path $z = h, -\ell \leq x \leq 0: \dfrac{M_0^2}{2E_0 I}\left(-\dfrac{3}{2}m^2\ell^2 + 3m\ell\right)$,

ii) along the path $x = -\ell, 0 \leq z \leq h: \dfrac{M_0^2}{2E_0 I}\left(\dfrac{3}{2}m^2\ell^2 - 3m\ell + 1\right)$.

Due to symmetry, the sum of these two expressions results in half the contribution to the J_e-integral, which is then given by

$$J_e = \frac{M_0^2}{E_0 I}, \tag{4.40}$$

leading to the following equation for the stress-intensity factor

$$K_I = \frac{M_0}{\sqrt{I}}. \tag{4.41}$$

Again K_I has the same dependence on the cross-sectional moment as in the homogeneous case. For arbitrary thickness, b say, J_e and K_I have to be multiplied by $1/b$ and $1/\sqrt{b}$, respectively, as in the previous example.

In both of these examples, a simple calculation leads to the energy-release rate $\mathcal{G}$ of the whole specimen per unit crack extension. In the first example, $\mathcal{G}$ is equal to J_e, whereas in the second example, $\mathcal{G}$ turns out to be

$$\mathcal{G} = \frac{M_0^2}{E_0 I}(1 - m\ell), \tag{4.42}$$

and does not coincide with J_e (equation (4.40)), as would be the case for a homogeneous body $(m = 0)$.

5 Conservation Laws in Elastodynamics

5.1 General Remarks

The homogeneous force equilibrium equations of elastostatics

$$\sigma_{ji,j} = 0$$

are extended to elastodynamics by addition of the inertia term $\rho u_i^{\bullet\bullet}$

$$\sigma_{ji,j} = \rho u_i^{\bullet\bullet} , \tag{5.1}$$

where σ_{ji} is the stress tensor, (2.42), ρ is the mass density and u_i, as before, the displacement vector. Dots indicate differentiation with respect to time.

In contradistinction to elastostatics, the physical stress σ_{ji} is now no longer divergence-free, i. e., we now have to deal with a source term in the stress formulation and an additional independent variable, the time t.

It appears that there exist three different possibilities in dealing with the inertia term of elastodynamics in the context of conservation laws.

1. The number of independent variables is increased by one, namely the time t, and conservation laws have now the form

$$P_{i,i} = 0 , \qquad \text{where} \qquad i = x_1, x_2, x_3 \text{ and } t \tag{5.2}$$

   and we find ourselves in the $(x_1, x_2, x_3\ t)$-domain.

2. One takes care of the time-dependence by means of an appropriate convolution.

3. The inertia term is dealt with as a true source term and the conservation laws of elastostatics become now balance laws. In global (i. e., integrated) form this implies that the path integrals of elastostatics have to be augmented by domain integrals, which together will be domain independent.

In this Section, we shall present only the first possibility, together with illustrative examples. The other two are discussed in Kienzler and Herrmann (2000).

5.2 Example: Elastic Bar

We discuss this procedure first with the aid of a spatially one-dimensional example of an elastic bar (Herrmann, 1980) of cross-section A, mass density ρ, mass per unit of length $\overline{m} = \rho A$, Young's Modulus E and axial stiffness EA.

The Lagrangian density (kinetic potential) per unit of length is given by

$$L = T - W\,, \tag{5.3}$$

where the kinetic energy T and the potential (strain) energy W are given as

$$T = \frac{1}{2}\overline{m}u^{\bullet 2} = \frac{1}{2}\overline{m}u_{,t}^2\,, \tag{5.4 a}$$

$$W = \frac{1}{2}EAu'^2 = \frac{1}{2}EAu_{,x}^2\,. \tag{5.4 b}$$

The Lagrangian L is obviously a function of u' and $u^{\bullet}$ and may depend explicitly on x and t if EA and $\overline{m}$ are not constant in space and time

$$L = L\left(x, t, u', u^{\bullet}\right). \tag{5.5}$$

In order to apply the Euler-Lagrange equation (2.39)

$$\mathbf{E}_{\beta}(L) = \frac{\partial L}{\partial v_{\beta}} - \frac{d}{dx_i}\left(\frac{\partial L}{\partial v_{\beta,i}}\right)$$

we have to identify the following quantities, with $\beta = 1$ (subscript omitted below) and $i = x, t$

$$v_{\beta} \rightarrow u\,, \tag{5.6 a}$$

$$\frac{d(\;)}{dx_1} \rightarrow \frac{d(\;)}{dx} = (\;)'\,, \tag{5.6 b}$$

$$\frac{d(\;)}{dx_2} \rightarrow \frac{d(\;)}{dt} = (\;)^{\bullet}\,, \tag{5.6 c}$$

$$v_{\beta,1} \rightarrow u'\,, \tag{5.6 d}$$

$$v_{\beta,2} \to u^{\bullet} , \tag{5.6 e}$$

$$\frac{\partial L}{\partial v_{\beta,1}} \to \frac{\partial L}{\partial u'} = -EAu' = -N , \tag{5.6 f}$$

$$\frac{\partial L}{\partial v_{\beta,2}} \to \frac{\partial L}{\partial u^{\bullet}} = \overline{m} u^{\bullet} = p . \tag{5.6 g}$$

Here $N = EAu'$ is the axial force and $p = \overline{m} u^{\bullet}$ is called the canonical momentum density (Morse and Feshbach, 1953) and corresponds to the momentum of a bar of unit length moving with velocity $u^{\bullet}$. The Euler-Lagrange equation (2.39) is rewritten as

$$\frac{d}{dt}\left(\frac{\partial L}{\partial u^{\bullet}}\right) + \frac{d}{dx}\left(\frac{\partial L}{\partial u'}\right) = 0 \tag{5.7}$$

and leads ($\overline{m}$ and EA not being functions of x and t) to the equation of motion

$$\overline{m} u^{\bullet\bullet} = EAu'' \tag{5.8}$$

or after division by $\overline{m}$, to the wave equation

$$u^{\bullet\bullet} - \alpha^2 u'' = 0 \qquad \text{with} \qquad \alpha = \sqrt{\frac{E}{\rho}} = \sqrt{\frac{EA}{\overline{m}}} , \tag{5.9}$$

α being the phase velocity.

We wish now to calculate the components of the material momentum tensor, here truly the elastic energy momentum tensor, because one of the independent variables is the time.

From the general equation for the current P_i (2.38),

$$P_i = \varphi_\beta \frac{\partial L}{\partial v_{\beta,i}} + \xi_j \left(L\delta_{ij} - v_{\beta,j} \frac{\partial L}{\partial v_{\beta,i}} \right)$$

we know that $-\partial L / \partial v_{\beta,i}$ must supply the components of physical momentum and $L\delta_{ij} - v_{\beta,j}\, \partial L / \partial v_{\beta,i}$ the negative components of material momentum. We already identified the components of physical momentum. Here it is a vector with components N and $-p$. We proceed now to calculate the components of material momentum (2.43)

$$b_{ij} = u_{,j} \frac{\partial L}{\partial u_{,i}} - L\delta_{ij}$$

with L given by

$$L = \frac{1}{2}\left(\overline{m}u^{\bullet 2} - EAu'^2\right). \tag{5.10}$$

The result is

$$b_{xx}: \quad b_{11} = u' \frac{\partial L}{\partial u'} - L = -\frac{1}{2}\left(\overline{m}u^{\bullet 2} + EAu'^2\right), \tag{5.11 a}$$

$$b_{xt}: \quad b_{12} = -Nu^{\bullet} = -EAu'u^{\bullet}, \tag{5.11 b}$$

$$b_{tx}: \quad b_{21} = pu' = \overline{m}u^{\bullet}u', \tag{5.11 c}$$

$$b_{tt}: \quad b_{22} = u^{\bullet} \frac{\partial L}{\partial u^{\bullet}} - L = \frac{1}{2}\left(\overline{m}u^{\bullet 2} + EAu'^2\right). \tag{5.11 d}$$

The quantity b_{xx} is the material force per unit of length, also called *wave stress* (cf. Morse and Feshbach, 1953), whereas b_{xt} is the stress rate of work or *rate of energy flow* along the bar, b_{tx} is called the *wave momentum* and finally, b_{tt} is the total energy per unit of length $H = T + W$ or the *Hamiltonian* per unit of length. In this one-dimensioanl example the material force b_{xx} happens to coincide with the expression for the total energy b_{tt} except for a minus sign; on the other hand, b_{xt} and b_{tx} differ merely by the factor $-\alpha^2$.

Equation (5.2) yields, with the implications $\varphi_\beta = \varphi = 0$ and $\xi_j = a_j =$ const.

$$b'_{xx} + b^{\bullet}_{tx} = 0 \qquad (a_1 = 1, a_2 = 0), \tag{5.12 a}$$

$$b'_{xt} + b^{\bullet}_{tt} = 0 \qquad (a_1 = 0, a_2 = 1), \tag{5.12 b}$$

or after substitution (Herrmann, 1980)

$$-\frac{1}{2}\left(Nu' + pu^{\bullet}\right)' + \left(pu'\right)^{\bullet} = 0, \tag{5.13 a}$$

$$-\left(Nu^{\bullet}\right)' + \frac{1}{2}\left(Nu' + pu^{\bullet}\right)^{\bullet} = 0. \tag{5.13 b}$$

The first equation above states that the spatial rate of change of the material force b_{xx} is balanced by the time rate of change of wave momentum b_{tx}. The second equation states that the spatial rate at which stresses do work in the velocity $u^{\bullet}$, b_{xt}, is balanced by the time rate of change of the internal energy density b_{tt}, or in other words, (5.13 b) corresponds to the equation of continuity for energy flow along the bar, or the principle of energy conservation (first law of thermodynamics).

The integral (global) form of $b_{ji,j}=0$ leads to the line integral

$$\oint_{\Gamma} b_{ji}\, n_j\, ds = 0\,.$$

To illustrate this form, let us consider a bar of length ℓ and a rectangular circuit in the $x-t$ plane sketched in Figure 5.1, where t^* is some arbitrarily fixed time.

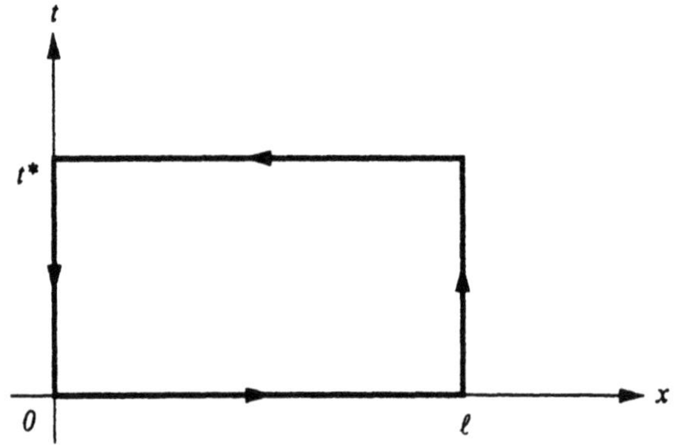

Fig. 5.1 Integration contour in the space-time domain

The closed line integral can then be evaluated as follows:

$$i=x:\quad \int_0^{\ell} pu'\,dx\Big|_{t=t^*} = \int_0^{\ell} pu'\,dx\Big|_{t=0} + \frac{1}{2}\int_0^{t^*}\left(pu^{\bullet}+Nu'\right)dt\Big|_{x=0}^{x=\ell}\,, \tag{5.14 a}$$

$$i=t:\quad \frac{1}{2}\int_0^{\ell}\left(pu^{\bullet}+Nu'\right)dx\Big|_{t=t^*} = \frac{1}{2}\int_0^{\ell}\left(pu^{\bullet}+Nu'\right)dx\Big|_{t=0} + \int_0^{t^*} Nu^{\bullet}\,dt\Big|_{x=0}^{x=\ell}\,. \tag{5.14 b}$$

The first term on the right-hand side of each equation above can be evaluated on the basis of prescribed initial conditions. Similarly, the second terms on the right-hand side can be evaluated on the basis of prescribed boundary conditions.

In the second equation (5.14 b) it can be ascertained that the second integral of the right-hand side vanishes, since either N (free displacement) or $u^{\bullet}$ (fixed displacement) are zero at the ends of a bar. This leads to the conclusion that the total energy of the system at time zero is equal to that at any arbitrary time t^*, i. e., the Hamiltonian is a constant of the motion, independent of time

$$\frac{d}{dt}\int_0^{\ell}\left(pu^{\bullet}+Nu'\right)dx=0\,, \tag{5.15}$$

or, in other words, the total energy is conserved.

By means of modal analysis it can be shown that the wave momentum $p=\overline{m}u^{\bullet}$ and the displacement gradient u' are orthogonal functions. Therefore, the first two integrals in (5.14 a) vanish, leading to the conclusion that, due to t^* being arbitrary, the material forces per unit of length b_{xx} at the two boundaries are equal

$$b_{xx}\Big|_{x=0}=b_{xx}\Big|_{x=\ell}$$

provided the bar is homogeneous. In other words, the material forces acting across the ends of the bar are in equilibrium with each other, because the resultant of the distributed material forces along the bar vanishes.

Having the conservation laws already established, it might be intriguing to determine, in a backward manner, the characteristic f employing the NA method. We know from Section 3 that the conservation laws are equal to the product of the equation of motion (5.8)

$$\Delta=\overline{m}u^{\bullet\bullet}-EAu''$$

multiplied by the characteristic f, i. e., an integrating factor. Replacing N and p by EAu' and, $\overline{m}u^{\bullet}$ respectively, it follows from (5.12) and (5.13)

$$b'_{xx}+b^{\bullet}_{tx}=-EAu'u''-\overline{m}u^{\bullet}u'^{\bullet}+\overline{m}u^{\bullet\bullet}u'+\overline{m}u^{\bullet}u'^{\bullet}=u'\left(\overline{m}u^{\bullet\bullet}-EAu''\right)=u'\Delta\,, \tag{5.16 a}$$

$$b'_{xt}+b^{\bullet}_{tt}=-EAu''u^{\bullet}-EAu'u'^{\bullet}+EAu'u'^{\bullet}+\overline{m}u^{\bullet}u^{\bullet\bullet}=u^{\bullet}\left(\overline{m}u^{\bullet\bullet}-EAu''\right)=u^{\bullet}\Delta\,. \tag{5.16 b}$$

Hence, the characteristic f is found as a sum of the displacement gradients in space and time

$$f=a_j u_{.j}=a_1u'+a_2u^{\bullet}\,, \tag{5.17}$$

such that the relation

$$P_{i,i} = -\left(\xi_j b_{ij}\right)_{,i} = -a_j b_{ij,i} = -a_j u_{,j}\left(\overline{m}u^{\bullet\bullet} - EAu''\right) = -f\Delta \qquad (5.18)$$

can be stated immediately.

It should be added that the characteristic Q (cf. equation 2.37) within the context of Neother's approach is given as

$$Q = \varphi - \xi_i u_{,i} = 0 - a_1 u' - a_2 u^{\bullet} \qquad (5.19)$$

and, as before, $Q = -f$, see equation (3.1) and (3.3).

Let us recall that the local conservation laws above are related to the translation in material space $\xi_j = a_j =$ const. and $\varphi_\beta = \varphi = 0$. The corresponding conservation law related to translation in physical space. $\varphi = a =$ const. and $\xi_j = 0$ is the equation of motion (5.8). Of course, further conservation laws exist and are discussed in Kienzler and Herrmann (2000).

5.3 The Energy-Momentum Tensor

Returning to the general domain (x_1, x_2, x_3, t) it might be instructive to represent the complete energy momentum tensor in the form of a 4 x 4 matrix (cf. Morse and Feshbach, 1953)

	x_1	x_2	x_3	t
x_1	b_{11}	b_{12}	b_{13}	b_{1t}
x_2	b_{21}	b_{22}	b_{23}	b_{2t}
x_3	b_{31}	b_{32}	b_{33}	b_{3t}
t	b_{t1}	b_{t2}	b_{t3}	b_{tt} .

(5.20)

The b_{ij} above are the components of the Eshelby tensor ($i, j, \beta = 1, 2, 3$)

$$b_{ij} = +u_{\beta,j}\frac{\partial L}{\partial u_{\beta,i}} - L\delta_{ij} = \qquad (5.21\text{ a})$$

$$= -\left(L\delta_{ij} + \sigma_{ik}\, u_{k,j}\right) \qquad (5.21\text{ b})$$

discussed in detail in Kienzler and Herrmann (2000)

The vector $\boldsymbol{R}$ with components b_{t1}, b_{t2}, b_{t3} is called the field- (or wave-) momentum density and is defined by

$$R_i = b_{ti} = +\rho u_{j,i}\, \dot{u}_j \,. \qquad (5.22)$$

The vector $\boldsymbol{S}$ with components b_{1t}, b_{2t}, b_{3t} is called the field intensity defined by

$$S_i = b_{it} = -\dot{u}_j\, \sigma_{ij} \qquad (5.23)$$

and represents the work of stresses along the rates of displacements, or the components of the energy-flow vector.

The element b_{tt} in the matrix above represents the total energy density $H = T + W$, i. e., the Hamiltonian.

The complete matrix is then written as

	x_1	x_2	x_3	t
x_1	b_{11}	b_{12}	b_{13}	$-\sigma_{1j}\dot{u}_j$
x_2	b_{21}	b_{22}	b_{23}	$-\sigma_{2j}\dot{u}_j$
x_3	b_{31}	b_{32}	b_{33}	$-\sigma_{3j}\dot{u}_j$
t	$\rho u_{j,1}\dot{u}_j$	$\rho u_{j,2}\dot{u}_j$	$\rho u_{j,3}\dot{u}_j$	H .

(5.24)

To facilitate a direct comparison with the development in Morse and Feshbach (1953), it should be mentioned that their matrix on p. 321 is the transpose of the matrix given above. It also should be mentioned that the representation (5.24) is identical to (5.21 a), provided, the indices i and j range from 1 to 4 $(x_4 = t)$, while the range of β remains 1 to 3.

As discussed in Section 1 the terms $b_{ti}(i = 1, 2, 3\, t)$ are densities and the terms $b_{ji}(j = 1, 2, 3; i = 1, 2, 3, t)$ are fluxes and the conservation laws indicate that time rate of a density is balanced by the divergence of the associated fluxes. The first three conservation laws (no summation over time t)

$$b_{jk,j} + b_{tk,t} = 0 \qquad j, k = 1, 2, 3 \qquad (5.25\text{ a})$$

or with (5.22)

$$b_{jk,j} + \dot{R}_k = 0 \qquad (5.25\text{ b})$$

express the balance between the net material force and the time rate of change of the field-momentum density. The fourth equation

$$b_{jt,j} + b_{tt,t} = 0 \tag{5.26 a}$$

or with (5.23)

$$S_{j,j} + H^{\bullet} = 0 \tag{5.26 b}$$

expresses the balance between the divergence of the energy-flow vector and the time rate of change of the total energy, i. e., the principle of energy conservation.

6 Concluding Remarks

In this contribution, the role of conservation laws in Configurational Mechanics has been described. Having defined the mathematical expression of a conservation law, two methods of establishing such laws for a given system have been presented. One methodology, based on Noether's theorem, is applicable for systems possessing a Lagrangian function, the other, called the Neutral Action Method and developed only recently, is applicable especially if the system is mathematically given by a set of differential equations and a Lagrangian function cannot be constructed. Applications of these methodologies in plane elastostatics and elastodynamics were presented.

Several special topics and some aspects of the general area discussed here could not be mentioned. These concern, among some others, trivial conservation laws and more general transformations (or symmetries) such as the so-called contact symmetries and generalized (or Lie-Bäcklund) symmetries, leading to further conservation laws. More general forms of the Lagrangian function, i. e., dependence on higher derivatives of dependent with respect to independent variables, as well as Olver's multi-index notion were also omitted.

Further, no mention of coupled fields, such as piezoelectricity and thermoelasticity, was included.

Some of these topics are detailed and referenced in the text by Kienzler and Herrmann (2000). As a final remark it should be emphasized that a great variety of conservation laws which can be established mathematically may not possess a ready physical interpretion.

References

Bessel-Hagen, E. (1921). Über die Erhaltungssätze der Elektrodynamk. *Mathematische Annalen* 84: 258-276.

Bluman, G.W., and Kumai, S. (1989). *Symmetries and Differential Equations*. Applied Mathematical Sciences No. 81, New York: Springer.

Chien, N. (1992). *Conservation laws in non-homogeneous and dissipative mechanical systems*. Ph.D. Thesis , Stanford University, Stanford CA.

Chien, N., Honein, T., and Herrmann, G. (1993). Conservation laws for linear viscoelasticity. *Zeitschrift für angewandte Mathematik und Physik (ZAMP)* 44: 44-52.

Delale, F., and Erdogan, F. (1983). The crack problem for a nonhomogeneous plane. *Journal of Applied Mechanics* 50: 609-614.

Eischen, J.W. (1987). Fracture of nonhomogeneous materials. *International Journal of Fracture* 34: 3-22.

Erdogan, F. (1983). Stress intensity factors. *Journal of Applied Mechanics* 50: 992-1002.

Euler, L. (1744). *Methodus inveniendi lineas curvas maximi minimive proprietate gaudentes sive solutio problematis isoperimetrici latissimo sensu accepti*, Additamentum I: De curvis elasticis. Lausanne: Bousquet, M.M. and Socios.

Gelfand, I.M., and Fomin, S.V. (1963). *Calculus of Variations*. Englewood Cliffs, N.J.: Prentice-Hall.

Günther, W. (1962). Über einige Randintegrale der Elastomechanik. *Abhandlung der Braunschweiger wissenschaftlichen Gesellschaft* 14: 53-72.

Herrmann, G. (1980). Some applications of invariant variational principles in mechanics of solids. In: Nemat-Nasser, S., ed., *Variational Methods in the Mechanics of Solids*. Oxford: Pergamon. 145-150.

Honein, T., Chien, N., and Herrmann, G.(1991). On conservation laws for dissipative systems. *Physsics Letters* A 155: 223-224.

Honein, T., and Herrmann, G. (1997). Conservation laws in non-homogeneous plane elastostatics. *Journal of Mechanics and Physics of Solids* 45: 789-805.

Kienzler, R. (1993). *Konzepte der Bruchmechanik*. Braunschweig: Vieweg.

Kienzler, R., and Herrmann, G. (2000). *Mechanics in Material Space*. Berlin: Springer.

Knowles, J.K., and Sternberg, E. (1972). On a class of conservation laws in linearized and finite elastostatics. *Archive of Rational Mechanics and Analysis* 44: 187-211.

Morse, P.M., and Feshbach, H. (1953). *Methods of Theoretical Physics*. New York: McGraw-Hill.

Noether, E. (1918). Invariante Variationsprobleme. *Nachrichten der Königlichen Gesellschaft der Wissenschaften zu Göttingen, Mathematisch-Physikalische Klasse* 2: 235-257.

Olver, P.J. (1993). *Applications of Lie Groups to Differential Equations*. Graduate Texts in Mathematics, No 107. New York: Springer, 2nd. edition.

Rice, J.R. (1968). A path indepenent integral and the approximate analysis of strain concentration by notches and cracks. *Journal of Applied Mechanics* 35: 379-386.

Williams, M.L. (1957). On the stress distribution at the base of a stationary crack. *Journal of Applied Mechanics* 24: 109-144.

Elements of Field Theory in Inhomogeneous and Defective Materials

Gérard A. Maugin[1] and Carmine Trimarco[2]

[1] Laboratoire de Modélisation en Mécanique,
Université Pierre et Marie Curie, Paris, France.
[2] Dipartimento di Matematica Applicata « U.Dini »,
Università di Pisa, Italy.

1 Introduction

So-called *configurational forces* , also called *material forces* in modern continuum mechanics, and more generally *energetic driving forces* , are those « forces » which are associated by duality to the displacement or motion of whatever may be considered a *defect* in a continuum field theory. Conceptually simple examples of such « defects » are dislocations in ordered crystals, disclinations in liquid crystals, vortices in fluid mechanics, cracks and cavities in materials science, propagating fronts in phase-transition problems, shock waves in continuum mechanics, domain walls in solid-state science, and more generally all manifestations, smooth or abrupt, of changes in material properties. In such a framework, the material symmetry of the physical system is broken by the presence of a field singularity of a given dimensionality (point, line, surface, volume). Until very recently all these domains were studied separately but a general framework emerged essentially through the works of the authors and co-workers, basing initially on inclusive ideas of J.D.Eshelby (deceased 1985) - hence the coinage of *Eshelbian mechanics* by the authors for the mechanics of such forces. In this framework which is developed in a somewhat synthetic form, all configurational forces appear as *forces of a non-Newtonian nature*, acting on the material manifold (the set of points building up the material whether discrete or continuous) and *not* in physical space which remains the realm of Newtonian forces and their more modern realizations which usually act per quantity of matter (mass or electric charge). That is, configurational forces act on spatial gradients of properties, on field singularities, *etc.* They acquire a true physical meaning only in so far as the associated expanded power is none other than a *dissipation* ; accordingly, configurational forces are essentially used to formulate *criteria of progress* of defects in accordance with the second law of thermodynamics. Within such a general vision, in fact, many irreversible properties of matter (e.g., damage, plasticity, magnetic hysteresis, phase transition, growth) are seen as *irreversible local rearrangements* of matter (material particles in an ordered crystal, spin layout in a ferromagnetic sample, director network in a liquid crystal) that are represented by pure material mappings. This is where some elements of modern differential geometry enter the picture following earlier works by Kröner, Noll, and others.

Having recognized the *material* (also called sometimes *Lagrangian* - as opposed to *Eulerian*) nature of configurational forces, the main progress was to identify (i) that the energetic driving forces on material inhomogeneities - whether *true inhomogeneities* like in recently developed *gradient-materials* and obviously in *composite materials*, or *quasi-inhomogeneities* , e.g., field singularities or gradients in the solution of non-mechanical fields, such as temperature in conductors of heat - are generated by a change in *particle* and not a change in the placement of the particle in physical space, (ii) that the accompanying *flux* is none other than the « energy-momentum » tensor introduced by J.D.Eshelby in defect mechanics in the early fifties (and obviously known in field theories since Hilbert) , and that, finally, (iiii) the associated *momentum* is a *material momentum*, also called *canonical momentum* in the nondissipative case, but which we prefer to call *pseudomomentum* (after the physicist Rudolph Peierls in optics). This kinetic quantity is a co-vector on the material manifold and it includes, a priori, contributions from all fields. Its conservation or non conservation relates to the fact that the physical system as a whole may exhibit properties which generally depend *explicitly* - and not through the field solution - on the considered material point. In many applications this dynamical entity is not so much relevant (e.g ., in fracture or phase transitions) but it plays an essential role in the linear dynamics of crystals (so-called *crystal momentum* of W.Brenig, 1955) and in the nonlinear dynamics of *solitonic structures* (wave momentum) and other nonlinear wave phenomena as was proved in the dynamics of nearly integrable systems by using the notion of *quasi-particles.*

Although configurational forces do not live, if we may say so, in the real world, they are « visible » through the duality they enjoy with their thermodynamic partners, material displacements. For instance, a macrocrack in the protecting vessel of a nuclear reactor is observable ; under changes of some external conditions (e.g ., periodic thermal heating of the structure), we « see » the unfortunate crack progress. According to the classical reasoning of mechanics, a *force* - the said *configurational force* - is associated with this progress. The theory allows one to *compute* this « force ». Similarly, lines of dislocations can be observed by some techniques (X-rays) . Under application of a system of physical forces to the specimen we see the dislocation lines « move ». The configurational force associated with this movement, the *Peach-Koehler* force, can be computed. Finally, the same holds true of the tip of a crack whose motion can be observed by infrared thermography in a transparent sample, or of domain walls in ferromagnets that can be observed by Bitter's technique, or still of the propagation of a phase-transformation front during quenching of a Japanese sword! Configurational forces do not live in our World because they are not the direct manifestation at a point of a classical cause at the same point. They are the results of complex , in general cooperative, and most often dissipative, phenomena and they can only be the result of a computation, analytical in some cases, numerical in most cases. The *mechanics of configurational forces* or *Eshelbian mechanics*, as we like to call it, provides the means to perform this computation, and gives the tools to exploit them in decisive circumstances. It is a *true* mechanics of forces in the sense that these forces contribute in dynamics to a balance of momentum in the general manner of Newton, and they also combine, add or subtract to one another, and they exhibit moments !

This contribution presents in a condensed manner the *field-theoretical* framework for *configurational forces* along the above-decribed line of thought. The power of the concepts , however, resides in the many applications which, in addition to the already mentioned cases, involve as well the study of the perturbative motion of solitonic structures (viewed as localized defects or *quasi-particles*) under the influence of external or coupled fields, and the accuracy of numerical computations in finite-difference schemes (FDS) and finite-element methods (FEM) where perturbating configurational forces appear if limit/boundary conditions are not exactly satisfied (FDS) or simply due to a bad design of the grid of computation in FEM. Although it is based initially on field-theoretic concepts and a somewhat complex geometric vision, the theory of configurational forces is thus endowed with computational interests in addition to the obvious understanding and convenient tools that it brings to fields of engineering (e.g ., in fracture, damage, plasticity) and condensed-matter physics. In our opinion, it constitutes one of the latest and most fruitful advances in macroscopic field theories that many considered a completely closed field of research offering no further progress, and therefore no true scientific interest, for quite a long time already. In this introduction it seems fitting to mention our elders who have been instrumental in uncovering the fundamental properties and representation of configurational forces, in addition to Eshelby (1951, 1970, 1975), D.Rogula (1970,1977), D.F.Nelson (1979), and P.Casal (1978). Working in close contact with M.Epstein since 1989 helped us to enrich our geometric vision. Pertinent remarks by P.Germain (1992, 1998) greatly enhanced the value of our contributions. Previous general presentations by Maugin (1993, 1995) are helpful as reading aids. A somewhat different viewpoint is expressed in the book of Gurtin (1999).

A progressive approach in complexity has been adopted in the present lecture notes. First (Section 2) we remind the reader of some elements of field theory (notion of *energy-momentum*, Noether's theorem, Noether's identity). Then (Section 3), as a simple application we consider the paradigmatic case of *finite-strain elasticity* - in the absence of any thermal and intrinsic dissipations at all material points - which helps one define the notions of material uniformity and inhomogeneity. Two new (to most readers) notions appear at that stage, those of *material force of true inhomogeneity* and *pseudomomentum* (canonical momentum). This involves some geometrical arguments. Forces driving singular sets, i.e., configurational forces *per se*, are the objects of Section 4, with particular examples from fracture. Then Section 5 introduces the notion of *material forces of quasi-inhomogeneity*, that is, material forces of dissipative origin in thermal conductors or other dissipative materials (with intrinsic dissipation) that act just like the material forces of true inhomogeneity. Generalizations to the cases of various materials with microstructure (e.g., polar elastic crystals or media endowed with so-called *internal variables of state*) are given in that section. This helps one formulate a general framework from which a canonical formulatioon of the mechanics of dissipative continua clearly emerges. In turn this allows one to present the theory of propagating interfaces (shock waves or phase-transition fronts) in a unified setting. More on smooth dynamics is given in Section 7 where we deal with the notion of *wave momentum* (part of the pseudomomentum) and its applications to linear elastic waves and nonlinear localized waves in elasticity-like physical systems of partial-differential equations. Finally more practical uses of configurational mechanics in three different types of numerical schemes are given in Section 8. A rich bibliography accompanies this contribution. Some proofs are offered by way of exercise.

2 Elements of field theory

We are concerned with simple general features of field theories in a continuum with space-time parametrization $\{\mathbf{X},t\}$, where $\mathbf{X}$ stands for *material coordinates* of classical continuum mechanics (e.g ., in Truesdell and Toupin, 1960), and t for a timelike scalar variable (Newton's absolute time). We consider *Hamiltonian actions* of the type

$$A(\phi;V)=\int_{V\times I} L\left(\phi^{\alpha},\partial_{\mu}\phi^{\alpha};X^{\mu}\right)d^{4}X\ , \tag{2.1}$$

where ϕ^{α}, $\alpha = 1,2,..,N$, denotes the ordered array of fields, say the independent components of a certain geometric object, and $d^4X = dV\,dt$. This is a Cartesian-Newtonian notation, with

$$\begin{aligned}&\left\{\partial_{\mu}=\partial/\partial X^{\mu};\mu=1,2,3,4\right\}\\&=\left\{\partial/\partial X^{K},K=1,2,3;\partial/\partial X^{4}\equiv\partial/\partial t\right\}\end{aligned}\ , \tag{2.2}$$

The summation over dummy indices (Einstein convention) is enforced. From expression (2.1) we can derive two types of equations: those relating to *each one* of the fields ϕ^{α}, and those which express a general conservation law of the system governing *all* fields simultaneously. The first group is obtained by imposing the requirement that the variation of the action A be zero when we perform a *small* variation $\delta\phi^{\alpha}$ of the field under well specified conditions at the boundary ∂V, of V (if V is not the whole of space), and at the end points of the time interval $I = [t_0,t_1]$ if such limitations are considered. However most field theories are developed for an infinite domain. The second group of equations are the result of the variation of the parametrization, and these results, on account of the former group, express the *invariance* or lack of invariance of the whole system under changes of this parametrization. In order to perform these variations we consider ε-parametrized families of transformations of *both* coordinates (parametrization) and fields such as

$$\left(X^{\mu},\phi^{\alpha}\right)\to\left(\overline{X}^{\mu},\overline{\phi}^{\alpha}\right), \tag{2.3}$$

with

$$\overline{X}^{\mu}=\kappa^{\mu}(\mathbf{X},\varepsilon),\ \overline{\phi}^{\alpha}(\overline{\mathbf{X}})=\Phi^{\alpha}\left(\phi^{\beta}(\mathbf{X}),\overline{\mathbf{X}},\varepsilon\right), \tag{2.4}$$

where ε is an infinitesimal parameter such that for $\varepsilon = 0$ we have identically $\kappa^{\mu}(\mathbf{X},0) = X^{\mu}$, $\Phi^{\alpha}(\phi^{\beta},\mathbf{X},0) = \phi^{\alpha}$. We assume that the quantity L is (2.1) transforms as a *scalar* quantity, i.e.,

$$L(\overline{\mathbf{X}},\varepsilon)=\det\left(\partial\mathbf{X}/\partial\overline{\mathbf{X}}\right)L(\mathbf{X}).$$

We note that derivations with respect $\mathbf{X}$ and ε commute, and the same holds true of integration in $\mathbf{X}$ space and derivation with respect to ε. The variation of a field ϕ^{α} is then defined by

$$\delta\phi^{\alpha} := \partial\Phi^{\alpha} / \partial\varepsilon\big|_{\varepsilon=0} . \tag{2.5}$$

With vanishing fields at infinity in space and vanishing variations at the ends of the time intervals, and applying an ε-parametrization to (2.1) we immediately have :

$$\delta A = \int d^4 X \left(\sum_{\alpha} \left\{ \frac{\partial L}{\partial \phi^{\alpha}} \delta\phi^{\alpha} + \frac{\partial L}{\partial\left(\partial_{\mu}\phi^{\alpha}\right)} \delta\left(\partial_{\mu}\phi^{\alpha}\right) \right\} + \frac{\partial L}{\partial X_{\mu}} \delta X^{\mu} \right) . \tag{2.6}$$

In order that δA vanish for all admissible $\delta\phi^{\alpha}(\mathbf{X})$, and any α, with **X** *fixed*, a classical computation yields the following *Euler-Lagrange equations* :

$$E_{\alpha} \equiv \frac{\delta L}{\delta \phi^{\alpha}} := \frac{\partial L}{\partial \phi^{\alpha}} - \partial_{\mu}\left(\frac{\partial L}{\partial\left(\partial_{\mu}\phi^{\alpha}\right)} \right) = 0 , \tag{2.7}$$

for each α = 1,2,..,N, at any space-time event **X**. We note that the result (2.7) is left unchanged if we substitute $L + \partial_{\mu}\Omega^{\mu}(\mathbf{X})$ to L. Equation (2.7) is a *strict* conservation law when L does not depend explicitly on ϕ^{α}.

The second group of equations that can derive from (2.1) by variation results from a simultaneous transformation of both the coordinates X^{μ} *and* the fields. We shall not repeat here the derivation of the resulting theorem known as *Noether's theorem* (Noether, 1918 ; Soper, 1976 ; Nelson, 1979). The latter states that to any *symmetry of the system there corresponds the conservation (or lack of strict conservation) of a current* (cf.Maugin, 1993, pp.99-103). For L given in (2.1), such a current generally reads

$$J^{\mu} = L \frac{\partial \bar{X}^{\mu}}{\partial \varepsilon} + \sum_{\alpha} \frac{\partial L}{\partial\left(\partial_{\mu}\phi^{\alpha}\right)} \left[\frac{\partial \bar{\phi}^{\alpha}}{\partial \varepsilon} - \partial_{\nu}\phi^{\alpha} \frac{\partial \bar{X}^{\nu}}{\partial \varepsilon} \right] , \tag{2.8}$$

where ε must be taken equal to zero. In spite of the notation J^{μ} is not always simply a four-vector. It all depends on the group of transformations considered. In particular, the most obvious symmetries in *Newtonian* space-time are represented by space-time translations, while there are no proper *variations* of the fields themselves, i.e., the general transformations (2.3) take on the following special form :

$$X^{\mu} \to \bar{X}^{\mu}(\mathbf{X}, \varepsilon) = X^{\mu} + \varepsilon \delta^{\mu}_{.\lambda} , \tag{2.9}$$

where δ^{μ}_{λ} is the Kronecker symbol ; i.e ., it equals one only when μ takes the fixed value granted to λ. On account of (2.9), we obtain the following current :

$$J^{\mu} \to J^{\mu}_{.\lambda} = T^{\mu}_{.\lambda} = L\delta^{\mu}_{.\lambda} - \sum_{\alpha} \partial_{\lambda}\phi^{\alpha} \frac{\partial L}{\partial\left(\partial_{\mu}\phi^{\alpha}\right)} \ . \tag{2.10}$$

This is a mixed second-order space-time tensor called the *energy-momentum tensor.* If L does not depend explicitly on **X**, then this tensor satisfies the following (four-dimensional) *strict conservation law* :

$$\partial_{\mu}T^{\mu}_{.\lambda} = 0, \lambda = 1,2,3,4 \ . \tag{2.11}$$

With the Cartesian notation introduced in (2.2) we see that the independence of L on t , $\lambda = 4$, yields a *scalar* conservation law in the explicit form

$$\frac{\partial H}{\partial t}\Big|_{\mathbf{X}} - \nabla_R . \mathbf{Q} = 0 \ , \tag{2.12}$$

while the independence of L on the spatial part of **X**, $\{X^K = 1,2,3\}$, yields the *material balance law* :

$$\frac{\partial P}{\partial t}\Big|_{\mathbf{X}} - div_R \mathbf{b} = \mathbf{0} \ , \tag{2.13}$$

where we defined the following canonical quantities :

- *energy (Hamiltonian density)* :

$$H := \sum_{\alpha} \dot{\phi}^{\alpha}\left(\partial L / \partial\dot{\phi}^{\alpha}\right) - L, \dot{\phi}^{\alpha} \equiv \partial\phi^{\alpha} / \partial t \ ; \tag{2.14}$$

- *energy flux vector* :

$$\mathbf{Q} = \left\{ Q^K := -\sum_{\alpha} \dot{\phi}^{\alpha} \frac{\partial L}{\partial\left(\partial_K \phi^{\alpha}\right)} \right\} ; \tag{2.15}$$

- *canonical (here material) momentum* :

$$P = \left\{ P_K := -\sum_{\alpha} \frac{\partial\phi^{\alpha}}{\partial X^K} \frac{\partial L}{\partial\left(\partial\phi^{\alpha} / \partial t\right)} \right\} ; \tag{2.16}$$

- *canonical stress tensor* :

$$\mathbf{b} = \left\{ b^K_{.L} := -\left(L\delta^K_{.L} - \sum_{\alpha} \frac{\partial\phi^{\alpha}}{\partial X^K} \frac{\partial L}{\partial\left(\partial\phi^{\alpha} / \partial X^K\right)} \right) \right\} . \tag{2.17}$$

The « explicit » independence of L on t - no right-hand side in (2.12) - stands for the *conservation of energy* (2.12). The « explicit » independence of L on X^K signifies that the material body is *materially homogeneous*. But this is not a fundamental requirement of physics, so that, in general, eqn.(2.13) may contain a nonzero right-hand denoted

$$\mathbf{f}^{inh} = \left\{ f_L^{inh} = \left(\partial L / \partial X^L \right)_{expl} \right. . \tag{2.18}$$

Had we considered a Lagrangian function L depending on higher-order space-time gradients, e.g., second-order one,

$$L = L\left(\phi^\alpha, \partial_\mu \phi^\alpha, \partial_\mu \partial_\nu \phi^\alpha; \mathbf{X}\right), \tag{2.19}$$

we would have obtained longer expressions for E_α, J^μ, and eqn.(2.11) ; e.g., in place of eqn.(2.7),

$$\begin{aligned} E_\alpha \equiv \frac{\delta L}{\delta \phi^\alpha} = \frac{\partial L}{\partial \phi^\alpha} - \partial_\mu \frac{\partial L}{\partial\left(\partial_\mu \phi^\alpha\right)} \\ + \partial_\mu \partial_\nu \frac{\partial L}{\partial\left(\partial_\nu \partial_\mu \phi^\alpha\right)} = 0 \end{aligned} . \tag{2.20}$$

But apart for some paradoxical theories with strange inertial terms (see, e.g., Maugin and Christov, 1997 , 2000), in classical theories L depends at most on the first- order time derivatives and the second-order gradients in (2.19) are purely spatial. When this is the case the formal definitions of both H and P are left unchanged while $\mathbf{Q}$ and $\mathbf{b}$ take on the following more general component form including second-order space-like (here material) gradients :

$$\begin{aligned} Q^K = -\sum_\alpha \left\{ \dot{\phi}^\alpha \left(\frac{\partial L}{\partial\left(\partial_K \phi^\alpha\right)} - \partial_L \left(\frac{\partial L}{\partial\left(\partial_L \partial_K \phi^\alpha\right)} \right) \right) \right\} \\ - \sum_\alpha \partial_L \dot{\phi}^\alpha \frac{\partial L}{\partial\left(\partial_L \partial_K \phi^\alpha\right)} \end{aligned} , \tag{2.21}$$

and

$$\begin{aligned} b_{.L}^K = -\left\{ L\delta_{.L}^K - \sum_\alpha \frac{\partial \phi^\alpha}{\partial X^L} \frac{\partial L}{\partial\left(\partial \phi^\alpha / \partial X^K\right)} \right\} \\ - \sum_\alpha 2 \frac{\partial^2 \phi^\alpha}{\partial X^L \partial X^M} \frac{\partial L}{\partial\left(\partial^2 \phi^\alpha / \partial X^M \partial X^K\right)} \\ + \sum_\alpha \frac{\partial}{\partial X^L} \left(\frac{\partial \phi^\alpha}{\partial X^M} \frac{\partial L}{\partial\left(\partial^2 \phi^\alpha / \partial X^M \partial X^K\right)} \right) \end{aligned} . \tag{2.22}$$

Other symmetries, and thus other consequences of Noether's theorem can be applied - e.g., rotations ; cf. Maugin, 1993, pp.110-12 - but these are not reminded here as they will not be exploited.

What must be principally gathered from the above is the essentially different nature of the Euler-Lagrange equations (2.7) or (2.20) for which one such equation is written for *each field* - or in a more mechanical jargon, *each degree of freedom* - and of the *canonical equations of energy and momentum* - e.g., eqns.(2.12) and (2.13) - which pertain to the whole physical system and by necessity consider all fields simultaneously [note the summations over α in the definitions (2.14) through (2.17)]. In particular, in establishing (2.11) we have made use of the celebrated *Noether's identity*

$$\left(\partial_\mu J^\mu\right)\delta\varepsilon + \sum_\alpha \left(E_\alpha \delta\phi^\alpha\right) = 0 , \tag{2.23}$$

which emphasizes the above-made remark , but naturally implies (2.11) whenever all *field* equations (2.7) are satisfied simultaneously. More is to be found on Noether's theorem and its relationship to the theory of *Lie groups* in Olver's book (1986). The above is quite sufficient for our purpose. Now it should be remarked that the above scheme applies only to *nondissipative systems*, but, in agreement with the motto of a noted physicist , we shall check that something can be done with dissipative systems when one knows the formulation for nondissipative ones.

3 Fundamental notions : The case of simple elasticity (hyperelasticity)

3.1 Direct application of field formalism

We consider here the purely nondissipative case of elasticity in finite strains. The finite-strain framework helps pedagogically in that it makes a clear distinction between *configurations.* The required elements of finite-strain kinematics are as follows. Suppose that the material body is an open, simply connected subset B of the *material manifold* M^3 of elements (so-called « particles » of continuum physics) $\mathbf{X}$ equipped with a scalar mass measure $\rho_0(\mathbf{X})$, which we assume to be sufficiently smooth (here no crack, no discontinuity, etc) in the reference configuration K_R of the body. The direct time-parametrized motion of $\mathbf{X}$ is given by the sufficiently regular function

$$\mathbf{x} = \chi(\mathbf{X},t), \tag{3.1}$$

where t and $\mathbf{X}$ have the same meaning as in Section 2. This provides the time-sequence of points occupied by $\mathbf{X}$ in the *actual configuration* K_t (at time t) in physical space E^3. The *physical velocity* $\mathbf{v}$ and *direct-motion gradient* $\mathbf{F}$ (usually called deformation gradient) are defined by

$$\mathbf{v} := \left.\frac{\partial \chi}{\partial t}\right|_X , \mathbf{F} := \left.\frac{\partial \chi}{\partial \mathbf{X}}\right|_t = \nabla_R \chi , \tag{3.2}$$

so that (T = transpose)

$$(\partial \mathbf{F}/\partial t)^T = \nabla_R \mathbf{v} , \tag{3.3}$$

where ∇_R (respectively, ∇) is the gradient operator at $\mathbf{X}$ on M^3 (respectively, at $\mathbf{x}$ in E^3). From $\mathbf{F}$, we can form the following most common measures of finite strain :

$$\mathbf{C} = \mathbf{F}^T.\mathbf{F}, \mathbf{E} = \frac{1}{2}(\mathbf{C} - \mathbf{1}_R). \tag{3.4}$$

Reciprocally, let $J_F = det\mathbf{F}>0$ always (at all times t). Then the *inverse motion*

$$\mathbf{X} = \chi^{-1}(\mathbf{x},t) , \tag{3.5}$$

is a well-behaved function, and we can define the *material velocity* $\mathbf{V}$ and the *inverse-motion* gradient $\mathbf{F}^{-1}$ such that

$$\mathbf{V} = \left.\frac{\partial \chi^{-1}}{\partial t}\right|_x , \mathbf{F}^{-1} = \left.\frac{\partial \chi^{-1}}{\partial \mathbf{x}}\right|_t . \tag{3.6}$$

By using the chain rule of differentiation, we can thus check that

$$\mathbf{v} + \mathbf{F}.\mathbf{V} = \mathbf{0}, \ \mathbf{V} + \mathbf{F}^{-1}.\mathbf{v} = \mathbf{0} , \tag{3.7}$$

and

$$\mathbf{F}^{-1}.\mathbf{F} = \mathbf{1}_R, \mathbf{F}.\mathbf{F}^{-1} = \mathbf{1} , \tag{3.8}$$

where $\mathbf{1}_R$ and $\mathbf{1}$ are unit dyadics.

For the bodies considered, the matter density ρ_0 at the reference configuration is at most a (smooth) function of the material point. Accordingly, the local expression of the conservation of mass reads :

$$\left.\frac{\partial \rho_0}{\partial t}\right|_X = 0 . \tag{3.9}$$

This means that the body presents a *smooth material inhomogeneity* from the point of view of *inertia.* Similarly for the elastic properties. Accordingly the Lagrangian density L of Section 2 reads

$$L = \bar{L}(\mathbf{v}, \mathbf{F}; \mathbf{X}) = K - W , \tag{3.10}$$

where kinetic (K) and potential (W) energies are given by

$$K = \frac{1}{2}\rho_0(\mathbf{X})\mathbf{v}^2, W = \overline{W}(\mathbf{F};\mathbf{X}), \tag{3.11}$$

where the former exhibits only *translational degrees of freedom.* The Lagrangian function $\overline{L}$ is a function of time and space derivatives of the basic field (3.1) - the *direct motion* (the latter does not intervene directly by virtue of the homogeneity of physical space) - and an explicit function of the point **X**. This is the print left by *material inhomogeneity.* In the absence of external body force and energy supply, it is readily shown that eqns. (2.7), (2.12) and (2.13) read :

$$\left.\frac{\partial \mathbf{p}}{\partial t}\right|_X - div_R \mathbf{T} = \mathbf{0} \ , \tag{3.12}$$

$$\left.\frac{\partial H}{\partial t}\right|_X - \nabla_R . \mathbf{Q}^{el} = 0 \ , \tag{3.13}$$

and

$$\left.\frac{\partial P}{\partial t}\right|_X - div_R \mathbf{b} = \mathbf{f}^{inh} , \tag{3.14}$$

wherein

$$\mathbf{p} = \rho_0 \mathbf{v}, \mathbf{T} = \partial \overline{W} / \partial \mathbf{F} , \tag{3.15}$$

$$H = K + W, \mathbf{Q}^{el} = \mathbf{G} = \mathbf{T}.\mathbf{v} \ , \tag{3.16}$$

$$P = -\mathbf{p}.\mathbf{F}, \mathbf{b} = -\left(\overline{L}\mathbf{1}_R + \mathbf{T}.\mathbf{F}\right), \mathbf{f}^{inh} = \left.\frac{\partial \overline{L}}{\partial \mathbf{X}}\right|_{expl} , \tag{3.17}$$

where the last notation means the *explicit material gradient* of $\overline{L}$, i.e., the material gradient taken at fixed fields **v** and **F**. The quantities defined in eqns.(3.15)-(3.16) are, respectively, the *physical linear momentum* **p** (components in E^3), the *first Piola-Kirchhoff stress* (a two-point tensor field with mixed « feet » in E^3 and M^3), the *Hamiltonian* (energy) density H per unit reference volume, the material « mechanical » *Poynting-Umov vector* (energy flux) **G**, the *canonical momentum* or *pseudomomentum* P (covariant components on M^3), the *Eshelby material stress* **b** (a mixed fully material tensor) , and the *material force of smooth inhomogeneities* $\mathbf{f}^{inh}$ (with covariant components on M^3). The latter reads more explicitly

$$\mathbf{f}^{inh} = \left(\frac{1}{2}\mathbf{v}^2\right)\left(\nabla_R \rho_0\right) - \left.\frac{\partial \overline{W}(\mathbf{F};\mathbf{X})}{\partial \mathbf{X}}\right|_{\mathbf{F} fixed} . \tag{3.18}$$

Exercis 3.1.. By multiplying eqn.(3.12) to the left by **F** and integrating by parts on account of (3.11) and (3.17), show that the following *Noether's identity* holds true :

$$\left(\frac{\partial \mathbf{p}}{\partial t} - div_R \mathbf{T}\right).\mathbf{F} + \left(\frac{\partial P}{\partial t} - \left(div_R \mathbf{b} + \mathbf{f}^{inh}\right)\right) = \mathbf{0} . \tag{3.19}$$

Remark 3.1. That is, had we known (3.12) *by any means*, we could have deduced (3.14) on account of the knowledge of the functional dependency of $\overline{W}$. This points to the way to be followed in a dissipative case.

In particular, in quasi-statics (neglect of inertial terms) and for a *materially homogeneous* elastic material, eqn.(3.19) reduces to *Ericksen's identity* (1977)

$$(div_R \mathbf{T}).\mathbf{F} + div_R \mathbf{b} = \mathbf{0} . \tag{3.20}$$

The general notion of « Ericksen-Noether identity » is dealt with in Maugin (1996) for mechanics and other problems (heat and electricity conductions).

Equation (3.13) is the equation of conservation of energy, a strict *conservation law* because $\overline{L}$ does not depend explicitly on time (eqn.(3.9) obviously is also a strict conservation law). Equation (3.12) is the balance of physical (linear) momentum in which the associated flux is the first Piola-Kirchhoff stress. Equation (3.14) is the balance equation of canonical momentum or *pseudomomentum* (or *material* (linear) *momentum*). While there exists only one representation of the local energy balance -eqn.(3.13) - equations (3.12) and (3.14) are, at all regular material points, but two statements, on different manifolds, of the *same* balance equation. The first one is often referred to as the *Lagrangian* representation, but this denomination is misleading because its components are in effect in the actual configuration (only its space-time parametrization is Lagrangian). The second one, which appears here as a simple pull back of the former onto M^3, is entirely material, being projected on M^3 and using a material space-time parametrization. It is *canonical*, because of this and not only because it derived from Noether's theorem. The Eshelby stress tensor **b** is fully material, and if there are no body couples acting on the material, satisfies the following symmetry condition (cf. Epstein and Maugin, 1990 ; this would be a natural consequence of the symmetry of the classical Cauchy stress) ; it is symmetric with respect to **C** :

$$\mathbf{C}.\mathbf{b} = \mathbf{b}^T.\mathbf{C} \quad . \tag{3.21}$$

If one introduces the *second Piola-Kirchhoff stress* **S** (this is a true contravariant material tensor) by

$$\mathbf{S} = \mathbf{T}.\mathbf{F}^{-T} , \tag{3.22}$$

it is immediately checked that **b** is given by

$$\mathbf{b} = -\left(\overline{L}\mathbf{1}_R + \mathbf{M}\right), \tag{3.23}$$

where

$$\mathbf{M} = \mathbf{S}.\mathbf{C} = \mathbf{T}.\mathbf{F} \ , \tag{3.24}$$

is the so-called *Mandel stress* (known in finite-strain elastoplasticity ; see below).

3.2 The notion of material inhomogeneity force

Equation (3.14) is not a true conservation law because of the presence of a source term in its right-hand side. We have naturally called « material inhomogeneity force » this right-hand side which exists by virtue of the material being both *inertially* and *elastically* inhomogeneous (properties depending on the material point itself). Such a « force » is not a physical one. It produces no power in the corresponding energy equation. As a matter of fact we have :

Exercise 3.2 : Show that eqn.(3.13) can also be written as :

$$\frac{\partial H}{\partial t} - \nabla_R.\left(\left(\mathbf{b} + L\mathbf{1}_R\right).\mathbf{V}\right) = 0 \ ; \tag{3.25}$$

and that this is also a consequence of the following identity :

$$\left(\frac{\partial H}{\partial t} - \nabla_R.\mathbf{G}\right) - \left(\frac{\partial P}{\partial t} - \left(div_R\mathbf{b} + \mathbf{f}^{inh}\right)\right).\mathbf{V} = 0 \ , \tag{3.26}$$

which is the timelike counterpart of the identity obtained by multiplying the identity (3.19) by $\mathbf{F}^{-1}$ to the right.

The advantage of eqn.(3.14) is that it introduces in a natural way the notion of *genuine material forces* and it renders explicit this smooth material inhomogeneity. The way such a « force » acts is illustrated by a simple example. Consider *quasi-statics* and a *continuously layered* material in the $X = X^1$ direction in such a way that we could write

$$\overline{W}(\mathbf{F};\mathbf{X}) = I(X)W_O(\mathbf{F}) \ , \tag{3.27}$$

where W_0 no longer depends explicitly on **X.** A typical example is one of exponential variation of the *layering function I*, i.e., $I(X) = exp(-X/\lambda)$. Then ($\mathbf{E}_1$= unit material vector along the X-direction)

$$\mathbf{f}^{inh} = -\left.\frac{\partial \overline{W}}{\partial \mathbf{X}}\right|_{expl} = \left(\lambda^{-1}\overline{W}\right)\mathbf{E}_1 \ . \tag{3.28}$$

With both λ and W_0 positive we see that for a fixed field solution the fictitious force $\mathbf{f}^{inh}$ is directed from the *harder* part of the material to the *softer* part. It is a *directional indicator* of the spatial variation of material properties. Other examples of this property are given in Maugin *et al* (1999). To be complete however, we should mention that, had we had a body force per unit volume (say the action of gravity) present in eqn.(3.12) - i.e., a force deriving from an external potential and acting per unit mass at actual placement **x** - this would create a material force by mere pull back, in the right-hand side of eqn.(3.14), but this would not be a true material force as it would depend on **X** only *through* the field (motion) (3.1), so that it remains a « physical or Newtonian » concept even after pull back. There are other material forces, which are closer to the true material meaning (see Section 5).

3.3 The notion of pseudomomentum

Because in the present case the relevant field happens to be the *motion* itself, we see that pseudomomentum - or canonical momentum - P simply is the pull back of physical momentum to M^3. This is not the case in the presence of additional fields (cf. definition (2.16) and other cases below). An interesting formula can be given for P on account of the definitions (3.4) and $(3.7)_2$. We immediately have (cf. Maugin, 1990a ; Maugin and Trimarco, 1991):

$$P = \rho_0 \mathbf{C}.\mathbf{V} \ . \tag{3.29}$$

This shows the interest of the presence of the minus sign in the definitition (2.16). Like in eqn.(3.21) , the strain tensor $\mathbf{C}$ acts here as a *deformed metric* on M^3 to build covariant quantities from contravectors (such as $\mathbf{V}$). More on the kinematic interpretation of $\mathbf{V}$ and P in Maugin (1993, pp.82-83). What may simply be noted is that we have the following

Exercise 3.3 : Prove that

$$K = \frac{1}{2}\rho_0(\mathbf{X})\mathbf{V}.\mathbf{C}.\mathbf{V}, P = \frac{\partial \overline{L}}{\partial \mathbf{V}}, H = P.\mathbf{V} - L \ . \tag{3.30}$$

The last two formulas, together with the definition of $\mathbf{V}$, hint at a canonical Hamiltonian formulation of inhomogeneous (and anisotropic) finite-strain elasticity. The last of (3.27) is none other than a Legendre transformation between Lagrangian and Hamiltonian densities. Equations (3.30) are based on the *inverse-motion* picture. The missing Hamilton's equation will read

$$\left.\frac{\partial \widetilde{P}}{\partial t}\right|_x = -\frac{\delta \widetilde{H}}{\delta \mathbf{X}}, \tag{3.31}$$

where $\widetilde{P}$ is the pseudomomentum per unit volume of actual volume, but still a material covector, the time derivative in the left-hand side is taken at fixed actual placement, the space derivative in the right-hand side is the *functional Euler derivative* of the inverse-motion Hamiltonian density per unit actual volume $\widetilde{H}(\mathbf{X} = \chi^{-1}, \mathbf{F}^{-1}, \mathbf{V})$. We let the reader prove that eqn.(3.30) is none other than a slightly transformed form of the pseudomomentum balance equation (3.14).

3.3 The notion of local structural rearrangement

In order to make ideas clear let us consider *the case of quasi-statics in the absence of body force* with an elastic energy density given by $(3.11)_2$. Then the field equation (3.12) - bulk equilibrium - reduces to

$$div_R \mathbf{T} = \mathbf{0} . \tag{3.32}$$

Following Epstein and Maugin (1990), we consider (thought experiment) the case where the material homogeneity can be artificially removed at each material point **X**, by effecting a change of reference configuration. This is done at each material point, and the reference change is therefore *local* and generally *not integrable* over the whole body. Such a change is called a *local structural rearrangement*, and this is conceived independently of any action in the actual configuration. Let **K(X)** this reference change so that accounting for the accompanying volume change $J_K = det\mathbf{K}$, it combines mutiplicatively to the right with **F** and, for energies, we can write

$$\overline{W}(\mathbf{F};\mathbf{X}) = J_K^{-1}\hat{W}(\mathbf{FK}(\mathbf{X})) = \tilde{W}(\mathbf{F},\mathbf{K}) . \tag{3.33}$$

Obviously we can compute the partial derivatives of the last mentioned function W, obtaining thus, as easily checked ,

$$\mathbf{T} = \frac{\partial \tilde{W}}{\partial \mathbf{F}} = \frac{\partial \overline{W}}{\partial \mathbf{F}}, \tilde{\mathbf{b}} = -\frac{\partial \tilde{W}}{\partial \mathbf{K}} = -(\mathbf{T}.\mathbf{F} - W\mathbf{1}_R).\mathbf{K}^{-T} . \tag{3.34}$$

Accordingly,

$$\mathbf{b} \equiv \tilde{\mathbf{b}}.\mathbf{K}^T = -\frac{\partial \tilde{W}}{\partial \mathbf{K}}.\mathbf{K}^T \equiv W\mathbf{1}_R - \mathbf{T}.\mathbf{F} . \tag{3.35}$$

This provides an elegant definition of the quasi-static Eshelby stress via the notion of local structural rearrangement. Assuming that we just know (3.32), we can then compute the material divergence of **b** from the result $(3.31)_2$ (***exercise***). The result is

$$div_R \mathbf{b} = \left. \frac{\partial \overline{W}}{\partial \mathbf{X}} \right|_{expl} , \tag{3.36}$$

which is nothing else but

$$div_R \mathbf{b} + \mathbf{f}^{inh} = \mathbf{0} . \tag{3.37}$$

But we can also compute the right-hand side of (3.36) through the following operation

$$\left. \frac{\partial W}{\partial \mathbf{X}} \right|_{expl} = (\nabla_R \mathbf{K}) : \frac{\partial \tilde{W}(\mathbf{F},\mathbf{K})}{\partial \mathbf{K}} = -(\nabla_R \mathbf{K}) : (\mathbf{b}.\mathbf{K}^{-T}) = \mathbf{b} : \Gamma , \tag{3.38}$$

where Γ is the (geometrical) *connection* based on the non-integrable mapping $\mathbf{K}$; that is, in components (to avoid any misunderstanding):

$$\Gamma^{A}_{B.K} := -\left(\mathbf{K}^{-1}\right)^{\alpha}_{.B} K^{A}_{.\alpha,K} . \tag{3.39}$$

Therefore, eqn.(3.37) also reads (Epstein and Maugin, 1990)

$$div_R \mathbf{b} = \mathbf{b} : \Gamma . \tag{3.40}$$

In some geometrical theories of *continuous distributions of dislocations*, the connection Γ is directly related to the *density of dislocations* (cf.Maugin, 1993, Chapter 6). Accordingly, we can say that in such « continuously dislocated » elastic bodies, dislocations create a material force density which is responsible for the non-divergence-free nature of the Eshelby stress tensor. Dislocations, which originally are discrete defects, act thus as a materially *distributed inhomogeneity force* in agreement with equation (3.40). We do not pursue further here this geometrical approach to continuously distributed defects (see more on differential geometry, the notions of material uniformity and homogeneity, the role of material symmetry groups, crystallographic basis, transplants, *G*-structure, and *G*-covariance in Epstein and Maugin, 1997).

4 Forces driving singularity sets

4.1 The notion of singularity set

Continuing with the purely elastic case of Section 3, it may happen that the elastic solution of a boundary-value problem based on integrating eqn.(3.12) and the associated boundary conditions over a material body, presents a singular behavior over a certain manifold. Working in three-dimensional physical space, these manifolds may only be of certain integer dimensions if we make abstraction of the appearance of fractals . Typically, one will observe the singularity of the strain and stress field along a *line* (case of dislocations, macroscopic cracks) or across a *two-dimensional surface* (case of a shock wave or a phase-transition front). Following Dascalu and Maugin (1994) - also Maugin (1996b) - , one must clearly distinguish between what may thus be considered an *elastic defect D*, and what is the *set of extension points*, *ED*, of the said elastic defect. For example, in the crack problem, the defect *D* is really two-dimensional (the crack, flat or not, itself with its two faces) while the set *ED* is the line representing the tip of the crack. This line itself can be obtained by a limiting process involving a family of open notches (see Paragraph 4.2). Similarly for dislocations, where *D* in fact is the missing half plane of atoms, while *ED* is the dislocation line. For shock waves and phase-transition fronts, the material on one side of the singular surface across which fields suffer finite discontinuities governed by the second law of thermodynamics, sees the other side as the *invading* or *progresssing* defect, which is therefore three-dimensional (a volume), *ED* being really the singular surface itself. The same holds true for the growth of a cavity where *D* is the cavity itself (therefore three-dimensional ; an open set of R^3) and *ED* is the surface of the cavity. In other problems such as the appearance of *damage, plasticity*, and *growth, ED* is potentially the whole body. This is particularly evident in the case of *volumetric growth* (cf. Epstein and Maugin, 1999, 2000) or elastoplasticity. This does not mean that there does not exist a surface separating the already damaged, plasticized or grown 3d-region from the rest of the body , and we can speak of the damage boundary, plastic boundary, *etc*, which is obviously 2d, and is an *unknown* in the evolution problem (free-boundary problem) resulting from the time evolution of physical (Newtonian) loads. However, in all cases the driving force causing the growth of the considered defect D is acting on the set of extension points ED, which is therefore one, two, and three-dimensional for dislocations and cracks, shock-waves and phase-transition fronts (and also walls between domains in ferroelectrics and ferromagnets), and damage, plasticity and growth, respectively. We illustrate this first with the case of *brittle fracture*. But the above discussion hints at introducing the *energy* of an elastic defect as a quantity defined over D (Dascalu and Maugin, 1994, and Paragraph 4.4 below) and this contradicts some classical vision ; for instance the energy of a defect such as a cavity is a bulk energy, and *not* a surface energy.

4.2 Example 1 : The analytical mechanics of brittle fracture

We consider general anisotropic finite-strain elasticity in *materially homogeneous* bodies. Outside singularity sets (at all regular material points $\mathbf{X}$ in B) we have locally the following reduced form of the momentum equation (3.12) and (3.14) :

$$\frac{\partial \mathbf{p}}{\partial t} - div_R \mathbf{T} = \mathbf{0}, \frac{\partial P}{\partial t} - div_R \mathbf{b} = \mathbf{0}. \tag{4.1}$$

On account of the assumed smoothness of all fields (in particular, $\mathbf{F}$), these two are entirely equivalent -redundant ; compare to Greg (1862) - according to Noether-Ericksen identity. Each of these can be integrated over an extended *regular* material region B, where the Green's divergence and Reynolds'transport theorems can be used without specific precaution. We have commutation rules for space integration and time differentiation since $(\mathbf{X},t)$ is a good set of independent space-time parameters from that viewpoint. We obtain thus

$$\frac{\partial}{\partial t} \int_B \mathbf{p} dV = \int_{\partial B} \mathbf{N}.\mathbf{T} dA \ , \tag{4.2}$$

$$\frac{\partial}{\partial t} \int_B P dV = \int_{\partial B} \mathbf{N}.\mathbf{b} dA \, . \tag{4.3}$$

Reasoning in the same way on the energy equation (3.13) , we have

$$\frac{\partial}{\partial t} \int_B H dV = \int_{\partial B} \mathbf{N}.\mathbf{T}.\mathbf{v} dA \, . \tag{4.4}$$

Note that eqns.(4.2) and (4.3) are no longer in direct correspondence since convection is lost in the space integration. Equations (4.2) and (4.3) are meaningful component-wise only.

Now if the fields $\mathbf{v}$ and $\mathbf{T}$, which are time and space derivatives of the basic field (motion), present a certain order of singularity over some subset of M^3, quantities like P , H, and $\mathbf{b}$ which are at least quadratic in these fields, will even be *more* singular. Thus, if the stress $\mathbf{T}$ is *not* a good indicator of the presence of field singularities (see Maugin, 1995, Sec.5), $\mathbf{b}$ may be a good indicator. This is exactly what happens in brittle fracture where the integration of $(4.1)_2$

and (3.13) over a subset of M^3 containing a field singularity, *captures this singularity* by making additional nonvanishing terms, characteristic of the singularity, appear in global equations of the type of (4.3) and (4.4). These terms are, respectively, the *material driving force* acting on the singularity and the so-called *energy-release rate* associated with the energy consumed in the irreversible progress of the defect, i.e., the evolution of ED ; the two must be related through the notion of power expended by the driving force. We follow Dascalu and Maugin (1994) in this illustration to fracture. Note that the present case already illustrates the different roles played *at singular manifolds* by eqns.$(4.1)_{1\text{-}2}$.

Following Rice (1968), a sharp-ended straight through crack C is viewed as the uniform limit of a family of end-rounded notches (cf.Figures 1 and 2). The flat faces of a notch and the rounded cylindrical end are assumed free of traction. We consider the body B of limiting surface ∂B excluding the notch , and the notch's end Γ_δ which is assumed to propagate inside the body with uniform material velocity $\overline{\mathbf{V}}$ parallel to the flat faces and to the axis $\mathbf{E}_1$. This region is regular but geometrically evolving in time (configurational change) due to the extension of the notch. In integrating $(4.1)_2$ and (3.13) over B we simply have to use generalized Green's and Reynolds' theorems. These yields (compare to (4.3) and (4.4))

$$\frac{\partial}{\partial t}\int_B P dV + F^{notch}(\delta) = \int_{\partial B} \mathbf{N}.\mathbf{b} dA \ , \tag{4.5}$$

$$\frac{\partial}{\partial t}\int_B H dV + G^{notch}(\delta) = \int_{\partial B} \mathbf{N}.\mathbf{T}.\mathbf{v} dA \, , \tag{4.6}$$

where we have defined a material force $F^{notch}(\delta)$ and an *energy-release rate* $G^{notch}(\delta)$ by

$$F^{notch}(\delta) = \mathbf{E}_1 \int_{\Gamma_\delta} \left(P_1\left(\overline{\mathbf{V}}.\mathbf{N}\right) - LN_1\right) dA \ , \tag{4.7}$$

and

$$G^{notch}(\delta) = \int_{\Gamma_\delta} H\left(\overline{\mathbf{V}}.\mathbf{N}\right) dA \, . \tag{4.8}$$

Assuming the convergence of the family of notches indexed δ towards the sharp crack and that the limit of the solutions of the corresponding sequence of elasticity problems converges towards the solution for the sharp crack and the first term in each of (4.5) and (4.6) converges to zero in this limit (this follows from the singularity of elasticity solutions and eqn.$(3.7)_1$) , we obtain for the driving force on the crack and the corresponding energy-release rate the following illuminating formulas :

$$F^{crack} = -\lim_{\delta \to 0} \mathbf{E}_1 \int_{\Gamma_\delta} \left(LN_1 - P_1\left(\overline{\mathbf{V}}.\mathbf{N}\right)\right) dA \ , \tag{4.9}$$

$$G^{crack} = \lim_{\delta \to 0} \int_{\Gamma_\delta} H\left(\overline{\mathbf{V}}.\mathbf{N}\right) dA \, . \tag{4.10}$$

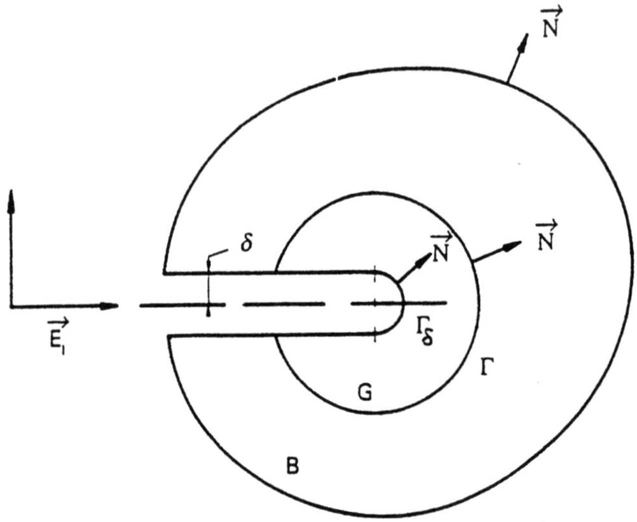

Figure 1. Notch δ in a nonlinear elastic body *B*

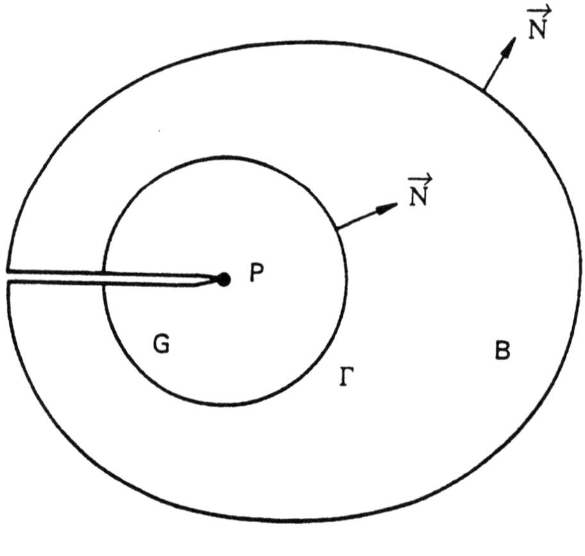

Figure 2. The straight-through crack as a limit of a notch

We have referred to these two basic formulas as those of the *analytical theory of brittle fracture*, because they involve the elasticity Lagrangian and Hamiltonian densities. The meaning of eqn.(4.10) is clear : it represents the energy consumed (« swallowed ») per unit time by the inward crack motion. The tip of the crack acts as an *energy sink*.

But we can equally well integrate eqns. $(4.1)_2$ and (3.13) over the regular material region of volume extension G of external (cylindrical) boundary Γ equipped with unit outward normal $\mathbf{N}$ and the notch as indicated in Figure 1, and moving inward with uniform material velocity $\overline{\mathbf{V}}$. Combining then the resulting expression with (4.7) and (4.8) , we obtain more classical (but less elegant) formulas for the quantities (4.9) and (4.10) :

$$F_1^{crack} = \int_\Gamma \left((\mathbf{N}.\mathbf{b})_1 + P_1 \left(\overline{\mathbf{V}}.\mathbf{N} \right) \right) dA - \frac{\partial}{\partial t} \int_G P_1 \, dV \tag{4.11}$$

and

$$G^{crack} = \int_\Gamma \left((H) \left(\overline{\mathbf{V}}.\mathbf{N} \right) + \mathbf{N}.\mathbf{T}.\mathbf{v} \right) dA - \frac{\partial}{\partial t} \int_G H \, dV \, . \tag{4.12}$$

It is readily checked that eqn.(4.11) reduces to the celebrated path-independent J-integral of brittle fracture in quasi-statics and small strains. Indeed, in these conditions, the pseudomomentum and kinetic-energy contributions are discarded, and

$$\mathbf{F} = \mathbf{1} + (\nabla_R \mathbf{u})^T \, , \tag{4.13}$$

where $\mathbf{u}(\mathbf{X},t)$ is the elastic displacement. On expanding the expression of $\mathbf{T}.\mathbf{F}$, we find that the first contribution in the stress contributes nothing because of its order of singularity and there remains the well-known expresssion

$$F^{crack} = J = \int_\Gamma \left(W N_1 - \mathbf{t}.\frac{\partial \mathbf{u}}{\partial X_1} \right) dL \tag{4.14}$$

per unit thickness of the body, where $\mathbf{t}$ is the traction at Γ.

More remarkably, a relationship can be established between (4.10) and (4.9) in the fully dynamical and finite-strain case. Let P a point of Γ_δ for the notch δ and $\mathbf{r}$ the vectorial poition of any point with respect to P. We can write $\chi(\mathbf{X},t) = \chi(\mathbf{X}_P + \mathbf{r},t)$, and through differentiation

$$\mathbf{v} = -\mathbf{F}.\overline{\mathbf{V}} + \frac{\partial \overline{\chi}}{\partial t}, \; \overline{\mathbf{V}} := \frac{\partial \mathbf{X}_P}{\partial t} \, . \tag{4.15}$$

On multiplication by $\mathbf{F}^{-1}$ to the left of $(4.15)_1$ we get

$$\mathbf{V} = \overline{\mathbf{V}} - \mathbf{F}^{-1}.\frac{\partial \overline{\chi}}{\partial t}, \overline{\mathbf{V}} // \mathbf{E}_1 \ . \tag{4.16}$$

Let all the points of Γ_δ be in uniform motion, then we can reasonably assume that the deformation is the same at all points of Γ_δ and set $\partial\overline{\chi}/\partial t = \mathbf{0}$. Since the only nonzero component of $\overline{\mathbf{V}}$ is along $\mathbf{E}_1$, multiplying the one-component of (4.7) by $\overline{V}_1$, and accounting for the simplified form of (4.16) and the identity $\mathbf{V.C.V} = \mathrm{v}^2$, we obtain for the notch

$$G^{notch}(\delta) = \overline{V}_1 F_1^{notch}(\delta). \tag{4.17}$$

Accordingly, for a uniform motion of the points of the notch, we find that the material force driving the notch coincides with the energy released during the notch progress of a unit length inside the body. Relation (4.17) is meaningful only if $\overline{\mathbf{V}}$ is a function of $\mathbf{X}$. This shows the intimate relationship between the notion of material configurational force with that of rigid-body (« en bloc ») motion.

In the limit case of the straight crack, we obtain

$$G^{crack} = \overline{V}_1 F_1^{crack}. \tag{4.18}$$

This is the *dissipation rate* due to the crack extension. It is in the classical bilinear form favored in irreversible thermodynamics. It is thus ready for the construction of crack-extension criteria (see Maugin, 1992, 1999a) for this.

We finally note that (4.18) happily complements eqns.(4.9) and (4.1) of the anlytical theory of brittle fracture for it can be established only thanks to the passing from the Lagrangian to the Hamiltonian density thanks to the pseudomomentum contribution and (4.16) which allows for the required Legendre tansformation between those two quantities, cf. eqn.$(3.30)_3$.

Case of smoothly materially inhomogeneous elastic materials : In this case eqns.$(4.1)_2$ and (4.3) contain a material inhomogeneity force contribution in their right-hand side. This modifies the expression (4.11) accordingly.

Exercise 4.1 **:** Show that for materially inhomogeneous elastic bodies eqn.(4.11) is replaced by

$$F_1^{crack} = \int_{\Gamma} \left((\mathbf{N.b})_1 + P_1 (\overline{\mathbf{V}}.\mathbf{N}) \right) dA - \left(\frac{\partial}{\partial t} \int_G P_1 dV - \int_G f_1^{inh} dV \right) . \tag{4.19}$$

Such an expression has been exploited in quasi-statics by Weichert and Haddi (1994). It shows that the crack driving force is a material force in its own right ; it plays the same role as a spatially integrated true inhomogeneity force.

4.3 Example 2 : The case of singular surfaces

The existence of a source term in the balance equation of pseudomomentum is witness of a loss of translational invariance on the material manifold M^3. Obviously, a material body B which is made of several assembled bodies B_α , $\alpha =$ 1,2,..,N, or a body made of a single material which suffers structural changes across an internal surface Σ - a so-called *interface* - is *globally materially inhomogeneous*, since translational invariance of material properties is globally lost. Accordingly, there should exist a patent manifestation of this fact through the existence of a density of *material forces* at the junction between bodies or phases. In the case of assembled materially homogeneous bodies with geometrically perfect contact between the constituting bodies, the notion of contact material body force which accounts for the jump in material properties (e.g ., the jump in the values of elasticity coefficients and the jump in density - whenever inertia is accounted for, thus in dynamical problems) has been examined by Maugin *et al* (1999). There indeed exist such material forces and in the case of perfect junction between bodies (continuity of elastic displacements and tractions), the corresponding material force has no energy content. It simply is a *directional indicator* of inhomogeneity, e.g., pointing from the harder to the softer material. This may be considered a limit case of (3.25) where the smooth variation over a short range of thickness d between two regions of practically uniform properties is taken as an abrupt change of finite value (a jump) in the limit as d goes to zero. To illustrate the obtaining of the expression of the said surface material force, we may follow Maugin and Trimarco (1995a) in their *quasi-static* variational formulation . Let the sufficiently regular surface Σ separate a body of material volume B into two regions B^+ and B^-. Let $\mathbf{N}$ the unit normal to Σ oriented from B^- to B^+. Let

$$[A] = A^+ - A^-, \tag{4.20}$$

denote the *finite jump* of the quantity A, A^+ and A^- being the uniform limits of A in approaching Σ along its normal on both sides of Σ. We shall note $\delta_{\mathbf{X}}\chi$ and $\delta_x\chi^{-1} = \delta_x\mathbf{X}$ the infinitesial variations of the direct motion (keeping the material point $\mathbf{X}$ fixed) and of the inverse motion (keeping the actual placement $\mathbf{x}$ fixed). These are sometimes referred to as the « Lagrangian » and « Eulerian » variations, respectively.

Exercise 4.2 : Prove that the two variations are related by

$$\delta_{\mathbf{X}}\chi + \mathbf{F}.\delta_x\chi^{-1} = \mathbf{0} \tag{4.21}$$

at all regular material points.

In order to generate the looked for equations we consider the following principle of potential energy :

$$\delta\Phi = -\delta_{\mathbf{X}} \int_B W(\mathbf{F})dV + \int_{\partial B} (\mathbf{N}.\mathbf{T}).\delta_{\mathbf{X}}\chi\, dA\,, \tag{4.22}$$

where, indeed (the interface Σ is assumed to carry no potential energy of its own),

$$\int_B W(\mathbf{F})dV = \int_{B^+} W^+(\mathbf{F})dV + \int_{B^-} W^-(\mathbf{F})dV , \tag{4.23}$$

where W^+ and W^- are the elastic energy densities per unit volume of the reference configuration for two essentially different anisotropic but homogeneous elastic materials. Noting $\mathbf{N}_\Sigma = \mathbf{N}^- = -\mathbf{N}^+$, a transport theorem yields

$$\delta_\mathrm{X} \int_{B^\pm} W^\pm(\mathbf{F})dV = \int_{B^\pm \, fixed} \delta_\mathrm{X} W^\pm dV - \int_{\Sigma^\pm} W^\pm\left(\mathbf{N}^\pm . \delta_x \mathbf{X}^\pm\right) d\Sigma , \tag{4.24}$$

so that

$$\delta_\mathrm{X} \int_B W(\mathbf{F})dV = \int_B \delta_\mathrm{X} W(\mathbf{F})dV + \int_\Sigma \mathbf{N}_\Sigma . \left[\mathrm{W} \delta_x \chi^{-1}\right] d\Sigma . \tag{4.25}$$

But in computing the first contribution in the right-hand side of this equation, we obtain (exercise)

$$\begin{aligned}\int_B \delta_\mathrm{X} W^\pm(\mathbf{F})dV = &-\int_{B-\Sigma} (div_R \mathbf{T}).\delta_\mathrm{X} \chi dV + \\ &\int_{\partial B-\Sigma} (\mathbf{N}.\mathbf{T}).\delta_\mathrm{X} \chi dV + \int_\Sigma \mathbf{N}_\Sigma . [\mathbf{T}.\delta_\mathrm{X} \chi] d\Sigma\end{aligned} \tag{4.26}$$

where the jump term comes from the application of the divergence theoremin both parts and the gluing back of the two resulting results.

The classical equilibrium equations read

$$\begin{aligned}&div_R \mathbf{T}^\pm = \mathbf{0} in B^\pm , \mathbf{T}^\pm = \partial W^\pm / \partial \mathbf{F}, \\ &\mathbf{N}.\mathbf{T} = \mathbf{T}^\mathrm{d} \, at \, \partial B - \Sigma, \, \mathbf{N}.[\mathbf{T}] = \mathbf{0} across \Sigma\end{aligned} \; . \tag{4.27}$$

Collecting the remaining contributions we have

$$\delta \Phi = -\int_\Sigma \mathbf{N}_\Sigma . \left[W \delta_x \chi^{-1} + \mathbf{T}.\delta_\mathrm{X} \chi\right] d\Sigma . \tag{4.28}$$

At this point we follow Hill (1986) by selecting the virtual variations in direct and inverse motions in the most realistic way. We cannot control the *direct* motion which, therefore, can suffer a jump in its first variation. But the situation is different for the inverse motion because in the absence of defects at Σ, we have continuity in the displacement, hence we assume continuity of the variation $\delta_x \mathbf{X}$. Accordingly, (4.21) renders

$$[\delta_\mathrm{X} \chi] = -[\mathbf{F}] \delta_x \mathbf{X} . \tag{4.29}$$

Exercise 4.3 : On account of the last of (4.22) and of the relations,

$$\langle A \rangle := \frac{1}{2}\left(A^+ + A^-\right), \tag{4.30}$$

$$[AB] \equiv \langle A \rangle [B] + [A] \langle B \rangle, \tag{4.31}$$

prove that

$$\delta\Phi = \int_\Sigma \mathbf{f}_\Sigma . \delta_x \mathbf{X} d\Sigma, \tag{4.32}$$

where

$$\mathbf{f}_\Sigma := -\mathbf{N}_\Sigma . \left[W(\mathbf{F}) \mathbf{1}_R - \langle \mathbf{T} \rangle . \mathbf{F} \right]. \tag{4.33}$$

By duality with the inverse-motion variation, this is *material force* per unit area of Σ. If potential energy is conserved (no dissipation), then

$$\mathbf{f}_\Sigma = \mathbf{0} \ at \ \Sigma. \tag{4.34}$$

If a potential energy localized at Σ exists (e.g., Σ constitutes a *third body* of infinitesimal thickness) then a surface energy term must be added in the right hand side of (4.23). This can be handled with the notion of *surface distribution* (introducing a *delta function* with compact support on Σ ; see Paragraph 4.4) ; As a result eqn.(4.34) will contain a source term akin to *surface tension* for a non-flat interface Σ.

Remark 4.1 : Notice that the quasi-static hypothesis (neglect of inertia) has played a role in the special writing of the expression of $\mathbf{f}_\Sigma$. A more general form is to be given in Section 6 in the fully dynamical case.

4.4 Use of distribution (generalized functions) theory

Here we just want to give a few ideas which would require much longer developments for an accurate mathematical exposition. Generalized functions or *distributions* were introduced in the 1940s by the French mathematician L.Schwartz following pioneering works by Hadamard and J.Leray in France, and S.L.Sobolev in Russia , respectively. By transferring operations of differentiation on sufficiently smooth *test functions* after integration by parts over a compact support, they allow for the *weak* formulation of solutions of partial differential equations which admit *singular* behaviors at points or jump discontinuities across surfaces. The principle of virtual power in continuum mechanics may itself be considered an exemplary use of a *distributional formulation* in which the role of test functions is played by virtual fields of displacements or velocities (see, e.g., Maugin, 1980). The theory of cracks in particular and, more generally, of « elastic defects » should have the theory of distributions for natural background. But few authors have worked along that line, probably by lack of sufficient mathematical education (one exception is H.D.Bui and co-workers). The argument defended by

Dascalu and Maugin (1994) is that the very structure of elasticity theory and the singularity order of elastic solutions dictate (i) the *expression of the energy balance law in presence of cracks* and (ii) that a *Griffith-like surface-energy distributions follows of necessity*. But a similar argument used for defects D such as cavities in quasi-static expansion shows that Griffith's argument must be carefully applied when the dimension of D is the *same* as that of the material body. Without repeating the proofs of Dascalu and Maugin, we think it appropriate to emphasize the following results as they bear direct relationship to the results exposed in previous sections.

First of all, while the distribution formalism allows one to directly incorporate the naturally associated boundary conditions in the *local* formulation of balance laws by using the notion of *delta* functions with hypersurfaces as support, only the *local energy balance* (as also the equation of pseudomomentum) is essentially modified to account simultaneously for boundary conditions and the singular behavior of mechanical fields at singular sets. But this generalized statement of the energy balance is not an *a priori* statement as in Griffith's original approach that accounts for surface energy along the crack faces : it is a *consequence* of the other local equations. Furthermore, if the additional term present in the local *distributional* balance of energy indeed represents the rate of energy released during the progress of the defect, then for a *steady* propagation criterion (à la Griffith), the distributional balance of energy takes the form of a true (distributional) *conservation law* for the total system « material plus defect ». Essentially, if one assumes (quasi-statics and no true material inhomogeneities) that eqns.(3.12), (3.13), (3.14) hold in the appropriately simplified form together with the accompanying balance of angular momentum, we have the following equations at all regular material points $\mathbf{X}$ in the absence of defects:

$$div_R \mathbf{T} = \mathbf{0}, \tag{4.35}$$

$$div_R (\mathbf{x} \times \mathbf{T}) = \mathbf{0}, \tag{4.36}$$

$$\frac{\partial W}{\partial t} - div_R (\mathbf{T}.\mathbf{v}) = 0 , \tag{4.37}$$

$$div_R \mathbf{b} = \mathbf{0} ; \tag{4.38}$$

then the first three of these take on the following *distributional form* over E^3 :

$$Div_R \mathbf{T} + \mathbf{T}^d \delta(\partial B) = \mathbf{0}, \tag{4.39}$$

$$Div_R (\mathbf{x} \times \mathbf{T}) + (\mathbf{x} \times \mathbf{T}^d) \delta(\partial B) = \mathbf{0}, \tag{4.40}$$

$$D_t W - Div_R (\mathbf{T}.\mathbf{v}) - (\mathbf{T}^d.\mathbf{v}) \delta(\partial B) = 0 , \tag{4.41}$$

where Div_R and D_t are material gradient and time-derivative distributional operators, $\mathbf{T}^d$ is the traction at ∂B, and $\delta(\partial B)$ is the delta distribution with support ∂B. In the presence of elastic defects (e.g., crack, cavity, dislocation), eqns.(4.39) and (4.40) are left unchanged while (4.41) takes on the following generalized form (cf. Dascalu and Maugin, 1994)

$$D_t W + \overline{V}\overline{G}\delta(S) = Div_R(\mathbf{T}.\mathbf{v}) + (\mathbf{T}^d.\mathbf{v})\delta(\partial B). \tag{4.42}$$

Here D is the *set of defect points*, B/C is the homogeneous elastic body, $ED = S$ is the *set of extension points* of the defect (the crack tip for a crack), $\overline{G}$ is the energy absorbed per unit area (or length) extension of the said defect, and $\overline{V}$ is the propagation speed. Equation (4.42) is valid at *all* points of B and ∂B. If, furthermore, there exists a steady-state propagation criterion of the Griffith type

$$\begin{aligned} &\overline{G} < G_c : \text{no extension} \\ &\overline{G} = G_c : \textit{possible} \text{ extension} \end{aligned} \tag{4.43}$$

then (4.42) becomes a *true* distributional *conservation law* in the form

$$D_t(W + G_c H(C)) = Div_R(\mathbf{T}.\mathbf{v}) + (\mathbf{T}^d.\mathbf{v})\delta(\partial B), \tag{4.44}$$

where $H(C)$ is a heaviside function and the quantity $G_c H(C)$ may truly be called the *energy of the defect.* Equation (4.44) holds true because we have the distributional result

$$D_t H(C) = \overline{V}\delta(S), G_c \text{ constant}. \tag{4.45}$$

Note that the « energy of the defect » depends on time through the time rate of change of the defect's size. Equation (4.39) was proved for a *straight* crack and for both screw and edge dislocations (cf.Dascalu and Maugin, 1994) by introducing regularizations of *nonintegrable* quantities (this is the case of $\partial W / \partial t$ in the case of the crack and of both $\partial W / \partial t$ and $\mathbf{T}.\mathbf{v}$ for dislocations , a more tedious case) and principal values in Cauchy-Hadamard-Schwartz sense. In both cases $\overline{G}$ is automatically found as the *J*-integral of fracture or the celebrated *Peach-Koehler force* of dislocation theory. These are cases where the set of defect points D has lower dimension than the body (line or plane versus plane or volume). The equivalent statement is somewhat different for a cavity where the dimension of D is the same as that of the material body, while the set ED is naturally of dimension two. In that case, contrary to what could be inferred through a hasty generalization of Griffith's idea, the « energy of the defect » has to be an *energy per unit volume*, just as W itself. Then eqn.(4.40) is shown to be replaced by the following *distributional conservation law*

$$D_t(W + W_c H(l - r)) = Div_R(\mathbf{T}.\mathbf{v}) + (\mathbf{T}^d.\mathbf{v})\delta(\partial B), \tag{4.46}$$

where l is the radius of the cavity and the steady-state expansion criterion reads $W = W_c$, and this is related to the M path-independent integral of Knowles and Sternberg (1972) by

$$M := \int_S \left((\mathbf{X}.\mathbf{N})W - \mathbf{N}.\mathbf{T}.(\nabla_R \mathbf{u})^T.\mathbf{X} \right) dS = \left(M_c = 2\pi l^2 W_c \right), \tag{4.47}$$

where the last expression within parentheses holds in the plane case. This formulation can be extended to *migrating intersurfaces* whre the expansion criterion involves the jump in W across the interface rather than W itself (cf. Maugin and Trimarco, 1994). Also, there exists a *distributional balance equation of pseudomomentum* which includes the driving force located at the set $ED = S$. This will be (to be proved by way of exercise)

$$Div_R \mathbf{b} + F^{crack} \mathbf{E}_1 \delta(S) + (\mathbf{N}.\mathbf{b})\delta(\partial B) = \mathbf{0}, \tag{4.48}$$

where F^{crack} is none other than the J-integral and S stands for the crack tip.

5 Material forces of quasi-inhomogeneity

5.1 The case of thermoelastic conductors

The thermoelasticity of conductors involves thermal dissipation and, therefore, cannot be deduced from a classical Lagrangian-Hamilotnian variational principle . This is where Remark 3.1 proves to be useful. We can deduce the missing pseudomomentum balance law from the classically known equations of finite strain thermoelasticity (cf. Epstein and Maugin, 1995a ; Maugin and Berezovski, 1999).

A. Reminder

For the nonlinear framework of thermoelasticity, we refer the reader to Suhubi (1975) and Maugin (1988). Here we simly list the relevant field equations and constitutive equations at any regular material point $\mathbf{X}$ in a material body. The Piola-Kirchhoff formulation is used from the start. Just as above, we consider no body forces for the sake of clarity. Then we have the following list of classical field equations :

- *Balance of mass :*

$$\left.\frac{\partial \rho_0}{\partial t}\right|_{\mathbf{X}} = 0, \tag{5.1}$$

- *Balance of linear (physical) momentum :*

$$\left.\frac{\partial \mathbf{p}}{\partial t}\right|_{\mathbf{X}} - div_R \mathbf{T} = \mathbf{0} \; ; \tag{5.2}$$

- *Balance of energy* (no body heat supply for the sake of simplicity):

$$\left.\frac{\partial H}{\partial t}\right|_{\mathbf{X}} - \nabla_R.(\mathbf{T}.\mathbf{v} - \mathbf{Q}) = 0 \; ; \tag{5.3}$$

- *Balance of entropy*

$$\left.\frac{\partial S}{\partial t}\right|_{\mathbf{X}} + \nabla_R.\mathbf{S} = \sigma_B \, . \tag{5.4}$$

The *second law of thermodynamics* at all regular material points **X** in the body imposes that

$$\sigma_B \geq 0 \; . \tag{5.5}$$

In the above set equations the total volume energy H, the volume kinetic energy K, the volume *internal energy* E, the volume entropy S, the material entropy flux **S** (not to be mistaken for the second Piola-Kirchhoff stress) and the bulk entropy source σ_B are given by

$$\begin{aligned} &H = K + E, K = \frac{1}{2}\rho_O \mathbf{v}^2, E = \rho_0 e, \\ &S = \rho_0 \eta, \mathbf{S} = \mathbf{Q}/\theta \, , \sigma_{\mathrm{B}} = -\theta^{-1}\mathbf{S}.\nabla_{\mathrm{R}}\theta \, , \end{aligned} \tag{5.6}$$

where θ is the absolute temperature , and e and η are the internal energy and entropy per unit mass in the actual configuration (at time t). In the classical *thermoelasticity of conductors*, the constitutive equations (laws of state) are deduced from the *free energy density* W by

$$W = W(\mathbf{F}, \theta; \mathbf{X}) = E - S\theta, \mathbf{T} = \frac{\partial W}{\partial \mathbf{F}}, S = -\frac{\partial W}{\partial \theta} \, . \tag{5.7}$$

The entropy flux should also satisfy a " continuity " requirement such that

$$\mathbf{S}(\mathbf{F}, \theta, \nabla_R \theta; \mathbf{X}) \to \mathbf{0} \;\; as \;\; \nabla_R \theta \to 0 \, . \tag{5.8}$$

Equations (5.1) through (5.89), together with a more precise expression for W and some of its mathematical properties (e.g., convexity), are those to be used in studying sufficiently regular *nonlinear dynamical processes in thermoelastic conductors.*

Exercise 5.1 : Account for (5.2) in the local statement (5.3) of the first law of thermodynamics to establish the so-called *internal-energy theorem* (*tr* = trace) :

$$\left.\frac{\partial E}{\partial t}\right|_{\mathbf{X}} = tr\left(\mathbf{T}.(\nabla_R \mathbf{v})^T\right) - \nabla_R.\mathbf{Q} \ .$$

Exercise 5.2 : Show that the *Clausius-Duhem inequality* reads

$$-\left(\frac{\partial W}{\partial t} + S\frac{\partial \theta}{\partial t}\right) + tr\left(\mathbf{T}.(\nabla_R \mathbf{v})^T\right) - \mathbf{S}.\nabla_R \theta \geq 0 \ .$$

B. Thermal material force

But for the explicit functional dependence of ρ_0 and W on $\mathbf{X}$, the above formalism does not reveal any material inhomogeneities, in particular the fact that thermoelastic properties may also be *particle dependent.* We need to write an equation of balance *directly on the material manifold* to see a manifestation of that inhomogeneity. Taking benefit of Remark 3.1 , we just pull back eqn.(5.2) to M^3 by applying $\mathbf{F}$ tp the right and integrate by parts, accounting for the indicated dependence of mass density and free energy, obtaining thus the following local balance of pseudomomentum :

$$\left.\frac{\partial P}{\partial t}\right|_{\mathbf{X}} - div_R \mathbf{b} = \mathbf{f}^{inh} + \mathbf{f}^{th} \ , \tag{5.9}$$

in which we have defined the (material) *pseudomomentum density* P, the Eshelby material stress tensor $\mathbf{b}$, the material force of true inhomogeneities $\mathbf{f}^{inh}$ and the material force of quasi-inhomogeneity due to thermal effects, $\mathbf{f}^{th}$, by

$$P = -\mathbf{p}.\mathbf{F} = \rho_0 \mathbf{C}.\mathbf{V} \ , \tag{5.10}$$

$$\mathbf{b} = -\left(L^{th}\mathbf{1}_R + \mathbf{T}.\mathbf{F}\right), \ \mathbf{f}^{inh} = \left.\frac{\partial L^{th}}{\partial \mathbf{X}}\right|_{expl} , \tag{5.11}$$

and

$$\mathbf{f}^{th} := S\nabla_R \theta \ , \tag{5.12}$$

together with an effective *Lagrangian density* for thermoelasticity L^{th} given by

$$L^{th} = K - W = \bar{L}^{th}(\mathbf{F}, \mathbf{v}, \theta; \mathbf{X}), \ K = \frac{1}{2}\rho_0 \mathbf{V}.\mathbf{C}.\mathbf{V} \ . \tag{5.13}$$

The fact that the " material force " $\mathbf{f}^{th}$ arises simultaneously with $\mathbf{f}^{inh}$, and the latter may have the interpretation of a continuous distribution of dislocations shows that, on the material manifold, a spatially nonuniform field of temperature may be viewed as a *quasi material inhomogeneity* or a *quasi-plastic phenomenon* amenable through a geometrical interpretation. Corresponding material forces are called *material forces of quasi-inhomogeneity*. This effect disappears once thermal equilibrium has been reached after complete diffusion of heat. Such an effect subsists in the absence of true material inhomogeneities, in which case eqn.(5.10) reduces to the simpler form,

$$\left.\frac{\partial P}{\partial t}\right|_{\mathbf{X}} - div_R \mathbf{b} = \mathbf{f}^{th} \ , \tag{5.14}$$

before thermal equilibrium is reached. In the present exposition (proof at any regular material point) eqn.(5.10) is nothing but an *identity* which expresses the *Noether-Ericksen identity* for the case of inhomogeneous, anisotropic thermoelasticity in finite strains. To our knowledge, the thermal material force - not with such a denomination - first appeared in small strains in a work of Bui (1977). We must emphasize that heat conduction does not modify here the formal expression of pseudomomentum, of the Eshelby stress, and of the true inhomogeneity force. This means that temperature, although a field in itself, is *not* an additional degree of freedom, at least in the present approach. There are other approaches where thermal effects alter the expression of both P and $\mathbf{b}$ (see Dascalu and Maugin, 1995) - these are, in essence, closer to the true spirit of field theory. We finally note that eqn.(5.4) may also be written as

$$\theta \left.\frac{\partial S}{\partial t}\right|_{\mathbf{X}} + \nabla_R . \mathbf{Q} = 0 \ , \tag{5.15}$$

which is the basic form of the *heat equation*. Although the right-hand side of this equation vanishes (here there is no intrinsic dissipation), eqn.(5.15) is *not* a strict conservation law because of the thermal factor in the first term. Equations (5.14) and (5.15) may frm the basis for a canonical formulation of dissipative continuum mechanics (see herein after). For further use in that direction, we note the following *double* Legendre transformation between H and L^{th} :

$$H = P.\mathbf{V} + S\theta - L^{th}, P = \frac{\partial L^{th}}{\partial \mathbf{V}}, S = \frac{\partial L^{th}}{\partial \theta} \ . \tag{5.16}$$

C. The problem of thermal brittle fracture

It should be again clear to the reader that at any *regular* material point $\mathbf{X}$ the local equations of momentum (5.2) and (5.14) are, through the Noether-Ericksen identity, direct consequences of one another. However, if we integrate each of these over a regular (simply connected) *homogeneous* material body B - of boundary ∂B equipped with unit outward normal $\mathbf{N}$ -, thanks to the trivial commuting of material integration and material time differentiation and the mathematically justified use of the divergence theorem, we obtain the following two global equations :

$$\frac{\partial}{\partial t}\int_B \mathbf{p} dV = \int_{\partial B} \mathbf{N}.\mathbf{T} dS \ , \tag{5.17}$$

$$\frac{\partial}{\partial t}\int_B P dV = \int_{\partial B} \mathbf{N}.\mathbf{b} dS + \int_B \mathbf{f}^{th} dV \ , \tag{5.18}$$

respectively, where the second must be understood *component-wise* (on the material manifold) because, in contrast to what happens in eqn.(5.17), the quantities involved in the integrands in eqn.(5.18) are material *covectors*, and the material manifold M^3 does not play the same neutral role as that played by the Euclidean physical manifold E^3 in (5.17). Obviously, in writing the global forms, we have again lost the convection property relating the two equations of linear momentum. This means that eqns.(5.17) and (5.18) can be used for different purposes, essentially (5.17) for *solving* the physical boundary-value, initial-value problem, and (5.18) for another purpose that becomes clear in two instances. The first of these is during numerical computations using a certain scheme. Not only does this scheme (e.g., *finite-difference scheme* in nonlinear wave propagation, or *finite-element method* in structural computations ; see Section 7 below) must be compatible with energy conservation, that is, the global equation obtained by summing eqn.(5.3) over the relevant material region, e.g.,

$$\frac{\partial}{\partial t}\int_B H dV = \int_{\partial B} \mathbf{N}.(\mathbf{T}.\mathbf{v} - \mathbf{Q}) dS \ , \tag{5.19}$$

but it must also be compatible with the additional balance law (5.18). Equations.(5.18) and (5.19) are the relevant equations to study the progress of fracture in the thermoelasticity of conductors because they capture thermo-mechanical singularities, i.e., they give rise to nonvanishing quantities that characterize this singularity : the driving force and the energy-release rate.We will not repeat here the analysis given in the purely elastic case. It suffices, accounting for the singularity order of temperature in the neighborhood of the crack tip (singularity of tthe Laplacian operator) to mention the thermal generalizations of equations (4.11) and (4.12) :

$$\frac{\partial}{\partial t}\int_B P_1 dV + F_1^{crack} = \int_{\partial B} (\mathbf{N}.\mathbf{b})_1 dS + \int_B \left(\mathbf{f}^{th}\right)_1 dV \ , \tag{5.20}$$

and

$$\frac{\partial}{\partial t}\int_B H dV + G^{crack} = \int_{\partial B} \mathbf{N}.(\mathbf{T}.\mathbf{v} - \mathbf{Q}) dS \ , \tag{5.21}$$

where we were cautious to write the first in component (*1*) form and we observe the appearance of source terms due to to the inward motion of the crack front. These are given the same formulas as those for pure elasticity except that L^{th} replaces the previous L. Accounting for these expressions, we finally obtain the global balance of pseudomomentum and energy over the subbody C with surrounding surface S as

$$F_1^{crack} = \int \left((\mathbf{N}.\mathbf{b})_1 + P_1 (\overline{\mathbf{V}}.\mathbf{N}) \right) dS + \int_G \left(\mathbf{f}^{th} \right)_1 dV - \frac{\partial}{\partial t} \int_G P_1 dV , \tag{5.22}$$

and

$$G^{crack} = \int \left(H (\overline{\mathbf{V}}.\mathbf{N}) + \mathbf{N}.(\mathbf{T}.\mathbf{v} - \mathbf{Q}) \right) dS - \frac{\partial}{\partial t} \int_G H dV . \tag{5.23}$$

We recognize in the last expression minus the change in *potential energy (dynamic thermoelastic, homogeneous case)* changed of sign, that is, the *dissipation rate* in the progress of the notch inside the material body. Accordingly, there must be a relationship between the latter quantity and the *power expanded* irreversibly by the *material force* of component F_1^{crack}. Because of the universal form of the expressions for F^{crack} and G^{crack}, just the same as before it is shown that

$$\overline{V}_1 F_1^{crack} = G^{crack} \geq 0 \quad , \tag{5.24}$$

where the inequality sign indicates the thermodynamical irreversibility of the crack growth phenomenon. This corresponds to the presence of a *hot heat source* at the crack tip, an effect that can be observed by means of infrared thermography.

Exercise 5.3 : If the temperature field in the neighborhood of the crack tip line satisfies the following condition

$$\overline{\mathbf{V}}.\nabla_R \theta = - \left. \frac{\partial \theta}{\partial t} \right|_{\mathbf{X}} + regularterm , \tag{5.25}$$

then the heat source localized at the crack tip is related to entropy and the material thermal force by (Maugin and Berezovski, 1999)

$$Q_{tip} \equiv \lim_{\Gamma \to 0} \int \mathbf{N}.\mathbf{Q} dS = \lim_{\Gamma \to 0} \left(\frac{\partial}{\partial t} \int_G (-S\theta) dV \right) - \lim_{\Gamma \to 0} \int_G \left(\overline{\mathbf{V}}.\mathbf{f}^{th} \right) dV . \tag{5.26}$$

Comments. The perspicacious reader will have noticed that expression (5.22) -component along the crack direction - is a dynamical thermoelastic generalization of the celebrated *Eshelby-Cherepanov-Rice J*-integral of brittle fracture if we remember the expression of the dynamical Eshelby tensor **b**. The formulation above is strictly based on thermomechanical arguments on the material manifold.

5.2 The case of dissipative media with internal variables

So far we have accounted only for thermal dissipation. In modern continuum mechanics, we account for a variety of microscopic phenomena responsible for macroscopic dissipation through the notion of *internal variables of state*. These variables have for essential property to be uncontrolable directly by external stimuli, so that they expend power only in the bulk. The theory of irreversible processes accounting for internal variables is documented in detail with many applications in a recent book (Maugin, 1999a) to which we refer the reader. We note collectively α these variables whose choice and tensorial nature depend on the physical accuteness of the theoretician helped by the experimentalist who should uncover those most representative variables (e.g., density of dislocation, plastic strain, work-hardening variables, etc). Such a thermodynamical framework is particularly powerful in the field of study of *anelastic behaviors* of solid-like materials (see, Maugin, 1992) and of the *non-Newtonian behavior* of complex fluids. Remarkably enough, compared to what was given in the preceding paragraph, the only initial change is the introduction of α in the list of functional arguments of the free energy W:

$$W = \overline{W}(\mathbf{F},\theta,\alpha;\mathbf{X}) \tag{5.27}$$

for an anisotropic, possibly anelastically inhomogeneous material in finite strains, whose basic behavior is elastic, but it may present combined anelasticity. Classically, $\overline{W}$ will depend on $\mathbf{F}$ only through another quantity such as an element of a multiplicative decomposition while α itself may contain another element of this decomposition (case of finite-strain elasto-plasticity and elasto-viscoplasticity). Just the same as before, but introducing the new thermodynamical force A associated with α, we have the following equations of state, definitions, and limit condition :

$$W = W(\mathbf{F},\theta,\alpha;\mathbf{X}), \mathbf{T} = \frac{\partial W}{\partial \mathbf{F}}, S = -\frac{\partial W}{\partial \theta}, A = -\frac{\partial W}{\partial \alpha} , \tag{5.28}$$

$$L^{th} = K - W; K = \frac{1}{2}\rho_0(\mathbf{X})\mathbf{v}^2 , \mathbf{f}^{inh} = \left.\frac{\partial L^{th}}{\partial \mathbf{X}}\right|_{expl} , \tag{5.29}$$

$$\mathbf{S}(\mathbf{F},\theta,\alpha;\nabla_R\theta;\mathbf{X}) \to \mathbf{0}\, as\, \nabla_R\theta \to \mathbf{0} . \tag{5.30}$$

Equations (5.1) through (5.3) are left formally unchanged. But, obviously, there exists now an *intrinsic dissipation* so that eqn.(5.15) is generalized to

$$\theta\frac{\partial S}{\partial t} + \nabla_R.\mathbf{Q} = \Phi^{int} \equiv A\left.\frac{\partial \alpha}{\partial t}\right|_{\mathbf{X}} . \tag{5.31}$$

The balance of pseudomomentum is accordingly modified. Indeed, performing on eqn. (5.2) the convection back to the reference configuration by applying **F** to the right and integrating by parts on account of eqns.(5.28)$_1$, we arrive at the following *inhomogeneous* balance of pseudomomentum on the material manifold :

$$\left.\frac{\partial P}{\partial t}\right|_{\mathbf{X}} - \left(div_R \mathbf{b} + \mathbf{f}^{inh}\right) = \mathbf{f}^{th} + \mathbf{f}^{int}, \mathbf{f}^{int} := A(\nabla_R \alpha)^T . \tag{5.32}$$

Clearly, insofar as this equation is concerned, internal variables produce a source term - a material force - that resembles $\mathbf{f}^{th}$ in both general expression and effects ! In particular, we may say that the microscopic irreversibility for which α accounts, manifests itself as a *quasi-inhomogeneity* in this equation, just the same as thermal effects. We also note that the new additional terms in (5.31) and (5.32) exhibit a consistency, one involving the time derivative, the other the space derivative, of the internal variable. This is more symmetric than the effect of temperature because there was nothing in the right-hand side of eqn.(5.15). But this is only apparent because we have the result of

Exercise 5.4 : Prove that (5.31) can also be written as (Maugin, 2000)

$$\left.\frac{\partial (S\theta)}{\partial t}\right|_{\mathbf{X}} + \nabla_R . \mathbf{Q} = \Phi^{th} + \Phi^{int}, \Phi^{\mathrm{th}} := S\frac{\partial \theta}{\partial t} . \tag{5.33}$$

Remark 5.1 : The similarity between variables α and θ is thus enhanced, but while the former is governed by the *heat equation*, the former has to be governed by a pure *evolution equation* subjected to the second law of thermodynamics (non-negative dissipation).

In the *thermo-anelastic fracture problem*, because of the already noticed similarity between source terms in the right-hand side of eqn.(5.32)$_1$ the proof of the consistency relation (5.24) will be facilitated. Indeed, for instance, we shall have

$$\bar{\mathbf{V}}. \lim_{\Gamma \to 0} \int_G \mathbf{f}^{int}\, dV \approx - \lim_{\Gamma \to 0} \int_G \left(A \left.\frac{\partial \alpha}{\partial t}\right|_{\mathbf{X}} \right) dV , \tag{5.34}$$

if α verifies a lemma such as in (5.25). We recognize in the right-hand side of (5.34), the limit, changed of sign, of the global intrinsic dissipation in the material domain G.

On may then ask if something is fundamentally changed in the formulation if the internal variable of state is such that its gradient must be introduced before hand in the functional form of W. This is a fashionable subject because it is related to the occurrence of *dissipative structures* (Maugin, 1990) and localization effects (Maugin, 1997a,b). The reader should consult the original papers for this as we can only give here a few elements. What is to be noted is that very often this additional dependence materializes in replacing a partial derivative by a functional *Euler-Lagrange* derivative in the definition of the thermodynamical force A, and that the Eshelby stress tensor contains additional terms due to the gradient of α as this tensor has for essential property to capture all gradients according to its canonical definition (cf.Maugin, 1993, 1995). Remarkably enough, to the price of weak assumptions, the simple but powerful consistency relation (5.24) still holds good in this complicated case (cf.Maugin, 1997a,b).

In effect, in this more general case where the notion of internal variable of state is questionable (because *diffusion* of the said variable requires an associated boundary condition, thus implying the possible direct action of an external stimulus on the variable at a boundary), we have :

$$W = \overline{W}(\mathbf{F},\alpha,\nabla_R\alpha,\theta;\mathbf{X}) \tag{5.35}$$

The dependence on $\nabla_R\alpha$ indicates a *weak nonlocality* in the α variable. No *inertia* of a classical type is here associated to α. We set

$$\begin{aligned} &\mathbf{T} = \partial\overline{W}/\partial\mathbf{F}, S = -\partial\overline{W}/\partial\theta,\\ &A = \overline{A} + \nabla_R.\mathbf{M} = -\delta\overline{W}/\delta\alpha,\\ &\overline{A} = -\partial\overline{W}/\partial\alpha, \mathbf{M} = \partial\overline{W}/\partial(\nabla_R\alpha), \end{aligned} \tag{5.36}$$

where $\delta/\delta\alpha$ indicates the Euler-Lagrange functional derivative with respect to α. The local balance of pseudomomentum is obtained following Remark 3.1 as

$$\frac{\partial P}{\partial t}\Big|_{\mathbf{X}} - \left(div_R\mathbf{b} + \mathbf{f}^{inh} + \mathbf{f}^{th} + \mathbf{f}^{int}\right) = \mathbf{0}, \tag{5.37}$$

wherein

$$P = -\mathbf{F}^{\mathrm{T}}.\mathbf{p}, \tag{5.38}$$

$$\mathbf{b} = -\left(L^{th}\mathbf{1}_{\mathrm{R}} + \mathbf{T}.\mathbf{F} + \mathbf{M}.(\nabla_R\alpha)^T\right), \tag{5.39}$$

$$\mathbf{f}^{inh} = \left(\frac{\partial L^{th}}{\partial\mathbf{X}}\right)_{\mathbf{expl}}, L^{th} := \frac{1}{2}\rho_0(X)\mathbf{v}^2 - \overline{\mathbf{W}}, \tag{5.40}$$

$$\mathbf{f}^{th} := S\nabla_R\theta, \mathbf{f}^{\alpha} := A\nabla_R\alpha \tag{5.41}$$

are, respectively, the *canonical* (material) *momentum*, the *Eshelby* material stress, the "material" force of the ***true*** material inhomogeneities, an effective Lagrangean density, the "material" thermal force, and a "material" force due to spatial disuniformities in the α variable. The latter, just like $\mathbf{f}^{th}$, may be referred to as a material force of *quasi-*inhomogeneitiy, *i.e.*, manifesting itself just like $\mathbf{f}^{inh}$ in the balance equation (5.37). Notice that the definition of P is left unchanged because the variable α does not carry inertia. In strict parallel to (5.37) we establish the equation of (mechanical) *energy* by taking the inner product of the (formally unchanged) physical momentum equation with $\mathbf{v}$ and integrating by parts to obtain

$$\frac{\partial H}{\partial t} - \nabla_R.(\mathbf{T}.\mathbf{v} + \mathbf{M}\dot{\alpha}) = \theta\frac{\partial S}{\partial t} - A\frac{\partial \alpha}{\partial t}, \tag{5.42}$$

where $H= K+E$ is the Hamiltonian density, $E=W+\theta S$ is the internal energy, and $K=\rho_0\mathbf{v}^2/2$ is the kinetic energy, all per unit reference volume. In the absence of bulk heat supply the *first law* of *thermodynamics* imposes that (5.42) take on a *strict conservative form* so that the r-h-s of (5.42) is necessarily the material divergence of a material vector, say $-\mathbf{Q}$. I.e., (5.42) transforms readily to a more classical form (compare to (5.3))

$$\frac{\partial H}{\partial t} - \nabla_R.(\mathbf{T}.\mathbf{v} + \mathbf{M}\dot{\alpha} - \mathbf{Q}) = 0 \tag{5.43}$$

where $\mathbf{Q}$ is called the material *heat-flux* vector, and the following *local entropy balance* holds

$$\theta\frac{\partial S}{\partial t}\Big|_{\mathbf{X}} + \nabla_R.\mathbf{Q} = \Phi^{int} , \tag{5.44}$$

where $\Phi^{int} = A\dot{\alpha}$ is the *intrinsic dissipation*, or

$$\frac{\partial S}{\partial t}\Big|_{\mathbf{X}} + \nabla_R.\mathbf{S} = \sigma_B \tag{5.45}$$

with

$$\begin{gathered} \mathbf{S} = \mathbf{Q}/\theta \\ \sigma_B = \theta^{-1}(A\dot{\alpha}) - \mathbf{S}.\nabla_R(ln\theta) \geq 0 \end{gathered}, \tag{5.46}$$

where the last inequality sign expresses the *second law of thermodynamics* at each regular material point in the body. The formulation (5.35) and (5.44) fits well in the formalism of *internal variables of state* in spite of the introduction of gradients of α and of the *diffusive* nature of this variable. We let the reader check that equation (5.43) is also obtained by computing the *power* of equation (5.37) in the material velocity fields $\mathbf{V} = -\mathbf{F}^{-1}.\mathbf{v}$ (in general a nontrivial computation).

Energy-release rate in fracture

We let the reader show that in fracture (neglecting thermal effects, and considering a materially homogeneous material) we can obtain the expression of the *energy-release rate* :

$$G^{crack} = \int_{\Gamma} \left(H(\overline{\mathbf{V}}.\mathbf{N}) + \mathbf{N}.\mathbf{T}.\mathbf{v} + \mathbf{N}.\mathbf{M}.\dot{\alpha} \right) dA$$

$$- \left(\frac{\partial}{\partial t} \int_{G} H dV + \int_{G} A \frac{\partial \alpha}{\partial t} \Big|_{\mathbf{X}} dV \right) \qquad (5.47)$$

while (5.37) yields, component-wise, a *global material force*, the driving force acting on the crack tip,

$$F_1^{crack} = \int_{\Gamma} \left((\mathbf{N}.\mathbf{b})_1 + P_1 (\overline{\mathbf{V}}\mathbf{N}) \right) dS$$

$$- \left(\frac{\partial}{\partial t} \int_{G} P_1 dV - \int_{G} f_1^{int} dV \right) \qquad (5.48)$$

with, in the limit as G shrinks to the crack tip , the *global dissipative inequality*

$$G^{crack} = F_1^{crack} \overline{V}_1 \geq 0 \qquad (5.49)$$

in which we have accounted for the fact that in these limits we have the following local behavior: $\mathbf{v} \approx -\overline{\mathbf{V}}.\mathbf{F}^T, \dot{\alpha} \approx -\overline{\mathbf{V}}.\nabla_R \alpha, A\dot{\alpha} \approx -\overline{\mathbf{V}}.\mathbf{f}^{int}$. The bilinear form (5.49) governs the irreversible thermodynamics of the extension of C independently of the precise physical *significance* of α.

5.3 The case of solids with a microstructure : Additional degrees of freedom

Firsr we look at the case where the *microstructure* is described by the fact that addditional kinetic properties are granted to any material point **X.** These may be properties of rotation, such as in so-called *polar* or *Cosserat* media, properties of affine deformation, such as in the *micromorphic* media of A.C.Eringen, a *gyroscopic magnetic spin* such as in ferromagnets at low temperature, etc. We still call generically α this additional degree of freedom, which may be represented by an orthogonal matrix, a general second order tensor, an axial vector, an n-vector in E^n, etc, depending on the case. The case of ferromagnets is examined in another contribution to this volume (cf. C.Trimarco and G.A.Maugin). A general abstract case has been examined in Maugin (1997a ; 2000a). Polar crystals have been dealt with in Maugin

(1998a). In these cases the self-interaction of the microstructure is taken into account through the presence of the gradient $\nabla_R \alpha$ in the free energy. This is reminiscent of the phenomenological representation of spatial ordering in materials with an order parameter and long-range interactions (Landau-Ginzburg). Accordingly, in the thermal and inhomogeneous framework the free energy still has expression (5.35), while the kinetic energy density of the continuum reads

$$K = \frac{1}{2}\rho_0(\mathbf{X})\mathbf{v}^2 + K_\alpha, K_\alpha = \frac{1}{2}\rho_0(\mathbf{X})\dot{\alpha}(\mathbf{X},t)\mathbf{I}_\alpha(\mathbf{X})\dot{\alpha}(\mathbf{X},t), \tag{5.50}$$

where $\mathbf{I}_\alpha$ is the inertia associated with α - this is a symmetric real-valued operator on the space of α's - and the superscript T represents the appropriate operation of transposition. There exists thus a *physical momentum* $\mathbf{p}_\alpha$ associated with α such that

$$\mathbf{p}_\alpha = -\rho_0(\mathbf{X})\mathbf{I}_\alpha(\mathbf{X})\left.\frac{\partial \alpha}{\partial t}\right|_{\mathbf{X}} . \tag{5.51}$$

In addition to eqns.(5.1) and (5.2) we therefore have the following two conservation laws (for α-inertia and physical momentum of the α-degree of freedom)

$$\left.\frac{\partial \mathbf{I}_\alpha}{\partial t}\right|_{\mathbf{X}} = \mathbf{0} \tag{5.52}$$

and

$$\left.\frac{\partial \mathbf{p}_\alpha}{\partial t}\right|_{\mathbf{X}} - \mathbf{f}_\alpha = \mathbf{0}, \mathbf{f}_\alpha = -\frac{\delta \overline{W}}{\delta \alpha}, \tag{5.53}$$

where we assume that the physical force acting on α is entirely potential. The first of (5.53) in fact reads

$$\left.\frac{\partial \mathbf{p}_\alpha}{\partial t}\right|_{\mathbf{X}} - div_R \mathbf{M} = \mathbf{g}_\alpha, \mathbf{g}_\alpha = -\frac{\partial \overline{W}}{\partial \alpha} . \tag{5.54}$$

The invariance of W , a scalar quantity, under rotations (here equivalent to the so-ccaled *objectivity* of W) imposes the following restriction

$$skew\left(\frac{\partial \overline{W}}{\partial \mathbf{F}}\mathbf{F}^T\right) + SO\left(\frac{\delta \overline{W}}{\delta \alpha}\right) = \mathbf{0} , \tag{5.55}$$

where the second contribution means the action of members of the proper orthogonal group *SO(3)* on W considered as a function of the tensorial object α and its first gradient. For instance, if α is a vector field $\mathbf{n}(\mathbf{X},t)$ in the actual configuration with values in E^3, eqn.(5.55) reads in components ($\mathbf{x}$ is referred to coordinates x^i and $\mathbf{X}$ to coordinates X^K, K=1,2,3)

$$\underset{[ij]}{skew}\left(\frac{\partial \overline{W}}{\partial F^i_{.K}} F^k_{.K}\delta_{kj} + \frac{\partial \overline{W}}{\partial n^i} n^k \delta_{kj} + \frac{\partial \overline{W}}{\partial n^i_{.,K}} n^k_{.,K}\delta_{kj}\right) = 0 . \tag{5.56}$$

this, like (5.55) , potentially contains the local *physical balance of moment of momentum.* Indeed, on account of eqns.(5.36) and $(5.54)_2$, eqn.(5.56) is none other than

$$T^{K[i}F^{j]}_{.K} = (\mathbf{g}_\alpha)^{[i} n^{j]} - M^{K[i} n^{j]}_{.,K} , \tag{5.57}$$

which reduces to the symmetry of the Cauchy stress $\sigma = J_F^{-1}\mathbf{F}.\mathbf{T}$, when the right-hand side of (5.57) vanishes, in particular when W does not depend on α and its gradient.

Exercise 5.5. Show that the equations of canonical momentum and energy are given by (Maugin, 19997a,b)

$$\left.\frac{\partial P}{\partial t}\right|_{\mathbf{X}} - \left(div_R + \mathbf{f}^{inh}\right) = \mathbf{f}^{th} + \mathbf{f}^{int} , \tag{5.58}$$

and

$$\left.\frac{\partial H}{\partial t}\right|_{\mathbf{X}} - \nabla_R . \left(\mathbf{T}.\mathbf{v} + \mathbf{M}.\left.\frac{\partial \alpha}{\partial t}\right|_{\mathbf{X}} - \mathbf{Q}\right) = 0 , \tag{5.59}$$

where

$$P = \rho_0 \mathbf{C}.\mathbf{V} - \mathbf{p}_\alpha . (\nabla_{\mathrm{R}}\alpha)^T , \tag{5.60}$$

$$H = K + W , L^{th} = K - W , \tag{5.61}$$

$$\mathbf{b} = -\left(L^{th}\mathbf{1}_R + \mathbf{T}.\mathbf{F} + \mathbf{M}.(\nabla_R \alpha)^T\right). \tag{5.62}$$

$$\mathbf{f}^{inh} = \left.\frac{\partial L^{th}}{\partial \mathbf{X}}\right|_{expl} , \mathbf{f}^{\mathrm{th}} = S\nabla_R\theta , \mathbf{f}^{int} = (\mathbf{g}_\alpha + div_R\mathbf{M}).(\nabla_{\mathrm{R}}\alpha)^T . \tag{5.63}$$

We have here an example where the microstructure contributes to *both* the canonical momentum P and the Eshelby-stress expression, as it contributes also to the energy equation in both the Hamiltonian density and the energy flux.

Note : In the nondissipative case, the expressions (5.60) and (5.62) are none other than the expressions (2.16) and (2.17). In the dissipative case eqn.(5.58) is obtained by exploiting the

Remark 3.1, i.e., forming the proper linear combination of eqns.(5.2) and $(5.53)_1$ and obtaining thus a generalized Ericksen-Noether identity.

The reader will find a complete application to the case of elastic polar crystals in Maugin (1998). A parallel application to liquid crystals (very special elastic liquids) is to be found in Maugin and Trimarco (1995d). Using the above-given results, it is not difficult to formulate the generalization of the *J*-integral of fracture to polar elastic solids (make the reasoning on eqns.(5.58-9)).

5.4 The case of solids with a microstructure : Gradient-theory of elasticity

Both the scheme briefly explored in the preceding paragraph and that dealing with *diffusive* internal variables exhibit *length scales*, a property not exhibited by classical elasticity and thermoelasticity, because only the deformation gradient is involved in the elastic energy, and not the placement itself). The above two cases exhibit a *length scale* which is related to the *microstructure* (*both* variables α and $\nabla_R \alpha$ are involved in W). Accordingly the microstructure will resonate with dynamically applied stimuli when the wavelength of the latter will match this characteristic or *internal length*. We will have the dynamical phenomenon of *dispersion*. There is another way a characteristic length or dispersion effects may be involved in elasticity. It is through the consideration of a *gradient theory* (cf.Maugin, 1980 for this general concept). This comes directly from the discrete description of crystal lattices in which one accounts for long-range interactions by accounting for interactions between second neighbors and higher-order ones (cf.Maugin, 1999b). For instance, in pure finite-strain elasticity we may consider an elastic potential energy density with the following dependence

$$W = W(\mathbf{F}, \nabla_R \mathbf{F}; \mathbf{X}). \tag{5.64}$$

This may be referred to as the *second-gradient theory of elasticity*, since $\nabla_R \mathbf{F}$ in truth is the second material gradient of the direct motion. With the same logic, it may also be referred to as a *strain-gradient theory*. This received some attention in the 1960s especially in works by R.A.Toupin (1962) and R.D.Mindlin and H.F.Tiersten (1962) - the latter for small strains. In the absence of dissipation, this corresponds exactly to the formal scheme given in eqns. (2.19) through (2.22), in which must be included a classical translational inertia and corresponding kinetic energy such as given by eqn.$(3.11)_1$. The reason for the last remark is that, although $\nabla_R \mathbf{F}$ is often considered as a variable independent of $\mathbf{F}$ in W for many purposes, there is still only one vectorial degree of freedom, the direct motion (3.1). In constructing specific field theories for representing, in a one-dimensional (space) framework, some ersatz model of three-dimensional or two-dimensional bodies, we may be led to introducing more complicated inertial terms (e.g., the laterial inertia of rods ; cf. Maugin and Christov, 2000, for theses exotic - from the point of view of analytical continuum mechanics - aspects). Reading off the formulas from Section 2, we shall have the following equations of physical momentum, canonical momentum and energy (see also Maugin and Trimarco, 1992 ; Maugin, 1993, Chapter 5)

$$\left.\frac{\partial \mathbf{p}}{\partial t}\right|_{\mathbf{X}} - div_R \mathbf{T}^{eff} = \mathbf{0}\,, \tag{5.65}$$

$$\left.\frac{\partial P}{\partial t}\right|_{\mathbf{X}} - div_R \mathbf{b}^{eff} = \mathbf{f}^{inh}\,, \tag{5.66}$$

$$\left.\frac{\partial H}{\partial t}\right|_{\mathbf{X}} - \nabla_R . \mathbf{G}^{\mathrm{eff}} = 0\,, \tag{5.67}$$

where we have defined the following quantities :

$$\mathbf{p} = \rho_0 \mathbf{v}\,,\ \mathbf{T}^{eff} = \mathbf{T} - div_R \mathbf{M}\ , \tag{5.68}$$

$$\mathbf{T} = \frac{\partial \overline{W}}{\partial \mathbf{F}}\,,\ \mathbf{M} = \frac{\partial W}{\partial (\nabla_R \mathbf{F})}\,, \tag{5.69}$$

$$P = -\,\mathbf{p}.\mathbf{F}\,,\ \mathbf{b}^{\mathrm{eff}} = \mathbf{b} - div_R(\mathbf{M}.\mathbf{F})\,, \tag{5.70}$$

$$\mathbf{b} = -\left(L\mathbf{1}_R + \mathbf{T}.\mathbf{F} + \mathbf{M} : (\nabla_{\mathrm{R}} \mathbf{F})^T\right), \tag{5.71}$$

$$H = K + W\,,\ L = K - W\,, \mathbf{G}^{eff} = \mathbf{G} - (div_R \mathbf{M}).\mathbf{v}\,,\ \mathbf{G} = \mathbf{T}.\mathbf{v} + \mathbf{M} : (\nabla_{\mathrm{R}} \mathbf{v})^T\,, \tag{5.72}$$

and

$$\mathbf{f}^{inh} = \left.\frac{\partial L}{\partial \mathbf{X}}\right|_{expl}. \tag{5.73}$$

In addition, we must express the rotational invariance of W, i.e., (compare to eqn.(5.55))

$$skew\left(\mathbf{T}\mathbf{F}^T + 2\mathbf{M} : (\nabla_{\mathrm{R}} \mathbf{F})^T\right) = 0 \tag{5.74}$$

which shows that the cauchy stress is *not* symmetric in this description since it is proportional to **F.T** .

Geometrical considerations generalizing those of Paragraph 3.4 may be found in works by M.Epstein and co-workers (Elzanowski and Epstein, 1992 ; Elzanowski *et al* , 1990 ; Epstein and de Leon, 1993, 1996))

5.5 Elastoplasticity as a structural rearrangement

The identification of the driving force of anelasticity is central to a thermomechanical approach to this type of irreversible behavior. Anelastic behavior is a macroscopic manifestation of microstructural material rearrangements. The corresponding time evolution in fact is the evolution of the distribution of microstructural material inhomogeneities, e.g., dislocation patterns. Prime importance is thus attributed to the *inhomogeneity mapping* **K** of Paragraph 3.4. and its time evolution. The inhomogeneity finite strain and the inhomogeneity velocity gradient can be defined by

$$\mathbf{C_K} := \mathbf{K}^{-T}.\mathbf{K}^{-1}, \mathbf{C}_\mathbf{K}^{-1} = \mathbf{K}.\mathbf{K}^T , \tag{5.75}$$

and

$$\mathbf{L_K} = \dot{\mathbf{K}}.\mathbf{K}^{-1} , \tag{5.76}$$

where a superimposed dot denotes partial time differentiation. The local configuration reached after the **K** mapping at **X** is called the *reference crystal* C_R by Epstein and Maugin (1990). But if **K** is taken to represent anelastic effects then by simple composition we see that

$$\mathbf{F}.\mathbf{K}(\mathbf{X}) = \mathbf{F}_e \tag{5.77}$$

is the *elastic deformation gradient.* But **K** being generally non-integrable, this is also the case of $\mathbf{F}_e$. Accordingly we can write the *Bilby-Bullough-Stroh-Kröner-Lee multiplicative decomposition*

$$\mathbf{F} = \mathbf{F}_e.\mathbf{F}_p , \mathbf{F}_p \equiv \mathbf{K}^{-1} , \tag{5.78}$$

where $\mathbf{F}_p$ may be referred to as the *plastic* (in general, *anelastic*) deformation gradient, in fact *not* a true gradient. The thermomechanics then goes as follows. The original *Clausius-Duhem inequality* per unit volume of the reference configuration K_R reads

$$-\left(\dot{W} + S\dot{\theta}\right) + tr\left(\mathbf{T}.\dot{\mathbf{F}}\right) - \theta^{-1}\mathbf{Q}.\nabla_R \theta \geq 0 . \tag{5.79}$$

We set

$$\begin{aligned} &J_e = det\mathbf{F}_e, J_p = det\mathbf{F}_p = J_K^{-1}, \\ &\mathbf{C}_e = \mathbf{F}_e^T.\mathbf{F}_e, \mathbf{C}_p = \mathbf{F}_p^T.\mathbf{F}_p = \mathbf{C_K} \end{aligned} , \tag{5.80}$$

and consider free energy densities of the following functional dependence

$$W = W(\mathbf{F}, \alpha, \theta) = J_K^{-1}\overline{W}(\mathbf{F}.\mathbf{K}(\mathbf{X}), \alpha) = J_p \overline{W}(\mathbf{F}_e, \alpha), \tag{5.81}$$

where α denotes the set of internal variables of state (e.g., hardening parameters, perhaps $\mathbf{F}_p$ itself). $\overline{W} = W_i$ now is the free energy density per unit volume of the so-called *intermediate configuration* K_i. The latter therefore is assimilated to our *reference crystal* at **X**. This is indeed in agreement with Mandel's (1971) concept of isoclinic intermediate configuration. There is in fact more connection to Mandel's theory of finite elastoplasticity than that. Remember that the

Mandel stress **M** is generally defined by eqn.(3.24). We can also define the fully covariant material stress tensor $\overline{\mathbf{S}}$ by

$$\overline{\mathbf{S}} = -\mathbf{C}.\mathbf{S}.\mathbf{C} = -\mathbf{C}.\mathbf{M} \ . \tag{5.82}$$

Then

$$\mathbf{B} := \mathbf{C}.\mathbf{b} = \mathbf{C}.\left(W\mathbf{1}_R - \mathbf{T}.\mathbf{F}\right) = \mathbf{C}.\left(W\mathbf{1}_R - \mathbf{S}.\mathbf{C}\right) = W\mathbf{C} + \overline{\mathbf{S}} = \mathbf{b}^T.\mathbf{C} \, , \tag{5.83}$$

where the last equality results from the symmetry of the Eshelby stress with respect to **C** -cf. eqn.(3.21). *Plastic incompressibility* imposed that $J_p = J_K = 1$ and

$$tr\mathbf{L}_K = tr\mathbf{L}_p = 0 \, , \tag{5.84}$$

where, in general,

$$\mathbf{L}_p = -\mathbf{K}^{-1}.\mathbf{L}_\mathbf{K}.\mathbf{K} \, . \tag{5.85}$$

We note that in plastic-incompressibility conditions the Mandel and Eshelby stresses are just the same but for a change of sign : **M** = - **b**. Furthermore, we can check in general that **b** is also symmetric with respect to $\mathbf{C_K}$, i.e.,

$$\mathbf{b}.\mathbf{C}_\mathbf{K} = \mathbf{C}_\mathbf{K}.\mathbf{b}^T \, . \tag{5.86}$$

From (5.79), the laws of state read

$$\mathbf{T} = \frac{\partial W}{\partial \mathbf{F}}, S = -\frac{\partial W}{\partial \theta}, A = -\frac{\partial W}{\partial \alpha} \ . \tag{5.87}$$

These can also be eventually written in terms of $\overline{W}$ (per unit volume of K_i). Then we let the reader prove the following expressions of the *intrinsic dissipation* :

- *per unit volume of the reference configuration* K_R :

$$\Phi_{R.} = -tr\left(\mathbf{b}.\mathbf{L}_\mathbf{K}\right) + A\dot{\alpha} = tr\left(\mathbf{M}.\mathbf{L}_\mathbf{K}\right) + A\dot{\alpha} \ , \tag{5.88}$$

or

$$\Phi_R = -tr\left(\mathbf{B}.\left(\mathbf{L}_\mathbf{K}.\mathbf{C}^{-1}\right)_S\right) + A\dot{\alpha} \ ; \tag{5.89}$$

- *per unit volume of the intermediate configuration* K_i :

$$\Phi_i = tr\left(\mathbf{S}_i.\left(\mathbf{C}_\mathbf{e}.\mathbf{L}_p\right)_S\right) + A_i\dot{\alpha} \ . \tag{5.90}$$

Here $\mathbf{S}_i$ is the second Piola-Kirchoff stress in the intermediate configuration, and a subscript s denotes the operation of symmetrization. Accordingly, the skewsymmetric parts of $\mathbf{L}_\mathbf{K}\mathbf{C}^{-1}$ and $\mathbf{C}_e.\mathbf{L}_p$ are left indeterminate in the formulations (5.89) and (5.90), respectively. We let the reader verify that

$$\overline{\mathbf{S}} = \mathbf{B} - W\mathbf{C} = \frac{\partial \widetilde{W}}{\partial \overline{\mathbf{E}}}, \widetilde{W} = \widetilde{W}(\overline{\mathbf{E}}, \alpha, \theta), \overline{\mathbf{E}} := \frac{1}{2}\left(\mathbf{1}_R - \mathbf{C}^{-1}\right), \tag{5.91}$$

in the first case, and

$$\mathbf{S}_i = 2\frac{\partial W_i}{\partial \mathbf{C}_e}, W_i = \overline{W}(\mathbf{C}_e, \alpha, \theta), \tag{5.92}$$

in the second case.

Following the principles enunciated in Maugin (1992), the thermomechanical formulations (5.89) and (5.90) are tantamount to writing the following *evolution laws* for the plastic behavior (regular plasticity surface):

$$\left(\mathbf{L}_\mathbf{K}.\mathbf{C}^{-1}\right)_S = \dot{\lambda}\frac{\partial f(\mathbf{B}, A, \theta)}{\partial \mathbf{B}}, \dot{\alpha} = \dot{\lambda}\frac{\partial f(\mathbf{B}, A, \theta)}{\partial A}, \tag{5.93}$$

and

$$\left(\mathbf{C}_e.\mathbf{L}_p\right)_S = \dot{\lambda}\frac{\partial g(\mathbf{S}_i, A_i, \theta)}{\partial \mathbf{S}_i}, \dot{\alpha} = \dot{\lambda}\frac{\partial g(\mathbf{S}_i, A_i, \theta)}{\partial A_i}, \tag{5.94}$$

with associated flow functions f and g, and plastic multiplier $\dot{\lambda}$; more on these developments in a recent work of Cleja-Tigoiu and Maugin (2000). The « classical » finite-strain elastoplasticity is presented in the books of Lubliner (1990) and Maugin (1992). Developments using the notion of Eshelby stress tensor as the *driving force* behind plasticity are given in Maugin (1994), Epstein and Maugin (1996, 1998). The geometrtical background using the notion of *evolving G-structures* is given in the same references.

5.6 Volumetric growth as a structural rearrangement

Basing on a recent work by Epstein and Maugin (2000), we view volumetric growth as a process that happens to a uniform body , which we describe in terms of a fixed reference configuration. As time goes on, more material of the *same* kind is « squished-in » smoothly into the body, but in such a way that material points preserve their identity. In other words, the process of growth can be seen as an evolution of material-point neighborhoods in a fixed reference configuration, an evolution that finds its expression in temporal changes of density and concomitant distorsions of material-point neighborhoods. While preserving the uniform material properties, the growth process will in general cause the developments of *inhomogeneities*, responsible for *residual stresses* in the body. Other types of growth, such as inclusions and surface growth (accretion) , are beyond the scope of this study. Accretive growth is dealt with by Gurtin (1993). Also, we do not deal here with the complicated physical or physiological mechanisms responsible for volumetric growth. Only a few elements of the continuum theory can be given here, the reader being referred to the original paper for details.
In contrast to previous sections, here mass density, even in the reference configuration, can evolve in time. Accordingly, the global mass balance equation reads

$$\frac{\partial}{\partial t}\int_B \rho_0 dV = \int_B \Pi dV + \int_{\partial B} \mathbf{N}.\mathbf{m}\, dS \ , \tag{5.95}$$

or, locally, replacing eqn.(3.9),

$$\left.\frac{\partial \rho_0}{\partial t}\right|_{\mathbf{X}} = \Pi + \nabla_R .\mathbf{m} \ . \tag{5.96}$$

The material vector $\mathbf{m}$ is the mass-flux vector, since the phenomenon is a priori considered as *diffusive*. Epstein and Maugin (2000) have shown that, for this theory, there exist an *effective* first Piola-Kirchhoff stress $\overline{\mathbf{T}}$, an *effective* Eshelby stress tensor $\overline{\mathbf{b}}$, a *material growth force* $\mathbf{f}^{gr}$, and a *growth dissipation density* Φ^{gr} which are given by the following set :

$$\overline{\mathbf{T}} = \mathbf{T} + \mathbf{m} \otimes \mathbf{v} \ , \tag{5.97}$$

$$\overline{\mathbf{b}} = -\left(L^{th}\mathbf{1}_R + \overline{\mathbf{T}}.\mathbf{F}\right) = -\left(L^{th}\mathbf{1}_R + \mathbf{T}.\mathbf{F} - \rho_0^{-1}\mathbf{m} \otimes P\right), \tag{5.98}$$

$$\mathbf{f}^{gr} := \Pi \mathbf{V} \ , \tag{5.99}$$

and

$$\Phi^{gr} = (\mathbf{m}.\nabla_R)E_0\big|_S = \mathbf{m}.\left(\nabla_R W + \mathbf{f}^{th}\right), \tag{5.100}$$

in such a way that the balance equation of pseudomomentum and the Clausius-Duhem inequality read

$$\left.\frac{\partial P}{\partial t}\right|_{\mathbf{X}} - \left(div_R \bar{\mathbf{b}} + \mathbf{f}^{inh}\right) = \mathbf{f}^{th} + \mathbf{f}^{int} + \mathbf{f}^{gr}, \tag{5.101}$$

and

$$-\left(\dot{W} + S\dot{\theta}\right) + tr\left(\dot{\mathbf{T}}.(\nabla_R \mathbf{v})^T\right) + \Phi^{gr} - \theta^{-1}\mathbf{Q}.\nabla_R \theta \geq 0. \tag{5.102}$$

In eqn.(5.100) the short-hand notation used in the first expression means a material gradient at fixed entropy of the internal energy density. This, in effect, yields the second expression. In eqn.(5.9) we have discarded any irreversible growth source of material momentum (cf. E.and M., 2000). Because of mass-flux, the Cauchy stress is *not* symmetric in the present theory. In terms of **T** this yields the condition

$$skew\left(\mathbf{T}.\mathbf{F}^T - \mathbf{v} \otimes (\mathbf{F}.\mathbf{m})\right) = 0. \tag{5.103}$$

Now we introduce the inhomogeneity mapping **K** called *transplant* in this growth theory with physiological applications. Let $J_K = det\, \mathbf{K}$, and ρ_c the truly invariant mass density in C_R so that

$$\rho_0 = J_K^{-1} \rho_c. \tag{5.104}$$

By time differentiation this shows that *growth* (increase in density) corresponds to $\frac{\partial}{\partial t}(det\mathbf{K}) < 0$ and *resorption* (decrease in density) corresponds to the opposite condition.

With a simplified functional dependence of W taken as

$$W = \overline{W}(\mathbf{F}, \theta, \mathbf{X}, t) \tag{5.105}$$

we can write

$$W = J_K^{-1} W_c(\mathbf{F}.\mathbf{K}(\mathbf{X}, t), \theta). \tag{5.106}$$

We also introduce the following two transplant velocity gradients :

$$\mathbf{L_K} = \dot{\mathbf{K}}.\mathbf{K}^{-1},\ \overline{\mathbf{L}}_\mathbf{K} = \mathbf{K}^{-1}.\dot{\mathbf{K}} = -\dot{\mathbf{K}}^{-1}.\mathbf{K}. \tag{5.107}$$

On account of (5.104) we have

$$\left.\frac{\partial \rho_0}{\partial t}\right|_{\mathbf{X}} = -\rho_0 tr \mathbf{L_K}. \tag{5.108}$$

Comparison of (5.108) and (5.96) yields

$$\Pi = -\left(\nabla_R.\mathbf{m} + \rho_0 tr \mathbf{L_K}\right). \tag{5.109}$$

This shows that there is no need of an evolution equation for the volumetric source, once we know the evolution of **K** and the diffusion **m.** These must be ruled by the second law of thermodynamics. As a matter of fact, $\overline{\mathbf{T}}$ and S receiving their classical form as state laws, the reduced dissipation reads as follows :

$$\Phi = tr\left(\overline{\mathbf{M}}.\mathbf{L}_{\mathbf{K}}\right)+\Phi^{gr} - \theta^{-1}\mathbf{Q}.\nabla_R\theta \geq 0\ ,. \tag{5.110}$$

where $\overline{\mathbf{M}}$ is the material Mandel stress based on $\overline{\mathbf{T}}$, and the second (growth) contribution can be shown to read (Epstein and Maugin, 2000)

$$\Phi^{gr} = \rho_0^{-1}\overline{\mathbf{T}} : \left(\nabla_\Gamma \mathbf{F}\right).\mathbf{m}\ , \tag{5.111}$$

where ∇_Γ denotes the *covariant* material gradient in the connection defined in eqn.(3.35). Expression (5.111) shows that a geometrically consistent theory including the effect of diffusion would require the consideration of second gradients of the motion, or *second-order transplants.* This is achieved (but not in all generality) in Epstein and Maugin (2000). In the simplifed (first-order) theory, therefore, we must impose $\mathbf{m} = \mathbf{0}$, so that the dissipation (5.110) and the mass-density evolution (5.96) take on the following forms :

$$\Phi = tr\left(\mathbf{M}.\mathbf{L}_K\right) - \theta^{-1}\mathbf{Q}.\nabla_R\theta \geq 0\ . \tag{5.112}$$

and

$$\left.\frac{\partial \rho_0}{\partial t}\right|_{\mathbf{X}} = \Pi = -\rho_0 tr \mathbf{L}_{\mathbf{K}}\ . \tag{5.113}$$

The system of constitutive equation must now be closed by a heat conduction law (e.g., Fourier's) and the evolution equation for the transplant **K**. The latter can be envisaged as the general relationship between various arguments :

$$\varphi\left(\mathbf{K}, \dot{\mathbf{K}}, \mathbf{M}\ or \mathbf{b}, \mathbf{F}, \dot{\mathbf{F}}\right) = 0\ , \tag{5.114}$$

reflecting thus some kind of *nonlinear viscoelasticity.* Using invariance arguments it can be shown (Epstein and Maugin, 2000), that this can be reduced to the following relationship :

$$\varphi\left(\overline{\mathbf{L}}_{\mathbf{K}}, \mathbf{b}_c, \mathbf{C}_e, \dot{\mathbf{C}}_e\right) = 0\ , \tag{5.115}$$

where $\mathbf{C}_e$ and $\dot{\mathbf{C}}_e$ come from $\mathbf{F}_e = \mathbf{F}.\mathbf{K}$ - the elastic deformation gradient , cf. eqn.(5.77)- , $\overline{\mathbf{L}}_{\mathbf{K}}$ comes from an invariant combination of **K** and $\dot{\mathbf{K}}$, and $\mathbf{b}_c$ is given by

$$\mathbf{b}_c = J_K \mathbf{K}^{-1}.\mathbf{b}.\mathbf{K}\ . \tag{5.116}$$

As a simple example of isotropic behavior for the symmetric part $\overline{\mathbf{L}}_S$ of $\overline{\mathbf{L}}_\mathbf{K}$ we have the evolution equation

$$\overline{\mathbf{L}}_S = \varphi_0 \mathbf{1} + \varphi_1 \mathbf{b}_c + \varphi_2 \mathbf{b}_c^2, \tag{5.117}$$

where the coefficients are functions of the fundamental invariants of $\mathbf{b}_c$. Such a law, akin to *nonlinear viscoelasticity*, is exploited in finite-element numerical computations of growth by Imatani and Maugin (2000).

6 Driving forces on singular surfaces

6.1 The case of coherent phase-transition fronts

The progress of phase-transition fronts in thermoelastic solids is essentially controlled by *temperature.* For instance the existence of several minimizers of the elastic potential depends on temperature in the Landau vision of the phenomenon. Consequently, thermal effects cannot be left out of the picture and they are essential in the following developments. We consider here the possible irreversible progress of a phase into another one, the separation between the two phases being idealized as a sharp, discontinuity surface Σ across which most of the fields suffer finite discontinuity jumps. It is only if we look at a smaller scale that we shall observe the structure of the necessarily nonzero-thickness interface between the two phases (cf. Maugin and Inoue, 1998 ; Maugin, 2000b).The thermodynamically irreversible processes occuring within that relatively « thick » interface are responsible for the irreversible motion of the idealized mathematical discontinuity (just the same as in shock-wave theory ; cf. Mandel, 1966 ; Germain, 1972). We consider the classical theory of Section 5.1. and acknowledge from the start that a phase transition consists in a local *material rearrangement* of the material. Therefore, a « material » description based on the exploitation of the Eshelby stress imposes itself. The presence of Σ breaks the material symmetry of the material body as a whole and manifests a *material inhomogeneity*. Consequently, the critical equation for the description of the phenomenon is not only that relating to dissipation, but also the equation associated with the *lack* of conservation of pseudomomentum across Σ, i.e., the associated *jump relation* that is generated by a change of « particle » on the material manifold : the driving force acting on the transition front will, therefore, be a « material force ».

The main problem here consists in writing down a consistent set of thermomechanical jump relations across the front Σ. Let $[A] = A^+ - A^-$, the jump of a discontinuous field A across Σ, the unit normal $\mathbf{N}$ to Σ being oriented from the « minus » to the « plus » side. Let $\overline{\mathbf{V}}$ the material velocity of the geometrical point of Σ. The phase transition fronts considered are *homothermal* (no jump in temperature ; the two phases coexist at the same temperature) and *coherent* (they present no defects such as dislocations). Consequently, we have the following continuity conditions :

$$[\theta]=0,[\mathbf{V}]=\mathbf{0} \ at \ \Sigma . \tag{6.1}$$

Jump relations associated with strict conservation laws in the bulk, are formulated according to the theory of *weak solutions* of hyperbolic systems. That is, we merely replace the operators $\partial/\partial t|_{\mathbf{X}}$ and ∇_R by the jump operators $-(\overline{\mathbf{V}}.\mathbf{N})[.]$ and $\mathbf{N}.[.]$, respectively. Thus the jump equations associated with the bulk equations (5.1), (5.2) and (5.3) that are strict conservation laws, read

$$\overline{V}_N[\rho_0]=0 \quad , \tag{6.2}$$

$$\overline{V}_N[\mathbf{p}]+\mathbf{N}.[\mathbf{T}]=\mathbf{0}, \tag{6.3}$$

$$\overline{V}_N[H]+\mathbf{N}.[\mathbf{T}.\mathbf{v}-\mathbf{Q}]=0 \ , \tag{6.4}$$

where $\overline{V}_N = \overline{\mathbf{V}}.\mathbf{N}$ is the normal speed of the points of Σ. The same pragmatic rule can be applied to the bulk equations (5.4) and (5.9) - that are *not* strict conservation laws - if we formally add *unknown* source terms, i.e., we *a priori* write the following two jump equations :

$$\overline{V}_N[S]-\mathbf{N}.[\mathbf{Q}/\theta]=\sigma_\Sigma \tag{6.5}$$

and

$$\overline{V}_N[P]+\mathbf{N}.[\mathbf{b}]+\mathbf{f}_\Sigma=\mathbf{0} \ . \tag{6.6}$$

The second law of thermodynamics at Σ reads

$$\sigma_\Sigma \geq 0 . \tag{6.7}$$

The problem consists in finding the expressions of σ_Σ and $\mathbf{f}_\Sigma$ that are compatible. We know the formal expression of both P and $\mathbf{b}$. The only manipulation we can do on eqn.(6.6) is to compute the power expended by $\mathbf{f}_\Sigma$ in the irreversible motion of Σ. To that purpose, we compute

$$p_\Sigma = \overline{\mathbf{V}}.\mathbf{f}_\Sigma \tag{6.8}$$

on account of the jump remaining equations (6.2) through (6.5). The result of these manipulations is

$$p_\Sigma = f_\Sigma \overline{V}_N = \theta_\Sigma \sigma_\Sigma \geq 0 \tag{6.9}$$

together with the fact that

$$q_\Sigma = p_\Sigma = \theta_\Sigma \sigma_\Sigma \geq 0 \tag{6.10}$$

is a *hot* heat source localized at Σ. In these equations θ_Σ is the temperature at Σ, f_Σ is the scalar driving force for which we need a *kinetic equation*, and we have at Σ the following *balance of « material » forces*

$$f_\Sigma + Hugo_{PT} = 0, \tag{6.11}$$

wherein (compare to eqn.(4.33) which is based on the quasi-static hypothesis)

$$Hugo_{PT} := [W - \langle \mathbf{N.T} \rangle.\mathbf{F.N}] \tag{6.12}$$

the symbolism <..> indicating the mean value at Σ. The surface « balance » equation (6.11) is written down just to emphasize the different roles of $Hugo_{PT}$ - a field quantity that is known once we know the field solution by any means on both sides of Σ - and the driving force f_Σ that is the thermodynamic conjugate of the normal speed $\bar{V}_N$. The expression of $\bar{V}_N$ in terms of f_Σ is the kinetic law for normal progress of which examples basing on a more microscopic approach can be found in Truskinowsky (1994). In the absence of microscopic justification, one simply applies the thermodynamical constraint (6.9) to formulate *a priori* an admissible kinetic law. The above-given thermomechanical approach finds its roots in a paper by Truskinowskii (1987), but the present formalism including the consideration of the pseudomomentum jump relation in material space is due to Maugin and Trimarco (1995a,b,c). What must be remarked concerning the above construct it that the normal « material force » $Hugo_{PT}$ is practically never zero since it must satisfy the inequality (6.9). However, one may *artificially* impose a vanishing $Hugo_{PT}$ for a progressing front (non zero $\bar{V}_N$). This means a *nondissipative* phase transition although we relate through Σ the phases which are in nonadiabatic evolution ($\mathbf{Q} \neq \mathbf{0}$). This is a *singular* situation for which one shows in one dimension that the transition then obeys *Maxwell's rule of equal area* (cf.Maugin,1998c, 1999b). For a fluid, this condition in fact materializes in the continuity of the *chemical potential*. The relationship of chemical potential and the Eshelby stress (sometimes called chemical potential tensor) was noticed by Grinfeld (1991). However, there is something more in the above results : it is that no quasi-static hypothesis was used, although it is exactly *shown* that no kinetic energy can enter the final expression (6.12). This agrees with the perspicacious view of Gibbs and Duhem who indeed foresaw that only the free enthalpy must govern the local matter rearrangement represented by a phase transition. Thus only the quasi-static part of the Eshelby stress finally contributes to the Hugoniot-Gibbs functional $Hugo_{PT}$ - in agreement with their vision.

The above result can be generalized to more complex material bodies such as electromagnetic ones (e.g., in Fomethe and Maugin, 1996) or polar thermoelasticity (Maugin, 1998a), as also thermo-anelastic materials. Furthermore, there is fundamntally nothing changed in the above-given argument in the presence of anelastic behaviors described by means of internal variables of state (cf.Maugin, 1998b).

6.2. The case of thermomechanical shock waves

In this case we do not satisfy the condition $(6.1)_1$. But we can keep the dependence on internal variables and therefore treat the general case of thermo-anelasticity, since the presence of α (without gradients) does not formally alter the presentation. The jump relations (6.2) through (6.6) constitute the general point of departure. However, if we kept the idea of a transition zone of small but finite thickness from one side of Σ to the other, then the source quantities in eqns.(6.5) and (6.6) would be known in terms of *excess quantities* computed throughout the thickness of the transition zonre. That is, without loss in generality in the absence of surface tension along Σ, imagine that this zone is locally a flat strip orthogonal to the coordinate axis X_1 and the problem is independent of the other two (transverse) coordinates X_2 and X_3. Then , assuming the transition zone of thickness δ we could write eqns.(6.5) and (6.6) as follows :

$$\bar{V}_N[S]-\mathbf{N}.[\mathbf{Q}/\theta]=\int_{-\delta/2}^{+\delta/2}\left(\sigma^{th}+\sigma^{\text{int}}\right)dX_1\geq 0 \tag{6.13}$$

and

$$\bar{V}_N[P]+\mathbf{N}.[\mathbf{b}]=-\int_{-\delta/2}^{+\delta/2}\left(\mathbf{f}^{th}+\mathbf{f}^{\text{int}}\right)dX_1 \tag{6.14}$$

where the jump now means the difference between the values a little bit before and beyond the ends of the interval δ , i.e ., outside the transition zone. This, which can be conceived as a zoom on the front, shows the remarkable role played by thermal and intrinsic entropy sources in the condition of entropy growth at Σ and the parallel role of the material forces $\mathbf{f}^{th}$ and $\mathbf{f}^{int}$ in writing the jump of the balance of pseudomomentum at Σ. Equations such as (6.13) are suggested by Mandel (1966,pp.268-9) and considered by some authors (Stolz, 1989,1994), but eqn.(6.14) is quite recent (Maugin, 1998b,c). It is the consistency between the source terms in eqns.(6.13)-(6.14) which provides the connection between entropy growth at Σ and the power expended by the material driving forces, even in the general theory of shock waves where the notion of driving force was heretofore ignored. But we shall consider the purely jump-like description (6.1)-(6.5), bearing (6.13)-(6.14) in mind as some « microscopic » justification. We then define the following scalar and vectorial quantities at Σ (such quantities are meaningful in themselves only if they are invariant across Σ. Otherwise, they become mathematically meaningful only once an operator such as [.] or <.>, attached to Σ, is applied to them) :

$$m:=\rho_0\bar{V}_N\ , \tag{6.15}$$

$$T_\Sigma:=m\mathbf{v}+\mathbf{N}.\mathbf{T}\ , \tag{6.16}$$

$$mQ_\Sigma:=m\bar{H}+\mathbf{N}.(\mathbf{T}.\mathbf{v}-\mathbf{Q})\ , \tag{6.17}$$

$$B_{\Sigma} := m\bar{P} + \mathbf{N}.\mathbf{b}\,, \tag{6.18}$$

$$mN := m\bar{S} - (\mathbf{N}.\mathbf{Q}/\theta), \tag{6.19}$$

where, apart from $\bar{V}_N$, quantities with overbars are per unit mass in the reference configuration. On account of definitions (6.15)-(6.19) the jump relations (6.1) through (6.5) now read

$$[m] = 0, [T_{\Sigma}] = \mathbf{0}, m[Q_{\Sigma}] = 0 \tag{6.20}$$

and

$$m[N] = \sigma_{\Sigma} \geq 0,\; [B_{\Sigma}] + \mathbf{f}_{\Sigma} = \mathbf{0}\;. \tag{6.21}$$

This is merely a rewriting, but it allows us to perform some simple manipulations and to express some remarkable results in an attractive manner. Define

$$\bar{L}^{th} = L^{th} / \rho_0 = \bar{L}^{th}(\mathbf{F}, \mathbf{v}, \theta, \alpha). \tag{6.22}$$

Then it is proved that the entropy source at Σ is given by (cf.Maugin, 1997b)

$$\sigma_{\Sigma} = m[\bar{L}^{th} / \theta] + mQ_{\Sigma}[\theta^{-1}] - T_{\Sigma}.[\mathbf{v}/\theta] \geq 0\,. \tag{6.23}$$

Guided by this, introduce the following *generating* or *Massieu* function M at Σ by

$$M := \theta^{-1}\left(m\left(Q_{\Sigma} + \bar{L}^{th}\right) - T_{\Sigma}.\mathbf{v}\right). \tag{6.24}$$

Then we prove the following two results (Maugin, 1997b)

$$\sigma_{\Sigma} = [M] \geq 0 \tag{6.25}$$

and

$$p_{\Sigma} = \bar{\mathbf{V}}.\mathbf{f}_{\Sigma} = [\theta M]. \tag{6.26}$$

That is, the generating function M - a notion already used in hydrodynamics and magnetohydrodynamics by P.Germain (1972) - provides *both* the entropy source and the power of the driving force at Σ, although we do not know yet the explicit form of the latter. From (6.25) and (6.26) there follows the following general relationship between p_{Σ}, σ_{Σ}, and the jump in temperature for a general shock wave relating two regions where adiabaticity is *not* supposed:

$$p_{\Sigma} = \sigma_{\Sigma}\langle\theta\rangle + \langle M - (mQ_{\Sigma} / \theta)\rangle[\theta]. \tag{6.27}$$

It is clear from the above results (6.25)-(6.26) than in the case of a phase-transition front across which temperature is constant we immediately have the result (6.9) since :

$$p_\Sigma = \theta_\Sigma [M] = \theta_\Sigma \sigma_\Sigma \geq 0 \ . \tag{6.28}$$

Otherwise, we have at hand a consistent theory of shock waves because, contrary to the simple theory in which the shock relates two regions in adiabatic situations, we here relate two regions in which dissipation of both thermal and intrinsic origins is taken into account so that we have a justification for the existence of σ_Σ of which the non-negativity is not imposed out of the blue. What about the celebrated *Hugoniot* equation of classical shock-wave theory (in which there are neither heat flux nor variables α present)? This is normally obtained by expanding the following identity on account of other (classical) jump relations :

$$[T_\Sigma]\langle \mathbf{v} \rangle \equiv 0 \ , \tag{6.29}$$

resulting in

$$Hugo_{SW} := [E(\mathbf{F}, S, \alpha) - \langle \mathbf{N}.\mathbf{T} \rangle . \mathbf{F}.\mathbf{N}] \equiv 0 \, at \, \Sigma \ , \tag{6.30}$$

where it is the *internal energy E* that is involved. This is the difference with the Hugoniot-Gibbs functional (6.12) of phase-transition fronts where it is the *free energy* that is involved. But in our formalism $Hugo_{SW}$ is a material driving force (more precisely, the double normal component of the jump of a material tensor) and the identity (6.30) clearly exhibits the inconsistency of classical shock-wave theory in which the driving force is set equal to zero (providing the Hugoniot equation) for a still propagating shock wave while the power expended by the driving force should be strictly related to the entropy growth, itself *not* supposed to be zero in general.

6.3 Thermal conditions at interfaces in thermoelastic composites

Now we can revisit the problem of interfaces in composites -mentioned at the beginning of Paragraph 4.1 - but when thermal effects are taken into account (cf. Maugin *et al*, 1999). As there exists already a nonzero *thermal material force* at all regular material points in the bulk of a thermoelastic material, it is quite normal that such a notion a fortiori exists at the crossing of an interface between two composite components having different thermoelastic properties. We shall again call $\mathbf{f}_\Sigma$ such a *surface material thermal force* when it exists at the junction of two thermoelastic conductors. From the viewpoint of mechanics we assume *perfect* matching between the two components so that we have

$$[\mathbf{X}] = \mathbf{0}, [\mathbf{V}] = \mathbf{0} \quad at \ \ \Sigma . \tag{6.31}$$

This supposes that we have neither dislocations nor delamination present at Σ. A variety of interface thermal conditions can be applied at Σ. We may have *perfect thermal contact*

$$[\theta]=0 \, at \Sigma, \tag{6.32}$$

or we can impose a heat source per unit area , q_Σ, for instance via Joule heating of a very thin layer of electric conductor placed in sandwich between the two components, or still, we may consider kind of mixed *adiabatic-isentropic* conditions

$$\mathbf{N}.\mathbf{Q}^{\pm}=0 \Rightarrow \mathbf{N}.[\mathbf{Q}]\equiv 0, [S]=0 \, at \Sigma. \tag{6.33}$$

A temperature difference

$$\vartheta=[\theta] \tag{6.34}$$

could be imposed by some kind of *thermocouple.* In any case we assume that there in general exists a heat source q_Σ and an entropy source σ_Σ in such a way that the jump relations for energy and entropy at Σ read (quasi-statics is sufficient for our purpose)

$$\mathbf{N}.\left[E\overline{\mathbf{V}}+\mathbf{T}.\mathbf{v}-\mathbf{Q}\right]+q_\Sigma=0 \tag{6.35}$$

and

$$\mathbf{N}.\left[S\overline{\mathbf{V}}-(\mathbf{Q}/\theta)\right]=\sigma_\Sigma \geq 0. \tag{6.36}$$

The corresponding jump of the (quasi-static) balance of pseudomomentum then reads

$$\mathbf{N}.[\mathbf{b}]+\mathbf{f}_\Sigma=\mathbf{0}, \tag{6.37}$$

where we emphasize that we do *not* know the expression of $\mathbf{f}_\Sigma$. The *inhomogeneous* equation (6.37) must be consistent with the statements (6.35) and (6.36). If the perfect thermal matching prevails, (6.32), then θ factorizes out in (6.36) and compatbility between the resulting equation and (6.35) imposes that the heat source at Σ be necessarily hot (or zero). But eqn.(6.37) must also be consistent. This is implemented by computing the power expended by the « force » $\mathbf{f}_\Sigma$ in a motion of material velocity $\overline{\mathbf{V}}$ such as in an oscillatory motion (at small frequency to respect the quasi-statics hypothesis, or else a virtual velocity field). On account of the mechanical matching condition $\mathbf{N}.[\mathbf{T}]=\mathbf{0}$, we obtain thus

$$p_\Sigma=\mathbf{f}_\Sigma.\overline{\mathbf{V}}=-\mathbf{N}.[\mathbf{b}].\overline{\mathbf{V}}=\theta[S]\overline{V}_N+q_\Sigma. \tag{6.38}$$

From this we obtain $\left(\overline{V}_N \neq 0, q_\Sigma=0\right)$

$$\mathbf{f}_\Sigma=\theta[S]\mathbf{N}. \tag{6.39}$$

In that case we see that the purely normal but *fictitious* thermal surface force is determined by the entropy source and is directed in the direction of entropy growth.

In the case where conditions such as those in (6.33) prevail, then a similar computation yields the dual formula

$$p_\Sigma = \mathbf{f}_\Sigma.\overline{\mathbf{V}} = S[\theta]\overline{V}_N\,, \tag{6.40}$$

so that the thermal driving force at the interface is given by

$$\mathbf{f}_\Sigma = \mathbf{f}_\Sigma^{th} := S[\theta]\mathbf{N}\,. \tag{6.41}$$

This « force » is purely normal and directed *from the colder to the warmer face of the interface*. Clearly, this is a limit case of eqn.(5.12). Indeed, the expression (6.41) appears to be the *singular jump form* associated with the bulk expression (5.12). To see this, one can compute the total thermal force (5.12) across a unit area of a layer of infinitesimally small thickness across which S is practically constant. This is in fact what is done in a different context when trying to justify the existence of $\mathbf{f}_\Sigma$ across singular surfaces (compare eqn.(6.14)). In the present case the temperature jump present in (6.41) could be imposed the same as in eqn .(6.34).

7 Application of configurational forces to computational schemes

It is usually, and correctly, remarked that at all regular material points, the balance of pseudo-momentum is « equivalent » to the balance of physical momentum. The reason for this is that at all such points, the former can be deduced from the latter by some algebraic and analytic operations allowed by the smoothness of the fields. That is, at such points the balance of pseudo-momentum is an *identity*. However, it is not exactly true that both equations of momentum are operationally equivalent, being then only two different « projections » on different manifolds, of the same equation, because they *cannot* play the same role in *problem solving*. This is due to the fact that, whether we like it or not, applied forces - which are real physical forces in the Newtonian-Euler-Cauchy sense - are prescribed in physical space (a world to which they belong by definition) so that they intervene in problem solving in the equations (field equations and boundary conditions) expressed in that space. Therefore, initial-value boundary-value problems are, of necessity, solved, analytically or numerically, in physical space. The question then arises of the usefulness of the balance of pseudo-momentum at regular material points, its role in capturing singularities being clear at singular points (cf. foregoing sections). This usefulness resides is the redundancy beween the two (essentially vectorial) equations as it can be exploited thus. While the balance of physical momentum - and other equations governing other fields than deformation -, are used in the direct problem solving, the balance of pseudo-momentum (or its « equilibrium » form where this applies) is to be exploited either in the form of a *criterion* to be satisfied by the already obtained point-wise field solution (eg., in fracture, phase-transition, anelasticity, etc), or as a way to checking globally the accuracy and quality of the obtained solution. This is very similar to the exploitation of the energy balance in the numerics of hyperbolic systems where the accurarcy of the adopted scheme is measured by a global energy criterion (i.e ., does the scheme conserve energy ?). Herein below we focus attention on such an application of the balance of pseudo-momentum in various numerical techniques, noting in passing that the notion of momentum is vectorial, and therefore « directional », and thus richer than that of energy (a scalar) which addresses only magnitude. Three numerical techniques are briefly examined in the light of recent progresses.

7.1 Finite-element method

We remind the reader of the basic equations of equilibrium, energy and pseudomentum in finite-strain elasticity (applied body force $\mathbf{f}^0$; corresponding material force $\mathbf{F}^0=-\mathbf{f}^0.\mathbf{F}$) :

$$div_R \mathbf{T} + \mathbf{f}^0 = \mathbf{0}, \tag{7.1}$$

$$\left.\frac{\partial W}{\partial t}\right|_{\mathbf{X}} = \nabla_R . (\mathbf{T}.\mathbf{v}) + \mathbf{f}^0.\mathbf{v} , \tag{7.2}$$

$$div_R \mathbf{b} + \mathbf{F}^0 = \mathbf{0}, \tag{7.3}$$

respectively. At all regular material points **X** we have the following *Ericksen-Noether identity*

$$\left(div_R \mathbf{T} + \mathbf{f}^0\right)\mathbf{F} + \left(div_R \mathbf{b} + \mathbf{F}^0\right) = \mathbf{0} \ . \tag{7.4}$$

Note that eqn.(7.1), valid at all regular points $\mathbf{X}$ *in* Ω_t, and the associated *natural boundary* condition at $\partial\Omega_t$,

$$\mathbf{N.T} = \mathbf{t}^0, \tag{7.5}$$

are equivalent to the following *weak formulation* (principle of virtual power)

$$P^*_{(i)} * + P^*_{(c)} + P^*_{(v)} = 0 \tag{7.6}$$

where

$$P^*_{(i)} = -\int_{\Omega_t} \mathbf{T} : (\nabla_R \mathbf{v}^*)^T d\Omega, P^*_{(c)} = \int_{\partial\Omega_t} \mathbf{t}^0.\mathbf{v}^* dS, P^*_{(v)} = \int_{\Omega_t} \mathbf{f}^0.\mathbf{v}^* d\Omega \ . \tag{7.7}$$

Equation (7.7) holds for all sufficiently smooth vectorial test functions (physical velocity fields) $\mathbf{v}^*$. In classical elastic engineering computations, when true material inhomogeneities or defects (field singularities) are absent, the *Finite-Element Method* (*FEM*) is based on a discretization of eqn.(7.6) by introducing interpolations of test functions. Imagine that the computation is also made in the absence of applied body forces ($\mathbf{f}^0 = \mathbf{0}$) . Then $div_R \mathbf{T} = \mathbf{0}$ is solved by the *FEM* and this yields a solution that depends on the location of nodes of the *FEM* net on the material manifold. Knowing such an *FEM* field solution the quantities $\mathbf{b}$ and $div_R \mathbf{b}$ can be evaluated, and this may yield

$$div_R \mathbf{b} \neq \mathbf{0}\left(say \mathbf{f}^{err}\right). \tag{7.8}$$

i.e., there may exist a spatial distribution of spurious material forces $\mathbf{f}^{err}$, where there should be none according to the Ericksen-Noether identity. Equation (7.8) says something about the faithfulness of the *FEM* net and the question naturally arises of the tailoring of a finite-element net in such a way that these spurious *configurational forces* be made to vanish. They are *configurational* forces as they do depend on the location of nodes on the material manifold. One may think to release (or make float) the inner material nodes so as to make these spurious material forces vanish, of for the least to minimize them. This idea belongs to M.Braun (1997) and P.Steimann (2000). In particular, the first author has shown that a distribution of practically vanishing configurational nodal forces could be obtained. Simultaneously, the total energy is reduced in the process and, therefore, becomes closer to the minimum attained by the exact solution. Unfortunately, this optimization procedure may be accompanied by the formation of badly shaped elements which may not improve the finite-element solution. In the process of the above *FEM*, since all required quantities are already computed in this optimization procedure, a particularly economical and rapid means of computation of the energy-release rate is offered. To see this multiply (7.4) by $\mathbf{F}^{-1}$ to the right, multiply the result scalarly by the test function **v***, and account for (7.6) to obtain the following expression :

$$-\int_{\Omega_t} \mathbf{b} : (\nabla_R \mathbf{V}^*)^T d\Omega + \int_{\partial\Omega_t} \mathbf{N}.\mathbf{b}.\mathbf{V}^* dS + \int_{\Omega_t} \mathbf{F}^0.\mathbf{V}^* d\Omega = 0, \tag{7.9}$$

for all *virtual material* velocity fields $\mathbf{V}^* = -\mathbf{F}^{-1}.\mathbf{v}^*$. This is formally equivalent to (7.6). But one should note that the second term in (7.9) does not correspond to a prescribed surface stress. Because of this, eqn.(7.9) is not applicable as such, i.e ., as a weak formulation of the original boundary-value problem. Rather, we note that for any velocity field we have

$$\int_{\partial\Omega_t} \mathbf{N}.\mathbf{b}.\mathbf{V}^* dS = \int_{\partial\Omega_t} (\mathbf{N}.\mathbf{V}^*) W dS + \int_{\Omega_t} \mathbf{t}^0.\mathbf{v}^* d\Omega \ , \tag{7.10}$$

so that we obtain the following original result (note that **N** is directed outward):

$$G(\Omega_t;\mathbf{V}^*) \equiv -\int_{\partial\Omega_t} W(\mathbf{N}.\mathbf{V}^*) dS = -\int_{\Omega_t} \mathbf{b} : (\nabla_R \mathbf{V}^*)^T d\Omega + \int_{\Omega_t} \mathbf{T} : (\nabla_R \mathbf{v}^*)^T d\Omega . \tag{7.11}$$

For a self-equilibrated (Piola-Kirchhoff) stress field ($\mathbf{f}^0 = \mathbf{0}, \mathbf{t}^0 = \mathbf{0}$), we have

$$\int_{\Omega_t} \mathbf{T} : (\nabla_R \mathbf{v}^*)^T d\Omega \equiv 0, \tag{7.12}$$

so that (7.11) reduces to the following formula for the *flux of elastic energy outward* the material volume Ω_t for a material velocity field **V*** :

$$G(\Omega_t;\mathbf{V}^*) = -\int_{\Omega_t} \mathbf{b} : (\nabla_R \mathbf{V}^*)^T d\Omega \ . \tag{7.13}$$

The quantity G is is usually called an *energy-release rate* (compare to the general definition involving the influx of the Hamiltonian density in eqn.(4.8)) .

Equation (7.9) can be generalized to more involved cases including additional degrees of freedom (e.g., a microstructure such as in micropolar elastic bodies, Maugin, 1998b) or *nonsimple* elastic media accounting for the second gradient of strain (so-called *second-gradient* theory ; cf. Maugin and Trimarco, 1992). This second case is particularly relevant to the contents of the forthcoming paragraph.

7.2 Finite-difference method

The *Finite-Difference Method* (*FDM*) is the numerical realm of nonlinear hyperbolic systems. This discretization method finds its origin in Analysis and the approximation of space derivatives of various orders by finite differences. The school of Courant (Lax, Wendroff, etc) in the USA and that of Godunov and Yanenko in Russia are responsible for its successes in treating problems including sharp field discontinuities such as shock waves in fluids. The accuracy of the devised *FDM* schemes is measured by their property to more or less conserve the global energy of the system. That is, in the present case, the numerical simulation of nonlinear-wave propagation is performed on the *FDM* discretized version of the *field* equation (3.12) - and possibly others such as $(5.53)_1$. For a conservative system it is checked whether the global (space integral) version of the energy equation - say (5.59) - holds good for the numerically obtained solution. The result of this check is expressed by announcing that e.g., « energy is conserved up to x% », values of the order of 10^{-3} being quite normal. From that view point, checking that the global version of the here redundant pseudomomentum balance law holds good for the numerical solution obtained is also a valid criterion for the accurarcy of the scheme of the *FDM* scheme, except that, because of the *vectorial nature* of that equation, the check is made along three different axes instead of that of the conservation of a single scalar entity as is the case for the energy equation. This places energy and pseudomomentum on equal footing or, as we like to say, on the same *ontological level.* In particular, for *systems* of field equations, both canonical equations and accuracy criteria concern the *global* physical system and not only one degree of freedom in spite of the 3D vectorial nature of pseudomomentum. Thus, following Christov and Maugin (1995), we may thus speak of FDM schemes that more or less conserve *both* energy and canonical momentum.

With the introduction of localized nonlinear waves of the solitary-wave type (kinks, humps, bound states, soliton complexes) in nonlinear *dispersive* systems of field equations, the exploitation of the pseudomomentum balance law is even more fruitful. This is easily understood by examining two specific questions. First, in a general way, for such systems , the field equations - eqn.(3.12) and $(5.53)_1$ - present both nonlinearities and high-order space or mixed derivatives. This is the case when treating an elastic crystal in which both nonconvexity of the strain energy and a weak nonlocality (gradient effects) are taken into account (e.g., in shape-memory alloys). An example of such systems is the following 1D generalized Boussinesq equation (cf. Maugin and Christov, 2000) where subscripts t and X stands for partial derivatives, s is a strain (eg., one shear component), $F(s)$ is a polynomial in s starting with second degree, and β is a positive or negative parameter :

$$s_{tt} - c_T^2 s_{XX} - \left[F(s) - \beta s_{XX} + s_{XXXX}\right]_{XX} = 0 . \qquad (7.14)$$

A *stiff* mathematical system such as this one, although one dimensional in space, requires devising a high-performance *FDM* scheme (see, e.g., Christov and Maugin, 1995). This makes that *FDM* is also an Art. We do not write here the pseudomomentum balance equation that corresponds to the field equation (7.14) because its formal expression suffices for our general purpose. Imagine that we have obtained analytically (with some luck) or numerically, strongly localized solitary-wave-like solutions of (7.14) thats we call « shapes » S. These corresponds usually to spatially uniform solutions at infinity. First the writing of the global form - by integration over the real line R in the absence of applied force - of the pseudomomentum equation provides theoretically an equation of the following type

$$\frac{d\mathbf{P}}{dt}=\mathbf{0};\ \mathbf{P}:=\int_{R} P dX\ . \tag{7.15}$$

This is a Newtonian-like equation of *inertial motion* for the « shape » S. There is no term in the right-hand side of eqn.$(7.15)_1$ because all field derivatives go to zero at infinity for these solutions S which, therefore, are steadilly progressing. As a matter of fact imposing the vanishing in the right-hand side of $(7.15)_1$ reveals the *natural limit* conditions on higher-order field derivatives, that classical types of boundary conditions do not consider. But in the numerical simulation of the steady propagation of « *shapes* » S, we have to work either on a periodic landscape arrangement or on a finite space interval. The satisfaction of these conditions is a necessity for the realization of the inertial motion globally governed by (7.15). Otherwise, the nonzero values at the interval boundaries will in fact create a perturbing *driving force* that will accelerate or slow down the supposedly steadilly moving shape, with a motion equation

$$\frac{d\mathbf{P}}{dt}=\mathbf{F}^{driving}\neq\mathbf{0} \tag{7.16}$$

replacing then $(7.15)_1$. Making $\mathbf{F}^{driving}$ zero or minimal is a justified endeavour. But it is clear that the same scheme - eqns.$(7.15)_1$ and (7.16) - can also be exploited to treat the influence of nonzero applied forces (or any term that can be viewed as a perturbation on the inertial motion of undeformed « shapes » present in the right-hand sides of the basic field equations to start with. Then the initial « shape » solutions are found - with some freedom left to some of their parameters - on the basis of the homogeneous form of these equations. Equation (7.16) where the right hand-side is then given by the space-integrated materially-convected form of this perturbing force, provides the time-evolution equation of these gross parameters (e.g., speed) in solutions (transient nonlinear localized wave forms) that cannot be obtained through any other known method.

One example of application of this technique are presented in Section 8 - see also Christov *et al* (1996), Maugin and Christov (2000) and , Maugin (1999b). Note that like in the previous section, and in accord with the general philosophy of this presention, the field equations, *per se*, serve one purpose, while the energy and/or pseudomomentum equations serve another one. Note by way of conclusion that (7.15) and (7.16) do not presuppose that the *point mechanics* (relationships between mass, momentum, velocity and energy) satisfied by the « shapes » in question is Newtonian. These relationships are part of the problem.

7.3 Finite-volume method - Continuous Cellular automata

The *Finite-Volume Method* (*FVM*) considers a fixed grid and cells, rather than discrete points, as discrete elements. It directly reflects the notion of conservation laws which apply to these cells. The theory presented in Sections 2-3 clearly emphasizes the notion of conservation laws. Therefore, the *FVM* seems, to the price of some adjustments, to be appropriate to the treatment of dynamical problems. This is all the more true that cells thus considered may also be viewed as the elementary blocks of a thermodynamics of so-called *discrete systems* in the manner of

Schottky (cf. Muschik, 1990). In this thermodynamics the state in one discrete system (e.g., one of the above cells) is defined in terms of its environment which may or may not be in thermodynamical equilibrium. *Contact thermodynamical quantities* (e.g., contact temperature, contact stresses) are introduced to characterize the state of the discrete system (in fact defined at the boundary surface of a cell in the *FVM*). This idea of making a cell's state depend on that of its neighbors is tantamount to introducing a *strategy* for the propagation of the thermodynamic state. This is akin to introducing the notion of *cellular automaton,* although discretization here is based on continuous balance laws, so that we may refer to this method as that of *continuous cellular automata.* The above referred to strategy is essential in some dynamical thermomechanical problems such as that of the propagation of a *phase-transition front.* It is along this line of thought that recent works develop.

In the new scheme developed starting with Berezovski's work of 1997, all thermomechanical balance laws are expressed for each cell, and the bulk quantities within each cell are related to the contact ones through the thermodynamics of discrete systems. A high-performance wave-propagation algorithm is exploited - using Lax-Wendroff and Godunov's ideas - that yields extremely good results in the simulation of the rapid progression of sharp wave fronts in 2D elasticity or thermoelasticity (cf. Berezovski and Maugin, 1999a,b). It is interesting to give a very few elements of that type of approach, for instance in linear isotropic, but materially inhomogeneous, thermoelasticity where the basic field equations are eqns.(5.2 and (5.15), the latter yielding in fact the equation of heat propagation [eqn.(5.9) is to be exploited later on]. More precisely, the balance of linear (physical) momentum, the time-rate of change of the Hooke-Duhamel constitutive equation, and the heat-propagation equation read as follows at any regular material point **x** in the absence of body force and heat source:

$$\frac{\partial(\bar{\rho}_0(\mathbf{x})v_i)}{\partial t}-\frac{\partial\sigma_{ij}}{\partial x_j}=0, \qquad (7.17)$$

$$\frac{\partial\sigma_{ij}}{\partial t}=\lambda(\mathbf{x})\frac{\partial v_k}{\partial x_k}\delta_{ij}+\mu(\mathbf{x})\left(\frac{\partial v_i}{\partial x_j}+\frac{\partial v_j}{\partial x_i}\right)+m(\mathbf{x})\frac{\partial T}{\partial t}\delta_{ij}\ , \qquad (7.18)$$

$$\frac{\partial(C(\mathbf{x})T)}{\partial t}=\frac{\partial}{\partial x_i}\left(k(\mathbf{x})\frac{\partial T}{\partial x_i}\right)+m(\mathbf{x})\frac{\partial v_k}{\partial x_k}. \qquad (7.19)$$

where ρ_0 is the density of matter, σ is Cauchy's stress, λ and μ are Lamé coefficients, $m = -\alpha(3\lambda+2\mu)$ is the thermoelastic coupling coefficient if α is the dilatation coefficient, C is the heat capacity, and k is the heat-conduction coefficient ; T denotes the small deviation of temperature. The material inhomogeneity is explicitly indicated by the **x**-dependence of some of these coefficients. By integration over a cell or finite-volume element V, we obtain the following system of FVM balance equations :

$$\frac{\partial}{\partial t}\int_V \rho_0 v_i\, dV=\int_{\partial V}\Sigma_{ij}n_j\, dA\ , \qquad (7.20)$$

$$\frac{\partial}{\partial t}\int_V \varepsilon_{ij}\,dV = \int_{\partial V} H_{ijk} n_k\,dA \ , \tag{7.21}$$

$$\frac{\partial}{\partial t}\int_V \sigma_{ij}\,dV = \int_{\partial V}\left(2\mu H_{ijk} n_k + \lambda\delta_{ij} V_k n_k\right) dA + \varphi_{ij} \ , \tag{7.22}$$

$$\frac{\partial}{\partial t}\int_V CT\,dV = \int_{\partial V}\left(k(\mathbf{n}.\nabla)T + mV_k n_k\right) dA + \varphi^{inh} \ , \tag{7.23}$$

where

$$H_{ijk} \equiv (1/2)\left(\delta_{ik}V_j + \delta_{jk}V_i\right) \tag{7.24}$$

and source terms due to material inhomogeneities (labelled « *inh* ») and thermoelastic couplings (labelled « *te* ») are given by

$$\varphi_{ij} = \varphi_{ij}^{te} + \varphi_{ij}^{inh} \ , \tag{7.25}$$

$$\varphi_{ij}^{te} = \int_V m\delta_{ij}\frac{\partial T}{\partial t} dV \ , \tag{7.26}$$

$$\varphi_{ij}^{inh} = -\int_V \left(v_k \frac{\partial \lambda}{\partial x_k}\delta_{ij} + v_i \frac{\partial \mu}{\partial x_j} + v_j \frac{\partial \mu}{\partial x_i} \right) dV \ , \tag{7.27}$$

$$\varphi^{inh} = -\int_V v_k \frac{\partial m}{\partial x_k} dV \ . \tag{7.28}$$

Herein above V_i and Σ_{ij} are the « contact » velocity and Cauchy stress defined at the *FV* element boundary of unit outward normal n_i. Θ denotes the « contact temperature ». Contact and bulk quantities are related by thermodynamic constraints such as the following one for the stress [labels (1) and (2) refer to two neighboring cells] , which express the continuity of partial derivatives of the internal energy :

$$\begin{aligned} &\sigma_{ij}^{(1)} + \Sigma_{ij}^{(1)} - T^{(1)}\left(\frac{\partial \sigma_{ij}^{(1)}}{\partial T}\right)_{\varepsilon} - \Theta^{(1)}\left(\frac{\partial \Sigma_{ij}^{(1)}}{\partial T}\right)_{\varepsilon} = \\ &= \sigma_{ij}^{(2)} + \Sigma_{ij}^{(2)} - T^{(2)}\left(\frac{\partial \sigma_{ij}^{(2)}}{\partial T}\right)_{\varepsilon} - \Theta^{(2)}\left(\frac{\partial \Sigma_{ij}^{(2)}}{\partial T}\right)_{\varepsilon} . \end{aligned} \tag{7.29}$$

The reader is referred to Berezosvki (1997) and Berezovski and Maugin (1999a,b) for the application to the numerics of smooth elastic and thermoelastic 2D wave propagation. What about the application to the propagation of phase-transition fronts where, of necessity, a criterion of progress (change of thermoelastic phase as the front progresses) is involved. This must necessarily exploits the *balance of pseudomomentum* in a form adapted to the continuous-cellular automaton formalism. As a matter of fact, in the present case, this additional balance law reads, at each regular material point :

$$\frac{\partial p_i^w}{\partial t} - \frac{\partial b_{ji}^w}{\partial x_j} = f_i^{inh} + f_i^{th} , \tag{7.30}$$

where we have set (W is the free energy per unit volume)

$$p_i^w = -\rho_0 v_j u_{j,i} , \; b_{ji}^w = -\left(L^{th}\delta_{ji} + \sigma_{jk} u_{k,i}\right) . \tag{7.31}$$

$$L^{th} = \frac{1}{2}\rho_0(\mathbf{x})\mathbf{v}^2 - W(\varepsilon_{ij}, T; \mathbf{x}), \; f_i^{inh} = \left(\frac{\partial L^{th}}{\partial x_i}\right)_{\exp l}, f_i^{th} = S\frac{\partial T}{\partial x_i}, \; S = -\frac{\partial W}{\partial T} . \tag{7.32}$$

The two source terms in eqn.(7.30) are due, respectively, to true material inhomogeneities and pseudo-inhomogeneities caused by thermal effets ; p_i^w, the purely quadratic part of pseudomomentum, is called *wave* or *crystal momentum* (see Section 8 below), and b_{ji}^w is the corresponding part of the Eshelby stress tensor, with u_i the infinitesimal displacement. The integral of (7.30) over a VE cell yields

$$\frac{\partial}{\partial t}\int_V p_i^w dV - \int_{\partial V} n_j B_{ji}^w dA = \Phi_i^w , \tag{7.33}$$

where

$$\Phi_i^w = \int_V \left(f_i^{inh} + f_i^{th}\right) dV , \tag{7.34}$$

and B_{ji}^w is the « contact » Eshelby stress tensor. Together with the Eshelby tensor of the neighboring cell, the Eshelby stress b_{ji}^w satisfies the following time-evolution equation and thermodynamical constraint :

$$\frac{\partial b_{ji}^w}{\partial t} = -\left(\frac{\partial L^{th}}{\partial t}\delta_{ji} + \frac{\partial \sigma_{jk}}{\partial t}\frac{\partial u_k}{\partial x_i} + \sigma_{jk}\frac{\partial v_k}{\partial x_i}\right), \tag{7.35}$$

and

$$-\frac{\partial}{\partial t}\left(\frac{\partial u_j}{\partial x_p}\right)\left(B_{pj}^w - b_{pj}^{w}{}^*\right) \geq 0 , \tag{7.36}$$

where the minus sign in (7.36) originates from the fact that the Eshelby stress expends power in the « inverse motion » velocity field. A short algebra allows one to show that the volume-element integral of (7.35) yields the following expression :

$$\frac{\partial}{\partial t}\int_V\left(b_{ji}^{w}+\frac{1}{2}\rho_0 \mathrm{v}^2\delta_{ji}\right)dV=-\int_{\partial V}\left(\Sigma_{jk}V_k n_i\right)dA+$$
$$+\int_V\left(\left(\sigma_{pq}\frac{\partial v_p}{\partial x_q}-C\frac{\partial T}{\partial t}\right)\delta_{ji}-\frac{\partial\sigma_{jk}}{\partial t}\frac{\partial u_k}{\partial x_i}+\frac{\partial\sigma_{jk}}{\partial x_i}v_k\right)dV \qquad (7.37)$$

The integrand in the left-hand volume integral is none other than the « *quasi-static* » Eshelby stress which indeed governs the phase transition as already seen in Section 6. The criterion of progress must, therefore, exploits eqns.(7.33) and (7.37) . Related works are in progress.

Conclusion : Referring to all above considered cases we must , by way of conclusion, emphasize the different roles played by the field equations, *per se*, and the canonical equation of momentum. The latter proves to be extremely useful, and in fact the only tool available, to devise criteria of progress. This applies in all numerical methods examined, whether of the finite-element, finite-difference, or finite-volume type. These applications succeed in granting a clearly engineering-like flavor to concepts, those of pseudomomentum and Eshelby stress, whose origin is to be found in abstract field theory and the geometry of the material manifold or material space. The next section brings us back to more theoretical and exotic matters.

8 Application of configurational forces to nonlinear-wave propagation

8.1 Wave-momentum

As a preliminary, consider eqns(3.12), the Ericksen-Noether identity, e.g., eqn.(3.19), and the decomposition (4.13) of the direct deformation gradient. If physical linear momentum is conserved, then for all regular displacement fields $\mathbf{u}(\mathbf{X},t)$, it is also true, by virtue of the Ericksen-Noether identity that, in the absence of true or quasi- material inhomogeneities, the following *field momentum*, is also conserved :

$$P^f = -\rho_0 \mathbf{v}.(\nabla_R \mathbf{u})^T \quad ; \tag{8.1}$$

the corresponding conservation equation reads

$$\left.\frac{\partial P^f}{\partial t}\right|_{\mathbf{X}} - div_R \mathbf{b}^f = \mathbf{0}, \tag{8.2}$$

where the *field* Eshelby stress tensor is given by

$$\mathbf{b}^f = -\left(L\mathbf{1}_R + \mathbf{T}.(\nabla_R \mathbf{u})^T\right). \tag{8.3}$$

Note that it is really $\mathbf{u}(\mathbf{X},t)$ that is the dependent field in continuum mechanics rather than $\chi(\mathbf{X},t)$, hence the adopted naming. In particular in *small-strain linear elasticity*, the above-stated equations yield

- *Physical (linear) momentum* :

$$\rho_0 \frac{\partial^2 \mathbf{u}}{\partial t^2} - div\sigma = \mathbf{0}, \sigma = \frac{\partial W}{\partial \mathbf{e}}, \mathbf{e} \equiv (\nabla \mathbf{u})_S \ , \tag{8.4}$$

where σ is the symmetric Cauchy stress tensor, $\mathbf{e}$ is the infinitesimal stress, and $W(\mathbf{e})$ is the strain energy per unit volume. We now longer distinguish between actual and reference configurations. This does not mean, however, that the balance of field momentum reduces to the same as eqn.(8.4) because the former is, of necessity, an equation with the same degree of nonlinearity (or singularity) as the energy equation. Therefore, it remains an equation that is quadratic in the fields σ and $\mathbf{e}$. Any « linearization » of this equation is a misconception of the role of that equation (for such a case, see Müller, 1999). Accordingly, the balance of field momentum remains eqn.(8.1) with

$$P^f = -\rho_0 (\nabla \mathbf{u}).\frac{\partial \mathbf{u}}{\partial t}, \tag{8.5}$$

$$\mathbf{b}^f = -\left(L\mathbf{1} + \sigma.(\nabla\mathbf{u})^T\right), \tag{8.6}$$

$$L = \frac{1}{2}\rho_0\left(\frac{\partial \mathbf{u}}{\partial t}\right)^2 - W(\mathbf{e}), \tag{8.7}$$

where $W(\mathbf{e})$ is quadratic in $\mathbf{e}$. It is in this framework that P^f is also referred to as the *crystal momentum* for an elastic crystal after the pioneering work of Brenig (1955). As a matter of fact, for quantified elastic waves under the name of *phonons* (these are special types of so-called *quasi-particles* that are quantum-mechanically associated with definite types, here elastic, of vibrations), this is the momentum that directly quantifies to *de Broglie's formula*

$$P^f = \hbar\mathbf{k}, \tag{8.8}$$

where $\hbar$ is Planck's reduced constant, and $\mathbf{k}$ is the wave vector. Notice that in the nonlinear framework, the wave vector is a *covariant vector* (since it is the dual of material position $\mathbf{X}$), and thus we understand the proportionality relation between the naturally covariant vector P or P^f and $\mathbf{k}$ (see pp. 35-37 and Chapter 9 in Maugin, 1993). Because of its role in *wave propagation* P^f is also referred to as the *wave momentum* P^w. Here we discussed this matter on the pure elastic case, but such considerations can be extended to more complicated cases, coupled systems, *etc.* Herebelow, this is considered by way of example in the framework of nonlinear wave propagation associated with a remarquable system of partial differential equations. But before closing this paragraph we note that because of the general definition (2.16), we have the following general relationship (Maugin, 1990a):

$$P^f = P + \mathbf{p}, \tag{8.9}$$

where $\mathbf{p}$ is the physical (linear) momentum associated with the theory. Such a relationship holds, for instance, in the pure electrodynamics of polarized and magnetized materials, even in the absence of deformations.

8.2 Example of application to nonlinear localized waves

Instead of considering a purely mechanical example, in order to show the somewhat universality of the framework, its power of application, and the general notion of wave-momentum in a system with mutliple and physically differing degrees of freedom, we shall consider a system of nonlinear dispersive partial differential equations, the so-called *sine-Gordon-d'Alembert system* or *SGdA system*. This is given by a minimum of two coupled equations:

$$\begin{aligned} &\phi_{tt} - \phi_{xx} - sin\phi - \eta u_x cos\phi = 0, \\ &u_{tt} - c_0^2 u_{xx} + \eta(sin\phi)_x = 0 \;. \end{aligned} \tag{8.10}$$

Here partial time and space derivatives are indicated by subscripts t and x. The basic set of fields is a set of two-scalar valued function $\{u(x,t)\ ,\ \phi(x,t)\}$; η is a numerical coupling coefficient, and c_0 is a characteristic speed. In fact there are two characteristic speeds in that system, c_0 and 1 (in the first of (8.9)), which is therefore globally dispersive. The system (8.9) has remarquable properties whic have been studied by Pouget and Maugin (1984) and Kivshar and Malomed (1989). It is also remarquable from the physical viewpoint because it encompasses a series of physical modellings from which it originates. Its first appearance is in Pouget and Maugin (1984) on the modelling of coupled nonlinear electroacoustic waves in elastic ferroeletrics. It was further shown to be a good model for coupled magnetoacoustic waves and moving domain walls of the Néel type in elastic ferromagnets (Maugin and Miled, 1986a). It is also applicable to the case of elastic micropolar materials (Maugin and Miled, 1986b ; Pouget and Maugin, 1989 ; Eringen , 1999) and elastic liquid crystals (Potapov et al, 1995). Finally, it seems to be essential in a phenomenological representation of acoustic emission (Trochidis and Polyzos, 1994) - cf. Maugin (1999b). In most of these cases c_0 is none other than the speed of transverse elastic waves of displacement u and ϕ is an angle representative of the orientation of an internal structure (an electric diplole, a magnetization vector, a director in liquid crystals, etc). The wave structure of the first of (8.10) originates in an approximation where the angular excursion of the internal structure is practically limited to a plane which is orthogonal to the x-axis.

The system (8.9) is conservative and admits both Hamiltonian and Lagrangian densities, e.g.,

$$H = K + W,\; L = K - W, \tag{8.11}$$

$$K = \frac{1}{2}\left(\phi_t^2 + u_t^2\right), W = \frac{1}{2}\left(\phi_x^2 + c_0^2 u_x^2\right) + \eta u_x \sin\phi + W_{ex}, \tag{8.12}$$

where

$$W_{ex} = -(1 + \cos\phi) \tag{8.13}$$

is a potential of external forces.

The equation of canonical or wave momentum associated with the system (8.10) can be obtained either by simple mathematical manipulations reproducing the construction of the Ericksen-Noether identity - i.e., multiplying $(8.10)_1$by ϕ_x and $(8.10)_2$ by u_x, integrating by parts, and combining the two results (cf.Maugin, 1992) - or by applying directly the canonical definitions of Section 2. The result is the *single* equation

$$\frac{\partial P^f}{\partial t} - \frac{\partial}{\partial x} b^f = 0, \tag{8.14}$$

where all quantities are scalars. In particular,

$$P^f = -\left(u_x u_t + \phi_x \phi_t\right), \tag{8.15}$$

$$b^f = -\left(L + u_x T + \phi_x M\right), \tag{8.16}$$

where we have introduced the « stresses »

$$T = \frac{\partial W}{\partial u_x} = c_0^2 u_x + \eta sin\phi , M = \frac{\partial W}{\partial \phi_x} = \phi_x . \tag{8.17}$$

According to the Ericksen-Noether identity, eqn.(8.14) is, at all regular points x, an identity that should be verified by the solutions, if any, of the system (8.10. This is conceivably a rather complex system. But it happens (Pouget and Maugin, 1984) that it possesses *exact analytic* travelling solutions $u(\xi = x - ct)$ and $\phi(\xi = x - ct)$, which are extremely localized in space, meaning by this that they correspond to a rapid change of the function ϕ from a spatially uniform value on one side to another spatially uniform value on the other side, being accompanied by a local hump for the function u. This is physically meaningful for the physical models for which (8.10) stands. These extremely localized solutions are close to *solitonic behavior*, but not completely in reason of the inevitable radiations allowed by eqn.$(8.10)_2$. But this is not the point. What is interesting is whether the canonical formalism developed is helpful in going further than the obtained analytical solutions. The answer is yes. Indeed, consider *perturbations* of the system (8.10) by several causes, e.g., a dissipative mechanism (rotational relaxation) associated with the degree of freedom ϕ , *a forcing* (externally applied electric or magnetic field) associated with the same degree of freedom, or a dissipative mechanism (classical viscosity) associated with the secondary degree of freedom u ; This degree is secondary because the whole nonlinearity and dispersion are associated with ϕ, and this is what causes the existence of strongly localized dynamical solutions via the phenomenon of *compensation* between *nonlinearity* (tendency to steepen a signal, e.g. formation of shock waves) and *dispersion* (tendency to spread a signal since various Fourier components would travel at different speeds in the linearized system). Let f_ϕ^{diss} ,$f_\phi^{forcing}$, and f_u^{diss} , the physical forces corresponding to the various perturbations. Equations (8.10) transform then to

$$\begin{aligned} &\phi_{tt} - \phi_{xx} - sin\phi - \eta u_x cos\phi = f_\phi^{diss} + f_\phi^{forcing} , \\ &u_{tt} - c_0^2 u_{xx} + \eta (sin\phi)_x = f_u^{diss} \quad . \end{aligned} \tag{8.18}$$

Finding directly a dynamical solution of this system becomes an insuperable, in fact impossible, task, even for simple given functions of (ϕ, u, x, t) for the introduced perturbations. It is relatively much simpler to perturb the exact analytical solutions found in the absence of perturbations , at least finding the alterations brought to the most essential characteristic properties of these solutions, e.g., their speed of propagation, yielding, for instance, acceleration (via forcing) or slow down (thanks to dissipation). To do so, we remind the reader that in the absence of perturbing forces, the analytical solutions obtained have space-time gradients that vanish rapidly to zero ouside a narrow interval, being thus nil at $\pm\infty$. Accordingly, for these solutions, eqn.(8.14) integrates over the real line to yield the *Newton-like* inertial motion

$$\frac{d}{dt}P^t(c) = 0,\ P^t(c) = \int_{-\infty}^{+\infty} P^f(\xi)d\xi \ . \tag{8.19}$$

In the presence of perturbations, remembering the construction of the Ericksen-Noether identity, the first of these will be replaced by

$$\frac{d}{dt}P^t(c) = -\int_{-\infty}^{+\infty}\left(\phi_x\left(f_\phi^{diss} + f_\phi^{forcing}\right) + u_x f_u^{diss}\right)dx \ . \tag{8.20}$$

The right-hand side of this equation can be expressed in terms of the speed c for solutions (ϕ, u) of the general form found in the absence of perturbations, leaving the parameter c free to vary in space and time. The result of this difficult computation is an ordinary time-differential equation for c (since x and t appear in a single variable) of the general type

$$\frac{d}{dt}P^t(c) = F(c) \ , \tag{8.21}$$

of which the integration, analytical or numerical, produces the time variation of the solitonic-like solution of the (ϕ , u) under the action of the given perturbations. The reader will find in Maugin (1999b) and the cited original papers the analytical and numerical results for several cases of perturbations of the system (8.10) or other nonlinear dispersive systems of mathematical physics, that perfectly illustrate the presented technique which is indeed based on the exploitation of the notion of configurational forces - this is what the right-hand side , of eqns.(8.20) and (8.21) is.

9. Conclusion : Toward a canonical mechanics of materials

Since the pioneering works of Casal (1978), and Rogula (1970, 1977) , and also the relativistic elasticity of Maugin (1971), it is clear that a *canonical mechanics of continua* in the classical framework is emerging - see also Germain (1992). What is really new, however, is the introduction of thermal and more general dissipative processes in a canonical framework, that of the « **material mechanics of materials** », for short $\boldsymbol{M}^3$. This is particularly evident on eqns. (5.32) and (5.33) which stand for (material) momentum and energy conservation in a formulation that renders these two equations structurally similar along space and time axes , respectively :

$$\left.\frac{\partial P}{\partial t}\right|_{\mathbf{X}} - \left(div_R \mathbf{b} + \mathbf{f}^{inh}\right) = \mathbf{f}^{th} + \mathbf{f}^{int} \ , \tag{9.1}$$

and

$$\left.\frac{\partial(S\theta)}{\partial t}\right|_{\mathbf{X}} + \nabla_R . \mathbf{Q} = \Phi^{th} + \Phi^{int} , \tag{9.2}$$

where , we remind the reader, the right-hand sides have exactly the same structure but for material-space and time derivatives, and the material force of true inhomogeneities $\mathbf{f}^{inh}$ has no energetic contents and, therefore, has no equivalent term in eqn.(9.2) if the system is holonomous (no explicit time dependence of an energy potential). The remarkable feature of the couple (9.1)-(9.2) of equations is that the relevant energy density per unit reference volume is neither the internal energy, nor the free energy, *but their difference.* In particular, in the absence of heat conduction - at spatially uniform and time-invariant temperature θ_0 - and of irreversible microscopic mechanisms accounted for through the internal-variable set α, and for a materially homogeneous material, eqns.(9.1)-(9.2) reduce to the following system of strict conservation laws :

$$\left.\frac{\partial P}{\partial t}\right|_{\mathbf{X}} - div_R \mathbf{b} = \mathbf{0}, \left.\frac{\partial(\theta_0 S)}{\partial t}\right|_{\mathbf{X}} = 0 , \tag{9.3}$$

a system that is Hamiltonian , and in fact corresponds to one considered by Kijowski and Magli (1999) in isothermal relativity . In the dissipative case (9.1)-(9.2) which includes exotic ones such as heat-conducting finite-strain elasticity and elasto-visco-plasticity (assuming material homogeneity for a more symmetric outlook), these equations read

$$\left.\frac{\partial P}{\partial t}\right|_{\mathbf{X}} + \nabla_R . (-\mathbf{b}) = -\left.\frac{\partial W}{\partial \mathbf{X}}\right|_{\mathbf{F}\,fixed} , \tag{9.4}$$

$$\left.\frac{\partial(\theta S)}{\partial t}\right|_{\mathbf{X}} + \nabla_R . \mathbf{Q} = -\left.\frac{\partial W}{\partial t}\right|_{\mathbf{F}\,fixed} , \tag{9.5}$$

a clearly (3+1)-dimensional formulation akin to to the space-time decomposition of a four-dimensional world-invariant formulation (compare to Truesdell and Toupin's World-invariant kinematics, 1960). We do not pursue further here this formalization that finds its best and esthetically pleasant expression in a recent work (Maugin, 2000a).

The reader will find in a companion contribution by Trimarco and Maugin (same volume) appropriate generalizations of many of the reasonings done in this chapter to the case of more involved continua, e.g., deformable continua in interaction with electromagnetic fields.

References

Berezovski A.,(1997), Continuous cellular Automata for Simulation of Thermoelasticity, *Proc.Est.Acad.Sci., Math.Phys.*, 46: 5-13.

Berezovski A. and Maugin G.A., (1999a), Application of High-resolution Wave-propagation Algorithm to 2D Elastic Propagation, *J.Comput.Phys.*, (to be published).

Berezovski A. and Maugin G.A., (1999b), Application of Wave-propagation Algorithm to 2D Thermoelastic wave Propagation in Inhomogeneous Media, in : *Proc.Intern.Conf.on Godunov' Method* (Oxford, October 1999), ed. E.F.Toro, Kluwer: Dordrecht (in press).

Braun M., (1997), Configurational Forces Induced by Finite-element Discretization, *Proc.Est.Acad.Sci., Math.Phys.*, 46: 24-31.

Brenig W., (1955), Bessitzen Schallwellen einen Impulz, *Zeit.Phys.*, 143: 168-172.

Bui H.D., (1977), *Mécanique de la rupture fragile*, Paris: Masson.

Casal P., (1978), Interpretation of the Rice Integral in Continuum Mechanics, *Lett.Appl. Engng.Sci.*, 16: 335-347.

Cleja-Tigoiu S. and Maugin G.A., (2000), Eshelby's Stress Tensors in Finite Elastoplasticity, *Acta Mechanica*, 139: 231-249.

Christov C.I. and Maugin G.A., (1995), An Implicit Difference Scheme for the Long-time Evolution of Localized Solutions of a Generalized Boussinesq System, *J.Comput.Phys.*, 116: 39-51.

Christov C.I., Maugin G.A. and Velarde M.G., (1996), Well-posed Boussinesq Paradigm with Purely Spatial Higher-order Derivatives, *Physical Review*, E54: 3621-3638.

Dascalu C. and Maugin G..A., (1993), Forces matérielles et taux de restitution de l'énergie dans les corps élastiques avec défauts, *C.R.Acad.Sci.Paris*, II-317: 1135-1140.

Dascalu C. and Maugin G.A., (1994), Energy of Elastic Defects ; A Distributional Approach, *Proc.Roy.Soc.Lond.*, A445: 23-37

Dascalu C. and Maugin G.A., (1995), Thermoelastic Material-momentum Equation, *J.Elasticity*, 39: 237-255.

Elzanowski M., Epstein M. and Sniatycki J., (1990), G-structures and Material Inhomogeneities, *J.Elasticity*, 23: 167-180.

Elzanowski M. and Epstein M., (1992), On the Symmetry Group of Second-grade Materials, *Int.J.Nonlinear Mech.*, 27: 638-645.

Epstein M. and de Leon M., (1993), On the Integrability of Second Order G-Structures with Applications to Continuous Theories of Dislocations, *Rep.Math.Phys.*, 33: 419-436.

Epstein M. and de Leon M., (1996), The Geometry of Uniformity in Second-grade Elasticity, *Acta Mechanica*, 114: 217-224.

Epstein M and Maugin G.A., (1990), The Energy-momentum Tensor and Material Uniformity in Finite Elasticity, *Acta Mechanica*, 83: 127-133.

Epstein M. and Maugin G.A., (1995a), Thermal Material Force : Definition and Geometric Aspects, *C.R.Acad.Sci.Paris*, II-320: 63-68.

Epstein M. and Maugin G.A., (1996), Geometrical Material Structure of Finite-strain Elasticity and Anelasticity, *Zeit.angew.Math.Mech.*, 76: S4, 125-128.

Epstein M. and Maugin G.A., (1997), Notions of Material Uniformity and Inhomogeneity, in :

Theoretical and Applied Mechanics (Proc.ICTAM, Kyoto, 1996), eds. Tatsumi T., Watanabe E., and Kambe T, Amsterdam: Elsevier. 201-215.

Epstein M. and Maugin G.A., (1999), Elements of a Theory of Growth , in : *Continuum Models and Discret Systems (CMDS9),* eds.E.Inan and K.Z.Markov, Singapore: World Scientific. 555-561.

Epstein M. and Maugin G.A., (2000), Thermomechanics of Volumetric Growth in Uniform Bodies, *Int.J.Plasticity*, 16:1-28.

Ericksen J.L., (1977), Special Topics in Elastostatics, in : *Advances in Applied Mechanics*, Vol.17, ed. C.S.Yih, Newn York: Academic Press. 189-244.

Eringen A.C., (1999), *Microcontinuum Field Theories, I. Foundations and Solids,* New York: Springer-Verlag.

Eshelby J.D., (1951), The Force on an Elastic Singularity, *Phil.Trans.Roy.Soc.Lond.,* A244: 87-112.

Eshelby J.D., (1970), Energy Relations and the Energy-momentum Tensor in Continuum Mechanics, in : *Inelastic Behavior of Solids*, eds.M.F.Kanninen, W.F.Adler, A.R.Rosenfeld and R.I.Jaffee, New York: McGraw Hill. 77-114.

Eshelby J.D., (1975), The Elastic Energy-momentum Tensor, *J.Elasticity*, 5: 321-335.

Fomethe A. and Maugin G.A., (1996), Material Forces in Thermoelastic Ferromagnets, *Cont.Mech.Thermodynam.*, 8: 275-292.

Germain P., (1972), Shock Waves, Jump relations and Structures, in : *Advances in Applied Mechanics*, ed.C.S.YihNew York: Academic Press. 131-194.

Germain P., (1992), Toward an Analytical Mechanics of materials, in : *Nonlinear Thermodynamical Processes in Continua*, Eds.W.Muschik and G.A.Maugin, Berlin: TUB-Dokumentation und Tagungen, Heft 61. 198-212.

Germain P., (1998), Functional Concepts in Continuum Mechanics, Meccanica, 33: 433-444.

Greg W.R., (1862), Why are Women Redundant ?, *National Review* (reprinted in Greg W.R., *Literary and Social Judgements*, London, 1863 ; also as a separate pamphlet, 1869).

Grinfeld M., (1991), *Thermodynamic Methods in the Theory of Heterogeneous Media,* ISIMM SeriesHarrow, Essex : Longman.

Gurtin M.E., (1993), The Dynamics of Solid-solid Phase Transitions.I. Coherent Interfaces, *Arch.Rat.Mech.Anal.,* 123: 305-335.

Gurtin M.E., (1999), *Configurational Forces as Basic Concepts of Continuum Physics,* New York: Springer-Verlag.

Haddi A. and Weichert D., (1995), On the Computation of the J-integral for Three-dimensional Geometries in Inhomogeneous Materials, *Comput.Mech.Sci.,* 5: 143-150.

Hill R., (1986), Energy-momentum Tensor in Elastostatics : General Reflections on the General Theory, *J.Mech.Phys.Sol.,* 34: 305-317.

Imatani S. and Maugin G.A., (2000) , A Constitutive Model for Growing Materials and Its Application to Finite-element Analysis, *Trans.ASME.J.Appl.Mech.*(submitted).

Kivshar Yu.S. and Malomed B.A., (1989), Dynamics of Solitons in Nearly Integrable Systems, *Rev. Mod. Phys.*, 61: 763-915.

Knowles J.K. and Sternberg E., (1972), Class of Conservation Laws in Linearized and Finite Elasticity, *Arch.Rat. Mech.Anal.*, 44: 187-211.

Lubliner J., (1990), *Plasticity Theory*, McMillan, New York.

Mandel J., (1966), *Cours de mécanique des milieux continus*, Vol.I, Paris: Gauthier-Villars.

Mandel J.,(1971), *Plasticité classique et viscoplasticité*, (CISM Lecture Notes, Udine, Italy),

Wien: Springer-Verlag.
Maugin G.A., (1980), Principle of Virtual Power in Continuum Mechanics : Application to Coupled Fields, *Acta Mechanica*, 35: 1-70.
Maugin G.A., (1988), *Continuum Mechanics of Electromagnetic Solids*, Amsterdam: North-Holland.
Maugin G.A., (1990a), Balance Law of Pseudomomentum in the Mechanics and Electrodynamics of Continua (in French), *C.R.Acad.Sci.Paris*, II-311: 763-768.
Maugin G.A., (1990b), Internal Variables and Dissipative Structures, *J.Non-Equilibr. Thermodynam.*, 15: 173-192.
Maugin G.A., (1992a), *The Thermomechanics of Plasticity and Fracture*, Cambridge University Press : Cambridge, U.K.
Maugin G.A., (1992b), Application of an Energy-momentum Tensor in Nonlinear Elastodynamics : Pseudomomentum and Eshelby Stress in Solitonic Elastic Systems, *J.Mech.Phys.Solids*, 40: 1543-1558.
Maugin G.A., (1993), *Material Inhomogeneities in Elasticity*, London: Chapman and Hall.
Maugin G.A., (1994), Eshelby Stress in Elastoplasticity and Ductile Fracture, *Int.J.Plasticity*, 10: 393-408.
Maugin G.A., (1995*), Material Forces : Concepts and Applications*, *ASME. Appl .Mech. Rev.*, 48: 213-245.
Maugin G.A., (1996a), On Ericksen's Identity and Material Balance Laws in Thermoelasticity and Akin Phenomona, in : *Contemporary Research in the Mechanics and Mathematics of Materials* (Ericksen's Anniversary Volume), eds. R.C.Batra and M.F.Beatty, Barcelona: C.I.M.N.E. 397-407.
Maugin G.A., (1996b), Variations on a Theme of Griffith, in : *Fracture : A Topical Encyclopedia of Current Knowledge Dedicated to Alan Arnold Griffith*, ed. G.P. Cherepanov, Melbourne, Florida : Krieger (published 1998). 517-536.
Maugin G.A., (1997a), Thermomechanics of Inhomogeneous-heterogeneous Systems : Application to the Irreversible Progress of Two- and Three-dimensional Defects, *ARI* (Springer-Verlag), 50: 41-56.
Maugin G.A., (1997b), On Shock Waves and Phase-transition Fronts in Continua, *ARI* (Springer-Verlag), 50: 141-150.
Maugin G.A., (1998a), On the Structure of the Theory of Polar Elasticity, *Phil.Trans. Roy.Soc.Lond.*, 356: 1367-1395.
Maugin G.A., (1998b), Thermomechanics of Forces Driving Singular Sets, *Arch.Mech.*, 50: 509-519.
Maugin G.A., (1998c), The Role of Canonical Balance Laws in the Study of the Progress of « Defects » in Microstructured Materials, *J.Phys.IV (France)*, 8: Pr8-231-Pr8-237.
Maugin G.A., (1999a), *The Thermomechanics of Nonlinear Irreversible Behaviors* , Singapore and River- Edge, N.J: World Scientific..
Maugin G.A., (1999b), *Nonlinear Waves in Elastic Crystals*, U.K.:Oxford University Press.
Maugin G.A., (2000a), On the Universality of the Thermomechanics of Forces Driving Singular Sets, *Arch.Appl.Mech.*, 70: 31-45.
Maugin G.A., (2000b), Multiscale Approach to a Basic Problem of Materials Mechanics (Propagation of Phase-transition Fronts) in : *Multifield Problems* (Proc. Intern.Conf. Stuttgart, Oct. 1999), ed. W.G.Wendland, Berlin: Springer-Verlag. 11-22.
Maugin G.A. and Berezovski A., (1999), Material Formulation of Finite-strain

Thermoelasticity and Applications , *J.Thermal Stresses*, 22: 421-449.
Maugin G.A. and Christov C.I., (1997), Nonlinear Duality Between Elastic waves and Quasi-Particles in Microstructured Solids, *Proc.Est .Acad.Sci.,* A46: 78-84 (Proc.EUROMECH Colloquium , Tallinn, Estonia).
Maugin G.A. and Christov C.I., (2000), Nonlinear waves and Conservation Laws (Nonlinear Duality Between Elastic waves and Quasi-particles), in : *Nonlinear Wave Phenomena*, eds. A.Guran and J.L.Wegner, Boston: Birkhauser (in press).
Maugin G.A. and Epstein M., (1998), Geometrical Material Structure of Elastoplasticity, *Int.J.Plasticity*, 14: 109-115.
Maugin G.A. and Inoue T.,(1998), Progress of Phase Boundaries and Walls in Thermo-deformable Solids : A Canonical Approach, *J.Phys.IV (France),* 9: Pr9-351-Pr9-360.
Maugin G.A., Inoue T. and Imatani S., (1999), Configurational or « Material Force of Inhomogeneity » at interfaces in Composites, *Trans.Jap.Soc.Mech.Engngn.* (pending publication)
Maugin G.A. and Miled A., (1986a), Solitary Waves in Elastic Ferromagnets, Physical *Review,* B33: 4830-4842.
Maugin G.A. and Miled A., (1986b), Solitary waves in Micropolar Elastic Crystals, *Int.J.Engng.Sci.,* 24: 1477-1499.
Maugin G.A. and Trimarco C., (1991), Pseudo-quantité de mouvement et milieux élastiques inhomogènes, *C.R.Acad.Sci.Paris*, II-313: 851-856.
Maugin G.A. and Trimarco C., (1992), Pseudo-momentum and Material Forces in Nonlinear Elasticity : Variational Approaches and Application to Brittle Fracture, *Acta Mechanica*, 94: 1-28.
Maugin G.A. and Trimarco C., (1995a), Dissipation of Configurational Forces in Defective Elastic Solids, *Arch.Mech.,* 47: 81-95.
Maugin G.A. and Trimarco C., (1995b), Dynamics of Configurational Forces at Phase-Transition Fronts, *Meccanica*, 30: 605-619.
Maugin G.A. and Trimarco C., (1995c), Configurational Forces and Coherent Phase-transition Fronts in Thermoelastic Solids, in : *Anisotropy, Inhomogeneity and Nonlinearity in Solid Mechanics* (IUTAM Symp. Nottinghma, 1994), eds. D.F.Parker and A.H.England, Dordrecht: Kluwer. 345-350.
Maugin G.A. and Trimarco C., (1995d), On Material and Physical Forces in Liquid Crystals, *Int.J.Engng.Sci.*, 33: 1663-1678.
Mindlin R.D. and Tiersten H.F., (1962), Effects of Couple Stresses in Linear Elasticity, *Arch.Rat.Mech.Anal.*, 11: 415-448.
Müller I., (1999), Eshelby Stress and Phase equilibrium, *Theoretical and Applied Mechanics*, 25: 77-89.
Muschik W., (1990), *Aspects of Non-Equilibrium Thermodynamics*, Singapore, and River Edge, N.J.: World Scientific.
Nelson D.F., (1979), *Electric, Optic and Acoustic Interactions in Dielectrics*, New York: Wiley-Interscience.
Noether (1918), Invariante Variationsproblem, *Klg.Ges.Wissen.Nach. Göttingen, Math.Phys.Kl.,* 2: 235.
Olver P.J., (1986), *Applications of Lie Groups to Differential Equations*, New York: Springer-Verlag.
Potapov A.I and Pavlov I.S., (1995), Nonlinear Wave in 1D Oriented media, *Acoustic*

Letters, 19: 110-115.

Pouget J. and Maugin G.A., (1984), Solitons and Electroacoustic Interactions in Ferroelectric Crystals-I , *Physical Review,* B30: 5306-5325.

Pouget J. and Maugin G.A., (1989), Nonlinear Dynamics of Oriented Elastic Solids I,II, *J.Elasticity,* 22: 135-55, 157-183.

Rice J.R., (1968), Path-independent Integrals and the Approximate Analysis of Strain Concentratiosn by Notches and Cracks, *Trans.ASME J.Appl.Mech.,* 33: 379-385.

Rogula D., (1970), Variational Principle for Material Coordinates as Dependent Variables. Application in Relativistic Continuum Mechanics, *Bull.Acad. Pol.Sci. Sér.Sci .Techn.,* XVII: 781-785.

Rogula D., (1977), Forces in Material Space, *Arch.Mech.,* 29: 705-715.

Soper D.E., (1976), *Classical Field Theory,* New York: J.Wiley.

Steinmann P., (2000), Application of Material Forces to Hyperelastic Fracture Mechanics, I-II, *Int .J.Solids Structures* (in the press).

Stolz C., (1989), Sur la propagation d'une ligne de discontinuité et la fonction génératrice de choc pour un solide anélastique, *C.R.Acad.Sci.Paris,* II-308: 1-3.

Stolz C., (1994), Sur le problème d'évolution thermomécanique des solides à changement brutal des caractéristiques, *C.R.Acad.Sci.Paris,* II-318: 1425-1428.

Suhubi E.S., (1975), Thermoelastic Solids, in : *Continuum Physics,* Vol.2, ed.A.C.Eringen, New York: Academic Press. 174-265.

Toupin R.A., (1962), Elastic Materials with Couple-stresses, *Arch.Rat.Mech.Anal.,* 11: 385-414.

Trochidis A. and Polyzos B., (1994), Dislocation Annihilation and Acoustic Emission During Plastic Deformations of Crystals, *J.Mech.Phys.Solids,* 42: 1933-1944.

Truesdell C.A. and Toupin R.A., (1960), The Classical Field Theories. In : *Handbuch der Physics,* Bd.III/a , ed.S.Flügge,Berlin: Springer-Verlag.

Truskinowskii L.M., (1987), Dynamics of Non-equilibrium Phase Boundaries in a Heat Conducting Nonlinearly Elastic Medium, *P.M.M.,* 51: 777-784.

Truskinowsky L.M., (1994), About the « Normal Growth » Approximation in the Dynamical Theory of Phase Transitions, *Cont.Mech.Thermodynam.,* 6: 185-208.

Weichert D. and Schultz M., (1993), J-integral Concept for Multi-phase Materials, *Comput.Mat.Sci.,* 1: 241-248. .

Material Mechanics of Electromagnetic Solids

Carmine Trimarco [1] and Gérard A. Maugin [2]

[1] Dipartimento di Matematica Applicata « U.Dini »,
Università di Pisa, Italy.
[2] Laboratoire de Modélisation en Mécanique,
Université Pierre et Marie Curie, Paris, France.

1 Introduction

Eshelby [1-5] introduced the notion (and the naming) of *Maxwell stress tensor of Elasticity* having in mind the Maxwell energy-stress of electromagnetism.

In a vacuum, the electromagnetic energy-stress tensor stems almost straightforwardly from the set of Maxwell equations. These equations entail an additional equation, which has the form of a vector balance equation and in which the Maxwell stress tensor $\mathbf{t}_M$ appears. In the electrostatic case and in a vacuum, the Maxwell stress tensor reads:

$$\mathbf{t}_M = \varepsilon_0 \mathbf{E} \otimes \mathbf{E} - \tfrac{1}{2}\, \varepsilon_0 E^2\, \mathbf{I} \tag{1.1}$$

where ε_0 is the electric permittivity of vacuum and $\mathbf{E}$ is the electric field. In a vacuum, the divergence of $\mathbf{t}_M$ vanishes identically. We can say that the total Maxwell force acting on any closed surface of the physical space vanishes though $\mathbf{t}_M$ itself may not vanish. This fact represents the main remarkable novelty of the Maxwell-Faraday theory of electromagnetism. Therefore, the field $\mathbf{E}$ (along with the electric displacement $\mathbf{D}$) such as conceived by Faraday and introduced by Maxwell, introduces a physical stress tensor at a point of the 'empty' space even in the absence of electric charges in that point. Similarly the field $\mathbf{H}$ and $\mathbf{B}$ (the magnetic field and the magnetic induction, respectively) introduce an analogous tensor which adds to the electrostatic one. The Maxwell balance of forces extends to electrodynamics. In the context of electrodynamics, the divergence of the stress tensor balances the time derivative of a quantity, which is interpreted as a momentum density, having in mind the mechanical description of a continuum body. As a result, one associates a mechanical momentum to the electromagnetic fields. Once more, what may be surprising is that this momentum survives in a vacuum as the electromagnetic quantities pervade the whole physical space.

However, severe difficulties arise in establishing univocally the contribution of the electromagnetic field to the momentum. The difficulties are strictly related to the problem of the proper expression for the Maxwell stress tensor in a material [6-11]. One of the main perplexing points for physicists has been that the Maxwell stress tensor may turn out to be not symmetric. As is known, the possible lack of symmetry for the stress tensor entails the failure of the balance of the moment of momentum. This occurrence would undermine the mechanical description. A thorough discussion of the question can be found in Nelson [6].

As the electric charge may be viewed as an inhomogeneity of the physical space, Eshelby suggested that, similarly, a material inhomogeneity or a material defect would produce a change in the elastic fields (and in the related stresses). According to Eshelby, the elastic field should vary in such a way that a balance law of the Maxwell kind would hold true, independently of the specific kind of defects.

Eshelby's arguments were initially proposed in linear elasticity. He was able to find the mentioned balance law and the energy-stress tensor that governs the behaviour of the material defects. Despite the close similarity of the Eshelby's tensor with the Maxwell electromagnetic stress tensor, the two tensors only share the form of an energy-stress, their physical meanings being different from one another. We will first discuss this point in finite elasticity by adopting a variational approach and two variational procedures. The latter are naturally related to the 'direct' and to the 'inverse' description, respectively. Within this framework, the physical nature of the classical Maxwell energy-stress emerges in a natural way. In fact, the Maxwell tensor is a Cauchy-like tensor in the mechanical interpretation, whereas the *Maxwell stress of elasticity*, such as introduced by Eshelby, is of different nature. The latter acts on the *material manifold* where the material points can re-arrange themselves.

Eventually, we are left with two balance laws for the energy-stress tensor. A first one, in which the energy-stress can be identified with the classical mechanical Cauchy-like stress that balances the momentum; a second one, in which the energy-stress is an Eshelby-like tensor and balances the total time derivative of a quantity, the so-called *pseudomomentum*. [7-9].

The proposed variational procedures provide a general criterion for establishing the physical nature of energy-stress tensors in different context. This is the case, for instance, of the Ericksen tensor that governs the disclinations in liquid crystals. The Ericksen tensor has the form of an energy-stress and the question may arise about its physical nature. Eshelby himself was involved in this controversy [12,15]. We will return to this point in Section (4.1).

We shall use here the same notation and definitions as in the contribution to the same volume by Maugin and Trimarco. Any definition not given here is to be found there.

2 Two variational principles in finite elasticity

2.1 Classical variation. The direct deformation.

The full set of equations for the traction problem of a hyperelastic body at equilibrium can be derived from a variational approach. The variational approach also provides successfully the set of equilibrium equations for the analogous problem in the presence of a stress energy density also known as the *complementary energy*. In the framework of finite elasticity the appropriate choice for the stress tensor, among the many that are currently employed in the literature, is of basic importance. For instance, the so-called first Piola-Kirchhoff stress tensor would not represent a convenient choice. We will return to this point shortly.

Let $W(\mathbf{F}, \mathbf{X})$ be the elastic energy density, per unit volume of the reference configuration, for an inhomogeneous body. By introducing the mass density ρ_0, one would write $W = \rho_0\, \Sigma$, Σ being the energy density per unit mass. However, we will keep the volume density in view of introducing the electromagnetic energy. The dependence on $\mathbf{F}$ of W (or Σ) has to be understood, as usual, through $\mathbf{C} = \mathbf{F}^T\mathbf{F}$. We refer to *Section* **3.1** of the notes by Maugin G.A. and Trimarco C., for notation. The total energy is given by the volume integral

$$\int_V W(\mathbf{F}, \mathbf{X})\, dX. \tag{2.1}$$

It is not difficult to show that the following equilibrium equations for the following traction problem

$$\begin{aligned} &div_R\, \mathbf{T} = \mathbf{f} && \text{in } V \\ &\mathbf{T}\mathbf{N} \;= \tau(\mathbf{X}) && \text{on } \partial V \end{aligned} \tag{2.2}$$

stems from the following variational equation:

$$\delta_X \mathcal{E} = \delta_X \left[\int_V W(\mathbf{F}, \mathbf{X}, \mathbf{x})\, dV + \int_{\partial V} \tau(\mathbf{X}).\mathbf{x}\, ds \right] = 0 \tag{2.3}$$

The subscript X stands for recalling that the reference volume is kept fixed in performing the variation. $\mathbf{N}$ is the outward unit normal to ∂V.

We may first note that, by performing the variation, the inhomogeneity, which is expressed by the explicit dependence of W on $\mathbf{X}$, does not affect the variation. Second, we note that $\tau(\mathbf{X})$ is also unaffected by the variation, which is explicitly performed as follows:

$$\delta_X\, (\tau(\mathbf{X}).\mathbf{x}) = \tau(\mathbf{X}).\, \delta_X\, \mathbf{x}\, . \tag{2.4}$$

$\tau(\mathbf{X})$ is known as a *dead-load*.

Expanding explicitly $\delta_X W$ in components:

$$\delta_X W = (\partial W / \partial \mathbf{F}) . \mathbf{F} = [\partial W / \partial (x^i_{,H})]\, \delta_X (x^i_{,H}) =$$
$$[\partial W / \partial (x^i_{,H})]\, (\delta_X x^i)_{,H} = [\partial W / \partial (x^i_{,H})\, (\delta_X x^i)]_{,H} +$$
$$- [\partial W / \partial (x^i_{,H})]_{,H} (\delta_X x^i)]. \tag{2.5}$$

(Here, capital cases are for components in the reference frame and lower cases are for components in the current frame.)

Hence, evaluate explicitly the equation (2.3) and write:

$$\delta_X \mathcal{E} = [\int_{\partial V} (\partial W / \partial \mathbf{F}).\mathbf{N} - \tau(\mathbf{X})]. (\delta_X \mathbf{x}) dV +$$

$$- \int_V [div_R (\partial W / \partial \mathbf{F}) - \partial W / \partial \mathbf{x}].(\delta_X \mathbf{x}) dV = 0, \tag{2.6}$$

$\forall\, (\delta_X \mathbf{x})$ which has the same regularity of $\mathbf{x} = \chi(\mathbf{X})$.

Equations (2.2) stem straightforwardly from (2.7) having assumed that $\mathbf{T} \equiv \partial W / \partial \mathbf{F}$ and that $\mathbf{f} = (\partial W / \partial \mathbf{x})$.

2.2 A twofold classical variation. The complementary energy

The introduction of the complementary energy density W_c through a Legendre transform, allows one to shift from strains to stresses as independent variables. As **T** can be viewed as the dual of **F**, one would be tempted to choose the first Piola-Kirchhoff tensor **T** as the natural variable for the stress. This choice would be additionally supported by the fact that **T** is naturally related to the traction. However, this choice would not be the appropriate one for several reasons. First, **T** is a two-field tensor. In Cartesian components it has one component in the reference frame, the second one in the current frame. Thus, **T** is not frame invariant. In addition, **T** is not positive (or negative) definite; thus, a polar decomposition theorem does not hold true as for **F**. As a result, basing on **T** as the only variable, the complementary energy W_c could not be made depending on Lagrangian quantities, as constitutive laws are requested. Second, a lack of uniqueness for the inversion with strains could emerge even in the presence of fixed traction at the boundary and unique solution in stresses. This lack of uniqueness is due to the indeterminacy of the finite rotation **R** [16]. For the appropriate choice of the stress tensor and for a detailed discussion of the problem, the reader is referred to [16-20].

Following Reissner [17], we can introduce the second Piola-Kirchhoff stress tensor and consider the following energy density for a hyperelastic material:

$$W(\mathbf{S}, \mathbf{E}) = \mathbf{S}.\mathbf{E} - W_c(\mathbf{S}) \tag{2.7}$$

where $\mathbf{E} = \frac{1}{2}(\mathbf{C} - \mathbf{I})$.

A variational principle can be stated for a placement-traction problem of equilibrium as follows:

$$\delta_X \mathfrak{I}_c(\mathbf{S}, \mathbf{x}) = \delta_X \Big[\int_V (\mathbf{S}.\mathbf{E} - W_c(\mathbf{S},\mathbf{X},\mathbf{x}))dV + \\ - \int_{\partial V1} (\mathbf{x} - \mathbf{x}_o).(\mathbf{F}\,\mathbf{S})\mathbf{N}\,dS - \int_{\partial V2} \tau(\mathbf{X}).\mathbf{x}\,dS\Big] = 0\,. \tag{2.8}$$

The variation is now a twofold variation as x and S can vary independently, while the material particle X is still kept fixed. The related Euler-Lagrange equations are:

$$\begin{aligned} &\mathit{div}_R\,\mathbf{FS} - \partial \mathbf{W}_c / \partial x = \mathbf{0} && \text{in } V \\ &\mathbf{FS} - \tau = 0 && \text{on } \partial V_2 \\ &\mathbf{x} - \mathbf{x}_0 = 0 && \text{on } \partial V_1. \\ &\partial W_c/\partial \mathbf{S} - \mathbf{E} = 0. && \end{aligned} \tag{2.9}$$

We could also introduce the so-called Biot stress $\mathbf{T_B}$ and the finite rotation $\mathbf{R}$ as alternate independent fields for the complementary energy, without referring to the elastic energy. In that case, the variation (once more a twofold variation with respect to $\mathbf{T_B}$ and $\mathbf{R}$) is performed under the constraint that these fields should satisfy the Cauchy equilibrium equation for the stresses. [19].

2.3 The inverse deformation. A novel variation.

We consider now the inverse mapping

$$\chi^{-1} : \mathbf{x} \rightarrow \mathbf{X} \tag{2.10}$$

and the quantities $\mathbf{C}^{-1} = \mathbf{F}^{-1}(\mathbf{F}^{-1})^T$ and $\mathbf{E} = \frac{1}{2}(\mathbf{I} - \mathbf{C}^{-1})$. $\chi^{-1} \in \mathcal{C}^3(V, \mathrm{V})$.

Introduce the elastic energy density per unit volume of the current configuration w $(\mathbf{F}^{-1}, \mathbf{X})$. The dependence of w with respect to $\mathbf{F}^{-1}$ has to be understood as through $\mathbf{C}^{-1}$ or through $\mathbf{E}$. Introduce also the stress tensor $\partial w / \partial \mathbf{F}^{-1}$, following the Piola procedure.

It is not difficult to check that $\partial w/\partial \mathbf{F}^{-1} = -J\, \boldsymbol{S}\, \mathbf{F}^{-1}$, where $J = \det \mathbf{F}$, $J\boldsymbol{S} = \partial w/\partial \mathbf{E} \equiv \mathcal{S}$. Note that the tensor $(\partial w/\partial \mathbf{F}^{-1})$ *is not naturally related to the external traction*, contrary to the case of the classical Piola-Kirchhoff stress $(\partial w/\partial \mathbf{F})$.

Introduce the following functional [18]:

$$\mathcal{E}^{i} = [\, \int_V [w(\mathbf{F}^{-1}, \mathbf{X}, \mathbf{x})\, dV + \int_{\partial V} \boldsymbol{T}^*(\mathbf{x}).\mathbf{X}\, ds\,] \tag{2.11}$$

which is apparently similar to the classical one. Consider the variation $\delta_x \mathcal{E}^{i}$, which is performed keeping the current configuration fixed, and the following variational equation:

$$\delta_x \mathcal{E}^{i} - \int_{\partial V} W\mathbf{N}.\, \delta_x\mathbf{X}\, ds = 0, \tag{2.12}$$

where

$$W = J\, w \quad \text{and} \qquad \delta_x\mathbf{X} = \partial \boldsymbol{X}(\mathbf{x}, \varepsilon)/\partial\varepsilon\,\big|_{\varepsilon=0}.$$

The equation (2.12) entails the following Euler-Lagrange equations:

$$\begin{aligned} &div\,(\partial w/\partial \mathbf{F}^{-1}) - (\partial w/\partial \mathbf{X})_F = 0 && \text{in } V \\ &[(\partial w/\partial \mathbf{F}^{-1})\mathbf{F}^{-T} - w\,\mathbf{I}]\mathbf{F}^{T}\,\mathbf{n} + \boldsymbol{T}^* = 0 && \text{on } \partial V \end{aligned} \tag{2.13}$$

$\mathbf{n}$ being the outward unit normal to ∂V.

This set of equations has no relationship, in principle, with the classical equilibrium problem. In fact, the equations found through the present variational principle state the balance among the *configurational forces,* which are represented by the expression $[div\,(\partial w/\partial \mathbf{F}^{-1})]$, and the *forces due to the inhomogeneities* $(\partial w/\partial \mathbf{X})_{expl.}$ We will see that the variational equation (2.12) is equivalent to the equation (2.6) provided that suitable conditions are introduced. These conditions will be discussed hereafter.

First we note that the spatial (body) force $(\partial w/\partial \mathbf{x})$ is no *captured* by the δ_x-variation. In fact, by performing the variation $(\delta_x\, w)$ one writes:

$$\delta_x\, w = (\partial w/\partial \mathbf{F}^{-1}).\delta_x\, \mathbf{F}^{-1} + (\partial w/\partial \mathbf{X})_{expl}\,.\,\delta_x\mathbf{X}\,. \tag{2.14}$$

The term $(\partial w/\partial \mathbf{x}).(\delta_x\mathbf{x})$ does not appear as $(\delta_x\mathbf{x}) \equiv 0$. Thus, we can conclude that he δ_x-variation is insensitive to the spatial forces. Contrary to what happens in the classical variation that is insensitive to the *force of inhomogeneities*, the δ_x-variation *does capture* the effects of these forces.

Second, we note that the following term

$$\delta_x\,(\boldsymbol{T}^*(\mathbf{x}).\mathbf{X}) = \boldsymbol{T}^*(\mathbf{x}).\delta_x\mathbf{X} \tag{2.15}$$

expresses a dead-loading condition, with respect to the δ_x-variation. Of course, $T^*(\mathbf{x})$ is not a classical dead load in the sense that it does change with the configuration.

Third, we note that the inhomogeneity force has a natural representation in the reference frame. In fact, in Lagrangian components, it reads:

$$(\partial w/\partial X^H) = f_H{}^{inh}. \qquad (2.16)$$

By contrast, the classical equilibrium equation is concerned with classical *physical forces* whose natural representation is in the current frame. The body force $(\partial w/\partial x^h) = f_h$ is the analogue (and the counterpart) of the force of inhomogeneity.

The variation $(\delta_X\, \mathbf{x})$ can be interpreted as the classical virtual displacement of a point in a continuum. Differently, $\delta_x \mathbf{X}$ has to be understood as a *re-placement or re-arrangement of material points* on the material manifold. This remark also suggests that the corresponding forces, which are related by coniugacy to $\delta_x \mathbf{X}$, be of thermodynamical nature. Thus, equations (2,13) express a balance law for these forces that are of different physical nature with respect to the usual ones.

Equation $(2.13)_1$ is nothing else than the Eshelby balance law and the tensor $(\partial w/\partial \mathbf{F}^{-1})$ is strictly related to the Eshelby tensor **b**. In fact, taking into account the equation $(2,13)_1$ and the identities in Appendix.2, one can write that

$$\begin{aligned}\int_{\partial V} (\partial w/\partial \mathbf{F}^{-1})\mathrm{n}\, ds &\equiv - \int_{\partial V} \mathbf{bN}\, dS = \\ &= \int_V (\partial w/\partial \mathbf{X})_F\, dv \equiv \int_V (\partial W/\partial \mathbf{X})_F\, dV. \end{aligned} \qquad (2.17)$$

Equation (2.17) represents the global *balance* of the *Eshelby 'traction'* (the material configurational forces) and the forces of inhomogeneity, either on a region V of the reference configuration or on the corresponding region V of the current configuration. A similar approach can be found in the papers by Golebiewska and Herrmann [21] and Pack and Herrmann [22].

A complementary energy principle can be also stated in the framework of the inverse deformation by introducing the functional

$$\begin{aligned} F_c(\mathbf{X}, \mathbf{x}, \mathcal{S}) = \int_V w(\mathcal{S}, \mathbf{E}, \mathbf{X}) + \\ + \int_{\partial V1} \mathcal{S}\mathbf{F}^{-1}\, \mathbf{n}.(\mathbf{X} - \mathbf{X}_0) ds + \int_{\partial V2} T^*(\mathbf{x}).\mathbf{X} ds \end{aligned} \qquad (2.18)$$

in analogy with the previous one $\Im_c$, and re-stating the variational principle (2.12) as follows:

$$\delta_x F_c - \int_{\partial V} W\, \mathbf{N}.\, \delta_x \mathbf{X}\ dS = 0. \qquad (2.19)$$

Once more, the fields $\mathbf{X}(\mathbf{x})$ and $\mathcal{S}$ are independently 'varied'.

2.4 Preserving the identity of the material point.

If one wishes to preserve *the identity of the material point* in the variational procedure, one can assume that

$$\delta_x X = \mathbf{0}, \tag{2.20}$$

or, more specifically, that

$$\partial/\partial\varepsilon\,[\chi^{-1}(\mathbf{x}(\mathbf{X},\varepsilon),\varepsilon)\,|_X = 0 \tag{2.21}$$

The conditions (2.20) or (2.21) entail the following relationship

$$\delta_x \mathbf{X} + \mathbf{F}^{-1}\,\delta_X\,\mathbf{x} = 0, \tag{2.22}$$

which is of primary importance in establishing the possible connections between the two variational procedures. If the relationship (2.22) holds true identically, we are able to state that the variational principles (2.6) and (2.12) are equivalent to one another, as well as the variational equations (2.8) and (2.19). The proof follows without difficulties by replacing the identity (2.22) in one of the aforementioned variational equations.

APPENDIX 2.

Useful identies.

I. $\partial W/\partial \mathbf{F}^{-1} = -\mathbf{F}^T(\partial W/\partial \mathbf{F})\,\mathbf{F}^{-1T}$

II. $\partial W/\partial \mathbf{F}^{-1} = \partial w/\partial \mathbf{F}^{-1} - wJ\,\mathbf{F}^T$

III. $w\,J = W$

IV. $\mathrm{div}\,(J^{-1}\,\mathbf{F}^T) = 0$ and $\mathrm{div}_R\,(J\,\mathbf{F}^{-1T}) = 0$

V. $\mathcal{S} = J^{-1}\,\mathbf{C\,S\,C} = J\,\mathbf{F}^T\,\mathbf{t}\,\mathbf{F}$

where **t** is the Cauchy stress tensor.

3 Electromagnetic fields and Maxwell equations

3.1 Preliminary comments

In a frame at rest the Maxwell equations read (in S.I. units ; *Notation*: *rot* $\mathbf{E} \equiv$ *curl* $\mathbf{E} \equiv \nabla \times \mathbf{E}$; *div* $\mathbf{D} = \nabla . \mathbf{D}$)

$$\begin{aligned} &div\, \mathbf{D} = \rho_e \\ &div\, \mathbf{B} = 0 \\ &\text{rot}\, \mathbf{E} = - \partial \mathbf{B}/\partial t \\ &\text{rot}\, \mathbf{H} = \mathbf{j}_e + \partial \mathbf{D}/\partial t \end{aligned} \tag{3.1}$$

The reader is referred to classical textbooks [23-30] for details and phenomenological interpretations. The equations (3.1) are assumed to hold true everywhere in a vacuum and in matter. It is worth noticing that all fields depend on the space variable **x** and on the time t. Conditions at infinity, conditions at the interface across the material boundary and initial conditions have to be stated for the fields in order that the problem is completely posed. The specific class of problems will suggest these conditions.

The quantities ρ_e and $\mathbf{j}_e$ are the free charge density and the free current density, respectively. The quantities **E, D, H** and **B** have been mentioned in the Introduction. The reader is warned to regard with caution the similarity between the fields **E** and **H**, on the one side, and between **D** and **B**, on the other side, in order not to be misled in the subsequent development and in order to avoid misunderstandings.

Apparently, the mentioned two couples of fields share similar symmetries in the Maxwell equations. In fact, **D** and **B** enter with the divergence operator in the Maxwell equations, whereas **E** and **H** with the rotation operator. Despite this apparent analogy, we shall rather point out the similarity between the fields **E** and **B** versus the similarity of the fields **D** and **H**. The main reason for adopting this view can be envisaged directly in the set of equations (3.1). In fact, the fields ρ_e and $\mathbf{j}_e$, which appear in $(3.1)_1$ and in $(3.1)_2$, respectively, represent the material *sources* for the field **D** and **H**, respectively. Hence, **D** and **H** are the fields intimately related with matter, although they are defined also in a vacuum. In order to specify their connection with matter and vacuum, **D** and **H** are currently split in two parts as follows:

$$\begin{aligned} &\mathbf{D} = \varepsilon_0 \mathbf{E} && \text{in } \mathcal{E}_3 - V \\ &\mathbf{D} = \varepsilon_0 \mathbf{E} + \mathbf{P} && \text{in } V \end{aligned} \tag{3.2}$$

$$\begin{aligned} &\mathbf{H} = (1/\mu_0)\, \mathbf{B} && \text{in } \mathcal{E}_3 - V \\ &\mathbf{H} = (1/\mu_0)\, \mathbf{B} - \mathbf{M} && \text{in } V \end{aligned} \tag{3.3}$$

$\mathcal{E}_3$ represents the Euclidean physical space.

P is the electric polarisation, **M** is the magnetisation. As **P** and **M** represent material properties, both identically vanish in a vacuum. Such material properties will be specified either by macroscopic constitutive equations or through a microscopic atomic model by suitable averaging.

As a final comment we note that the vector fields **B, H** and **M** (the latter also appears now in the Maxwell equations) do not behave as proper vectors with respect to inversion of a spatial axis. These vectors, also known as *axial vectors*, are better regarded as skew symmetric second order tensors.

3.2 Maxwell equations at the microscopic level.

In the microscopic framework one deals with atoms and ions. The electromagnetic fields of physical interest are generally assumed to act out of the region occupied by these particles. In this view, one renounces to face the very difficult problem of electromagnetism: to find the fields inside the atoms or inside the ion of the possible crystal lattice. Thus, one is only concerned with the fields **E** and **B** and these fields, following the view of Lorentz, will be retained as the basic electromagnetic fields.

These fields will be denoted by $\mathbf{e}_m$ and $\mathbf{b}_m$ for reminding us that they represent microscopic fields. Their possible relationship with the macroscopic quantities is found by suitable statistical averaging procedures. In the microscopic framework, the Maxwell equations read:

$$\begin{aligned} \varepsilon_0\, div\, \mathbf{e}_m &= \rho_m \\ rot\, \mathbf{e}_m &= -\partial\, \mathbf{b}_m / \partial\, t \\ div\, \mathbf{b}_m &= 0 \\ (\mu_0^{-1})\, rot\, \mathbf{b}_m &= \varepsilon_0\, \partial\, \mathbf{e}_m / \partial\, t + \mathbf{j}_m + \rho_m \mathbf{v}. \end{aligned} \tag{3.4}$$

The subscript '$_m$' stands for microscopic.

It is worth remarking that it is not obvious that the *identity of each microscopic field is preserved* in the macroscopic description, after the averaging process [31]. Although, there is a general agreement among physicists that the polarisation **P** and the magnetisation **M** arise as the result of averaging the fields ρ_m, $\mathbf{j}_m$ and $\rho_m\mathbf{v}$. We will not discuss here these interesting points and refer to the proper literature.

Fundamental fields and potentials

Along with the above introduced Maxwellian fields, alternate additional fields are usually introduced. This is the case of the vector potential **A**, which is related to **B** as follows: *rot* **A** = **B**. It is easy to check that such a statement is consistent with the second Maxwell equation $(3.1)_2$. The field **A** is not uniquely defined as long as its divergence is not specified. The specific condition for (*div***A**) is prescribed according to the specific problems or class of problems. For instance, in order that waves propagates and that the propagation equation possess the simple standard form, the following condition is introduced

$$div\,\mathbf{A} + \partial\,\Phi\,/\partial\,t = 0, \quad \text{(Lorentz condition)}, \tag{3.5}$$

where Φ, known as the scalar potential, is defined here below by the equation (3.8).

In the quasi-static case, a more appropriate condition is the following :

$$div\,\mathbf{A} = 0 \quad \text{(Coulomb condition)}. \tag{3.6}$$

By summarising the results, we report the following equations that Φ and $\mathbf{A}$ satisfy

$$rot\,\mathbf{A} = \mathbf{B}, \tag{3.7}$$

$$(\mathbf{E} + \partial\,\mathbf{A}\,/\partial\,t) = -\nabla\,\Phi. \tag{3.8}$$

along with the above mentioned conditions (3.5) and (3.6).

Equation (3.8) stems straightforwardly from equation $(3.1)_2$. In fact, as all fields depend on $\mathbf{x}$ and t, the spatial operators (the *rot* operator in this case) commute with the time derivative, which has to be understood at a fixed $\mathbf{x}$. Φ is also not uniquely defined. Its indeterminacy is related to $\mathbf{A}$. The discussion of this point addresses to the so-called gauge condition and to the problem of gauge invariance. In electrostatics or in quasi-statics the indeterminacy can be removed by suitable conditions at infinity and/or at the boundary of conductors. Once more, the specific class of problems will suggest the appropriate conditions. Having introduced the fields Φ and $\mathbf{A}$, one can note that the unknowns in the Maxwell equations (once they are specified in the Lorentz form) are so reduced to four scalar unknowns (i.e. Φ and the three components of the vector potential) instead of the six components of the fields $\mathbf{E}$ and $\mathbf{B}$.

3.3 Polarisation and magnetisation

In order to state the problem for macroscopic materials, all electromagnetic fields must be taken into account. Polarisation and magnetisation express the main feature of matter and one is led to regard them as the fundamental electromagnetic properties of materials. In the Lorentz view, the $-$ (div $\mathbf{P}$) $\equiv \rho_P$ is interpreted as a *polarisation charge* density that adds up to ρ_e, having taken into account the equations (3.2). One may think of introducing likewise the *magnetisation charge* density as $-$ (div $\mathbf{M}$). However, the latter notion may be considered disturbing, as there is not evidence for the existence of the magnetic charge. In this respect, the general attitude is rather to accept the *Ampèrian model.* Basing on the Ampère conjecture, one assumes that **M** is due to microscopic magnetisation currents. Hence, **M** and **H** share the same physical nature and the same units. In this view, the relevant physical quantity is (*rot* **M**) rather than (*div* **M**). However, nothing prevents us to interpret **M** as a volume density of magnetic dipoles just as one does with **P**.

In terms of forces, different interpretations for **M** lead to different alternate expressions for the magnetic force density. A first one, which is reminding of the electric force acting upon a polarisation charge, reads

$$\mathbf{f}_v = -(div\,\mathbf{M})\,\mathbf{B}_0 \quad \text{and} \quad \mathbf{f}_S = (\mathbf{M}.\mathbf{n})\,\mathbf{B}_0, \tag{3.9}$$

for volume and surface densities, respectively. $\mathbf{B}_0$ represents the known magnetic induction. Notice that, in this context, (**M.n**) represents a surface magnetisation charge. Should we look for the magnetic field in a cavity within a material, this quantity would be useful for the solution of the problem. The second expression, which bases on the *Ampérian* view, reads:

$$\mathbf{f}_v = -(rot\,\mathbf{M}) \wedge \mathbf{B}_0 \qquad \text{and} \quad \mathbf{f}_S = -(\mathbf{M} \wedge \mathbf{n}) \wedge \mathbf{B}_0. \tag{3.10}$$

A third expression is related to the interpretation of **M** as dipole volume density and reads

$$\mathbf{f} = (\nabla\mathbf{M}).\,\mathbf{B}_0. \tag{3.11}$$

The related Maxwell magnetic tensor would be different in the three mentioned cases. As a result, one would be led to different expressions for the traction at the material boundary or at the interfaces [29,30,32], although the expression for the resultant force does not differ in the various formulations.

A similar problem may arise for the electric forces. We will no enter into the details of this very involved problem. The interested reader is addressed to the proper literature that is quoted in the bibliography. However, we will return to this point in Section **3.7**.

3.4 Polarisation and magnetisation in moving frames

Maxwell equations do not hold good in the form (3.1) in moving frames, even in uniformly translating moving frames. This lack of invariance with respect to Galilean transformations represents the main difficulty in reconciling the Maxwell's electromagnetism with classical mechanics. Maxwell equations are invariant under the so-called Lorentz transformations and the reconciliation with mechanics is possible only within the framework of the Relativity theory. Here we will consider the approximation of small velocities with respect the velocity of light and look for the Maxwell equations in this approximation. There is not a general agreement on the proper form that Maxwell equations should have in moving frames. Here we follow the Lorentz's arguments and investigate how the fields should be modified in order to preserve the Maxwell equations in uniformly moving frames. As said, the velocity is small with respect to that of light. As first, the attention will be focused on the electromagnetic sources: the charges and the currents.

Along with the Maxwell equations one introduces the notion of force acting upon a moving charge q in a vacuum, such as given by Lorentz:

$$\mathbf{F} = q\,\mathcal{E}, \quad \mathcal{E} = \mathbf{E} + \mathbf{v} \wedge \mathbf{B}, \tag{3.12}$$

where **v** is the velocity of the moving charge.

The assumption for that the electric field acting on the moving charge, $\mathcal{E}$ is different from **E**, is of fundamental importance for the subsequent discussion on the Maxwell equations in a moving frame. In this view, the polarisation of moving media has to be related to the field $\mathcal{E}$, rather than to the field **E**. Hence, the constitutive relationship in the simplest case of a linear, rigid, homogeneous and isotropic dielectric would read:

$$\mathbf{P} = \varepsilon_0(\varepsilon-1)\,[\mathbf{E} + \mathbf{v} \wedge \mathbf{B}]. \tag{3.13}$$

Accordingly, the electric displacement would be:

$$\mathbf{D} = \varepsilon_0\,\mathbf{E} + \varepsilon_0(\varepsilon-1)\,[\mathbf{E} + \mathbf{v} \wedge \mathbf{B}] = \varepsilon_0\mathbf{E} + \varepsilon_0(\varepsilon-1)\,\mathbf{v} \wedge \mathbf{B}. \tag{3.14}$$

that the electric displacement in the formula (3.14) only changes its constitutive relationship but does not change its form, such as given by (3.2).

A second assumption is based on the important remark that a moving polarised material produces a magnetic field, even in the absence of initial magnetisation or in the absence of electric currents. This magnetisation has the following form

$$\mathbf{M}_d = \mathbf{P} \wedge \mathbf{v}. \tag{3.15}$$

The subscript 'd' stands for dielectric. The assumption (3.15) affects the fourth of Maxwell equations in which the electric currents enter as sources. The underlying assumption is that the contribution of polarisation to the total current is given by a *convection term* $\rho_P\mathbf{v}$ and by a *polarisation current* $(\partial\,\mathbf{P}/\partial\,t\,)_{conv}$, in the following fashion:

$$\mathbf{J} = \mathbf{j} + \rho_e\mathbf{v} + \mathbf{j}_P \equiv \mathbf{j} + (\rho_e + \rho_P)\mathbf{v} + (\partial\,\mathbf{P}/\partial\,t\,)_{conv}. \tag{3.16}$$

where $\rho_P \equiv -(\mathit{div}\mathbf{P})$; Here $(\partial\,\mathbf{P}/\partial\,t\,)_{conv}$ denotes the *convected time derivative* of **P**. It explicitly reads:

$$(\partial\,\mathbf{P}/\partial\,t\,)_{conv} = (\partial\,\mathbf{P}/\partial\,t\,) + \mathit{rot}\,(\mathbf{P} \wedge \mathbf{v}) + \mathbf{v}(\mathit{div}\mathbf{P}). \tag{3.17}$$

There is a connection between the *time convected derivative* and the time derivative of the global fluxes of the quantities of interest through a moving surface. This connection can be found in the Appendix, at the end of the present Section.

The fields **D** and **H** are regarded as secondary fields in the Lorentz form of Maxwell equations. In this view, one would find that the field *rot* $(\mathbf{P}\wedge\mathbf{v})$ adds up to the field *rot* $\mathbf{H}_0$, where $\mathbf{H}_0 \equiv (\mu_0)^{-1}\mathbf{B}$ represents the field in a vacuum. Hence, one is encouraged to identify the field $(\mathbf{P}\wedge\mathbf{v})$, that appears in the formula (3.17), with a magnetisation. Classical experimental evidences seem to support this interpretation [24].

The second and third Maxwell equations $(3.1)_{2,3}$ are not influenced by these arguments as no sources enter these equations. Nonetheless, equation $(3.1)_3$ can be expressed in terms of the field $\mathcal{E}$

3.5 Maxwell equations in Eulerian form

Having introduced the field

$$\mathcal{H} = \mathbf{H} - \mathbf{v} \wedge \mathbf{D}, \tag{3.18}$$

the set of Maxwell equations in the moving frame can be written as follows:

$$\begin{aligned} &div\,\mathbf{D} = \rho_e \\ &div\,\mathbf{B} = 0 \\ &rot\,\mathcal{E} + (\partial\mathbf{B}/\partial t)_{conv} = 0 \\ &rot\,\mathcal{H} - (\partial\mathbf{D}/\partial t)_{conv} = \mathbf{j} + \rho_e\mathbf{v}, \end{aligned} \tag{3.19}$$

by taking into account the equations (3.12), (3.13) and (3.17).

With reference to the equations (3.16) and (3.17), the time convected derivatives for **B** and **D** explicitly read:

$$(\partial\mathbf{B}/\partial t)_{conv} = (\partial\mathbf{B}/\partial t) + rot\,(\mathbf{B}\wedge\mathbf{v}) \tag{3.20}$$

and

$$(\partial\mathbf{D}/\partial t)_{conv} = (\partial\mathbf{D}/\partial t) + rot\,(\mathbf{D}\wedge\mathbf{v}) + \rho_e\,\mathbf{v}. \tag{3.21}$$

As we wish to recover decomposition for $\mathcal{H}$, in analogy with the formulas (3.3), we note that $\mathcal{H}$ also reads:

$$\mathcal{H} = (\mu_0)^{-1}\,\mathcal{B} - \mathcal{M}, \tag{3.22}$$

where

$$\mathcal{B} = [\mathbf{B} - (1/c^2)(\mathbf{v}\wedge\mathbf{E})] \tag{3.23}$$

and

$$\mathcal{M} = \mathbf{M} + \mathbf{v} \wedge \mathbf{P}. \tag{3.24}$$

$\mathcal{B}$ and $\mathcal{M}$ are the fields in the moving frame. $(1/c^2) \equiv \varepsilon_0\mu_0$.

Notice that $\mathcal{M}$ is invariant under Galilean transformations, whereas **M** is not. This remark is consistent with the microscopic interpretation.

Equations (3.19) differ from the classical Maxwell equations only by the presence of the convection current $\rho_e\mathbf{v}$. Notice that equation $(3.19)_4$ can be also written as follows:

$$rot\ \mathcal{H}' - (\partial \mathbf{D} / \partial t) = \mathbf{j} + \rho_e \mathbf{v}, \tag{3.19.a$_4$}$$

where $\mathcal{H}' \equiv \mathcal{H} + \varepsilon_0 \mathbf{v} \wedge \mathbf{E} \equiv \mathbf{H} - \mathbf{v} \wedge \mathbf{P}$; in the absence of free charges, equation (3.19.a$_4$) is formally invariant with respect to equation $(3.1)_4$.

An alternate way to derive the set of equations (3.19) is by stating the basic electrodynamical laws in global form for a moving region [see Appendix 3]. This approach will be also useful in order to derive the Maxwell equations in the material framework. In the current configuration these laws can be written as follows:

$$\begin{gathered}
\int_{\partial V} \mathbf{D}.\mathbf{n}\, dv = \int_V \rho_e\, dv\ , \\
\int_{\partial V} \mathbf{B}.\mathbf{n}\, dv = 0\ , \\
\oint_C (\mathbf{E} + \mathbf{v} \wedge \mathbf{B}).\mathbf{t}\, dl = d/dt \int_{Sc} \mathbf{B}.\mathbf{n}\, ds\ , \\
\oint_C (\mathbf{H} - \mathbf{v} \wedge \mathbf{D}).\mathbf{t}\, dl = d/dt \left[\int_{Sc} \mathbf{D}.\mathbf{n}\, ds\right] + \int_{Sc} \mathbf{j}.\mathbf{n}\, ds\ ,
\end{gathered} \tag{3.25}$$

and are supplemented by the equation

$$\int_{Sc} \mathbf{j}.\mathbf{n}\, ds + d/dt \int_V \rho_e\, dv = 0, \tag{3.26}$$

which expresses the conservation of the electric charge. Here V , S_c and C represent volume, surfaces and closed lines, respectively. (3-dimensional, a 2-dimensional and a 1-dimensional measurable sets).The equivalence of equations (3.25) with the set of equations (3.19) stems without difficulties, by taking into account the result of Appendix 3.I and 3.II. The supplementary equation for the conservation of charge needs to be added. This reads:

$$div\, \mathbf{j} + (\partial \rho_e / \partial t)\big|_{\mathbf{x}} + div\, (\rho_e \mathbf{v}) = 0. \tag{3.27}$$

Should the free charges and currents vanish, the equation (3.26) would be identically satisfied. In the following, we will often disregard the presence of free charges and currents, as the main concern will be on the remaining Maxwell fields.

3.6 Deformable materials. Material formulation of Maxwell equations

In order that Maxwell equations preserve their form in deformable media, the electromagnetic quantities need to change their form with respect to the original expressions, once more. In this respect, the electromagnetic quantities need to be properly convected in the referential framework from the current one. The transformation should also ensure that the conservation of the electric charge would be preserved in time.
Notice that in the present context the fields will depend on $\mathbf{x}(\mathbf{X}, t)$ and on the time t explicitly.

By transforming equations (3.19) in the reference frame, they can be expressed in the following full material form:

$$div_R\, \mathfrak{D} = J\rho_e$$

$$div_R\, \mathfrak{B} = 0$$

$$rot_R\, \mathfrak{E} + (\partial \mathfrak{B}/\partial t)|_X = 0$$

$$rot_R\, \mathfrak{H} - (\partial \mathfrak{D}/\partial t)|_X = \mathcal{j}. \qquad (3.28)$$

The fields introduced above are defined as follows:

$$\mathfrak{D} = J\mathbf{F}^{-1}\,\mathbf{D}\,; \qquad \mathfrak{B} = J\mathbf{F}^{-1}\,\mathbf{B}\,;$$

$$\mathfrak{E} = \mathbf{F}^{T}\,\mathcal{E}; \quad \mathfrak{H} = \mathbf{F}^{T}\,\mathcal{H} \equiv \mathbf{F}^{T}\,\mathbf{H} + \mathbf{V} \wedge \mathfrak{D}\,;$$

$$\mathcal{j} = J\mathbf{F}^{-1}\,\mathbf{j} + J\rho_e \mathbf{V}\,. \qquad (3.29)$$

The following auxiliary fields are also worthy to be introduced:

$$\mathcal{P} = J\mathbf{F}^{-1}\,\mathbf{P}; \quad \mathfrak{M} = \mathbf{F}^{T}\,\mathbf{M};$$

$$\mathfrak{E}^* = \mathbf{F}^{T}\,\mathbf{E} \equiv \mathfrak{E} + \mathbf{V} \wedge \mathfrak{B};$$

$$\mathfrak{B}^* = J\mathbf{F}^{-1}\,\mathcal{B} \equiv \mathfrak{B} + (1/c^2)\,\mathbf{C}^{-1}(\mathbf{V} \wedge \varepsilon_0\, J\, \mathbf{C}^{-1}\,\mathfrak{E}^*);$$

$$\mathfrak{M}^* = \mathbf{F}^{T}\,\mathcal{M} \equiv \mathfrak{M} - (\mathbf{V} \wedge \mathcal{P}). \qquad (3.30)$$

Notice that the contravariant nature of the components of **B** and ***D*** are naturally preserved by the equations $(3.29)_{1,2}$. We recall that the electric charge in a volume is given by the surface integral ($\int_{\partial V}\mathbf{D.n}$), which is an invariant scalar quantity. The same argument applies to **B.** By contrast, the fields **E** and **H** have been pulled-back in the reference configuration by preserving their covariant nature. The fields **M** and $\mathcal{M}$ can alternatively be understood as densities of dipoles per unit volume, like **P**.
In this case, their pull-back in the reference frame would be the following: $J\mathbf{F}^{-1}\,\mathcal{M} = \boldsymbol{M}$, (Eringen, Maugin, v.I, p. 82 [10]).

Remarks.

(i) For future use, we note that the field $\mathfrak{H}$ can be also written as follows:

$$\mathfrak{H} = (\mu_0 J)^{-1} \mathbf{C}\mathfrak{B} + \mathbf{V} \wedge \varepsilon_0 J \mathbf{C}^{-1} \mathfrak{E}^* - \mathfrak{M}^*, \tag{3.31}$$

or, equivalently:

$$\mathfrak{H} = (\mu_0 J)^{-1} \mathbf{C}\mathfrak{B}^* - \mathfrak{M}^*, \tag{3.31a}$$

by taking into account the equations (3.22), (3.23), (3.24), (3.29), (3.30) and the subsequent remark.

(ii) For any two vector fields **a** and **b** the following identity yields:

$$\mathbf{F}^T (\mathbf{a} \wedge \mathbf{b}) = J \left(\mathbf{F}^{-1}\mathbf{a}\right) \wedge \left(\mathbf{F}^{-1}\mathbf{b}\right). \tag{3.32}$$

The proof relies on the following identities, (see the Appendix 3):

$$e_{ijh}\, x^i_{,I}\, x^j_{,J}\, x^h_{,H}\, J = e_{IJH} \;;\; \text{and } e_{IJH}\, X^I_{,i}\, X^J_{,j}\, X^I_{,i}\, J^{-1} = e_{ijh}. \tag{3.33}$$

e_{ijh} and e_{IJH} represent the Ricci's alternating tensors in the two frames, respectively.

CHECK that:

I. $(-\mathbf{V} \wedge \mathfrak{B}) = \mathbf{F}^T (\mathbf{v} \wedge \mathbf{B}).$ (3.34a)

In components:

$$\mathbf{F}^T (\mathbf{v} \wedge \mathbf{B})_I = x^i_{,I} (e_{ijh} v^i B^j) = x^i_{,I}\, e_{ijh} (-x^j_{,J} V^J)(J^{-1} x^h_{,H} \mathfrak{B}^H) = -(\mathbf{V} \wedge \mathfrak{B})_I.$$

II. $\mathbf{F}^T (\mathbf{v} \wedge \mathbf{E}) = -J^{-1}(\mathbf{V} \wedge \mathbf{C}^{-1} \mathfrak{E}^*).$ (3.34b)

The proof is as above, having noted that $\mathbf{F}^{-1}\mathbf{E} = \mathbf{C}^{-1}\mathfrak{E}^*$.

III. $\mathbf{F}^{-1}(\mathbf{v} \wedge \mathbf{E}) = -\mathbf{C}^{-1}(\mathbf{V} \wedge \varepsilon_0 J \mathbf{C}^{-1} \mathfrak{E}^*).$ (3.34c)

In components:

$$[\mathbf{F}^{-1}(\mathbf{v} \wedge \mathbf{E})]^H = X^H_{,i}\, e_{ijh}\, v^i E^h = X^H_{,i}\, e_{IJH}\, X^I_{,i}\, X^J_{,j}\, X^I_{,i}\, J\, v^i E^h = -[\mathbf{C}^{-1}(\mathbf{V} \wedge \varepsilon_0 J \mathbf{C}^{-1} \mathfrak{E}^*]^H.$$

With reference to the equation $(3.28)_3$, can note that the material time derivative $(\partial/\partial t)|_X$ commutes with the spatial operators that now act in the reference configuration. Basing on this remark and on the equation $(3.28)_2$, one introduces the *material scalar and vector potentials*, ϕ and $\mathfrak{A}$, respectively, in analogy with the case of the Maxwell equations in the frame at rest:

$$\mathfrak{B} = rot_R\, \mathfrak{A} \quad \text{and} \quad \mathfrak{E} = -[\nabla_R \phi + (\partial \mathfrak{A} / \partial t)\,|_X]. \tag{3.35}$$

ϕ and $\mathfrak{A}$ are related to Φ and **A**, respectively, as follows:

$$\phi = \Phi - \mathbf{A}.\mathbf{v}\,; \qquad \mathfrak{A} = \mathbf{F}^{-1T}\,\mathbf{A}$$

$$\Phi = \phi + \mathbf{V}.\,\mathfrak{A}. \tag{3.36}$$

Remark.

Note that : $\mathbf{A}.\mathbf{v} = -(\mathbf{F}^{-1T}\,\mathfrak{A}).\,\mathbf{F}\,\mathbf{V} = -\,\mathfrak{A}.\mathbf{V}$.

3.7 Forces and stresses

The proper expression for the volume density of the electromagnetic force in polarised and magnetised materials is a controversial matter. As mentioned earlier, slightly different and apparently very similar viewpoints lead to different expressions. Some of them are mathematically more elegant, others are more consistent with specific class of problems, others seem to adhere better to the physical microscopic picture. All different approaches certainly enlighten the topic by providing new interesting views and unexpected results. The interested reader is addressed to the quoted literature where these fascinating problems are expounded thoroughly.

Here we assume that such forces, known as *ponderomotive* forces, are given by the following expression, in which the effects of free currents are omitted [10,11]:

$$\mathbf{f}^{em} = \rho_e\,\mathcal{E} + (\partial \mathbf{P}/\partial t)_{conv} \wedge \mathbf{B} + (\mathbf{P}.\nabla)\mathcal{E} + (\nabla \mathbf{B}).\,\mathcal{M} \tag{3.37}$$

he expression (3.37) reduces to the Lorentz force for charged moving particles. It is worth remarking that the arguments expounded hereby are independent of the choice of the specific expression of $\mathbf{f}^{em}$.

Taking into account the expression (3.37), the following identity yields [10,11]:

$$\mathbf{f}^{em} = div\,\mathbf{t}^{em} - (\partial\,\mathbf{p}^{em}/\partial t\,). \tag{3.38}$$

$\mathbf{t}^{em}$ can be understood as the stress tensor and $\mathbf{p}^{em}$ as the electromagnetic momentum. This interpretation allows for a combination with the mechanical balance of momentum. The following explicit identities yield:

$$\mathbf{t}^{em} = \varepsilon_0\, \mathbf{E} \otimes \mathbf{E} - (\mu_0)^{-1}\, \mathbf{B} \otimes \mathbf{B} - \tfrac{1}{2}\,[\varepsilon_0 E^2 + (\mu_0)^{-1} B^2]\, \mathbf{I} +$$

$$+ \mathcal{E} \otimes \mathbf{P} - \mathcal{M} \otimes \mathbf{B} + (\mathbf{B}.\mathcal{M})\mathbf{I}, \tag{3.39}$$

and

$$\mathbf{p}^{em} = \varepsilon_0\, \mathbf{E} \wedge \mathbf{B}. \tag{3.40}$$

$\mathbf{t}^{em}$ reduces to $\mathbf{t_M}$ in a vacuum. In addition, the expression (3.39) accounts for the validity of the superposition principle for the stresses, as the following relationship yields:

$$\mathbf{t}^{em} = \mathbf{t_M} + \mathbf{t}^{int}, \tag{3.41}$$

where

$$\mathbf{t}^{int} = \mathcal{E} \otimes \mathbf{P} - \mathcal{M} \otimes \mathbf{B} + (\mathbf{B}\,.\,\mathcal{M})\, \mathbf{I} \tag{3.42}$$

represents the interaction tensor. Both, $\mathbf{t}^{em}$ and $\mathbf{t}^{int}$, are clearly *non-symmetric* and accounts for the presence of an electromagnetic torque.

Note that $\mathbf{p}^{em}$ survives also in the absence of matter. In fact, the expression (3.38) reduces to:

$$\mathit{div}\, \mathbf{t}_M - (\partial\, \mathbf{p}^{em} / \partial\, t) = 0, \tag{3.43}$$

in a vacuum. This remark shows in evidence that $\mathbf{p}^{em}$ can be interpreted as the momentum of the pure field and that it is unaffected by the presence of the material. We are entitled now to write the full balance of momentum in full Eulerian form as follows:

$$\mathit{div}\, (\mathbf{t}^* - \mathbf{p}^t \otimes \mathbf{v}) = \rho\, \mathbf{f} + (\partial \mathbf{p}^t / \partial\, t)\,|_X, \tag{3.44}$$

where

$$\mathbf{p}^t = \mathbf{p}^{em} + \mathbf{p}_{mech} \tag{3.45}$$

is the total momentum, $\mathbf{p}_{mech}$ is the mechanical momentum, and

$$\mathbf{t}^* = \mathbf{t_M} + \mathbf{t}^{int} + \mathbf{t}_{mech} \equiv \mathbf{t_M} + \mathbf{t}^E \tag{3.46}$$

is the combined electromechanical stress tensor, which includes the mechanical contribution. Equation (3.44) can be treated and transformed as a mechanical balance equation and, thus, pulled-back in a reference configuration, whenever the introduction of the latter is meaningful. In such a reference configuration the Cauchy-like stresses, such as introduced above, will assume the form of Piola-Kirchhoff stresses. The related mechanical and electromagnetic momenta are found by multiplying the various $\mathbf{p}$'s by J, respectively.

From the mathematical standpoint, the introduction of the reference configuration in mechanics allows for stating the problem in a known domain, being the current domain unknown. Boundary conditions (and/or interface conditions) need to be specified, as usual, and the behaviour of the electromagnetic fields at infinity as well.

APPENDIX 3.

I. The convected time derivative.

Compute the time derivative of a global flux of a vector field **a** across a moving surface. We write:

$(\partial/\partial t)_X\ [\int_{\partial V}(\mathbf{a}.\mathbf{n})ds\ ]\ ds = (\partial/\partial t)|_X\ [\int_{\partial V} J\ (\mathbf{F}^{-1T}\mathbf{a}).\mathbf{N}ds\ ]dS =$

$= \int_{\partial V}\ (\partial/\partial t)|_X\ [J\ (\mathbf{F}^{-1T}\mathbf{a}).\mathbf{N}]\ dS = \int_{\partial V}\ [J\ (\mathbf{F}^{-1T}\mathbf{a}).\ \mathbf{N}\ (div\mathbf{v}) + J\ \mathbf{N}.(\partial/\partial t)|_X\ (\mathbf{F}^{-1}\mathbf{a})dS =$

$= \int_{\partial V}[(div\mathbf{v})\ \mathbf{F}^{-1}\mathbf{a} + (\partial/\partial t)|_X\ (\mathbf{F}^{-1}\mathbf{a})]\ .\ \mathbf{F}^T\mathbf{n}\ ds = \int_{\partial V}[\mathbf{a}(\mathrm{div}\mathbf{v}) + \mathbf{F}.(\partial/\partial t)|_X\ \mathbf{F}^{-1}\mathbf{a}].\mathbf{n}\ ds =$

$=\int_{\partial V}[\mathbf{a}(\mathrm{div}\mathbf{v}) - (\partial\mathbf{a}/\partial t)|_X + [(\partial/\partial t)|_X\ \mathbf{F}]\ \mathbf{F}^{-1}\mathbf{a}].\mathbf{n}\ ds =$

$=\int_{\partial V}\{(\partial\mathbf{a}/\partial t)|_X - (\nabla\mathbf{v})\mathbf{a} + \mathbf{a}\ (div\mathbf{v})\}.\mathbf{n}\ ds =$

$=\int_{\partial V}\{(\partial\mathbf{a}/\partial t)|_x + div\ [\mathbf{a}\otimes\mathbf{v} - \mathbf{v}\otimes\mathbf{a}] + \mathbf{v}(div\ \mathbf{a}) + \mathbf{a}\ (div\mathbf{v}) - \mathbf{a}\ (div\mathbf{v})\ \}.\mathbf{n}\ ds =$

$=\int_{\partial V}\{[(\partial\mathbf{a}/\partial t)|_x + rot\ (\mathbf{a}\wedge\mathbf{v}) + \mathbf{v}(div\mathbf{a})\}.\mathbf{n}\ ds =$

$=\int_{\partial V}(\partial\mathbf{a}/\partial t)_{conv}\ .\mathbf{n}\ ds.$

II. Maxwell equations.

From Eulerian to material form.

$(\#)\int_V div\ \mathbf{B}\ dv = \int_{\partial V}\mathbf{B}.\mathbf{n}\ ds = \int_{\partial V} J\ \mathbf{F}^{-1}\mathbf{B}.\mathbf{N}\ dS = \int_V (div_R\ J\ \mathbf{F}^{-1}.\mathbf{B})dV = \int_V (div_R\ \mathfrak{B})dV = 0.$

$(\#\#)\ \int_V div\ \mathbf{D}\ dv\ = \int_V (div_R\ \mathfrak{D})\ dV = \int_V \rho_e\ dv = \int_V J\rho_e\ dV.$

$(\#\#\#)\ \int_S\ [rot\ \mathbf{E} + (\partial\mathbf{B}/\partial t)|_x\ ].\mathbf{n}\ ds = 0.$

We first examine the equality (###) and remark the following :

$(\partial/\partial t)|_X\ [\int_S \mathbf{B}.\mathbf{n}\ ds] = (\partial/\partial t)|_X\ [\int_S \mathfrak{B}.\mathbf{N}\ dS] = \int_S (\partial\mathfrak{B}/\partial t)|_X.\mathbf{N}\ dS =$

$$= \int_S [(\partial \mathbf{B}/\partial t)|_x + \mathbf{v}\,(div\,\mathbf{B}) + rot\,(\mathbf{B}\wedge\mathbf{v})].\mathbf{n}\,dS =$$

$$= \int_S [(\partial \mathbf{B}/\partial t)|_x + rot\,(\mathbf{B}\wedge\mathbf{v})].\mathbf{n}\,dS = \int_S [(\partial \mathbf{B}/\partial t)_{conv}..\mathbf{n}\,ds,$$

whence:

$$\int_S [(\partial \mathbf{B}/\partial t)|_x\,ds = \int_S (\partial \mathcal{B}/\partial t)|_X.\mathbf{N}\,dS + \int_S rot\,(\mathbf{B}\wedge\mathbf{v})].\mathbf{n}\,ds.$$

Then remark that :

$$\int_S [rot\,\mathbf{E} + (\partial\,\mathbf{B}/\partial t)|_x].\mathbf{n}\,ds = \int_S [rot\,(\mathbf{E} + \mathbf{v}\wedge\mathbf{B})].\mathbf{n}\,ds + \int_S (\partial \mathcal{B}/\partial t)|_X.\mathbf{N}\,dS$$

$$\int_S [rot\,(\mathbf{E} + \mathbf{v}\wedge\mathbf{B})].\mathbf{n}\,ds = \oint_{CS} \mathcal{E}.\,\boldsymbol{t}_l dl = \oint_{CS} \mathbf{F}^T \mathcal{E}.\,\boldsymbol{t}_L dL = \oint_{CS} \boldsymbol{\mathfrak{E}}.\,\boldsymbol{t}_L dL =$$

$$\int_S [rot_R\,\boldsymbol{\mathfrak{E}} + (\partial \mathcal{B}/\partial t)|_X].\mathbf{N}\,dS.$$

$\boldsymbol{t}_l$ and $\boldsymbol{t}_L$ are the unit tangent vectors to *CS* and to CS, respectively. The same arguments apply to the last of Maxwell equations.

From the frame at rest to the moving frame.

- $\int_S [rot\,\mathbf{E} + (\partial\,\mathbf{B}/\partial t)|_x].\mathbf{n}\,ds = \int_S [rot\,(\mathbf{E} + \mathbf{v}\wedge\mathbf{B}) + (\partial \mathbf{B}/\partial t)_{conv}].\mathbf{n}\,ds =$

$$= \int_S [rot\,\mathcal{E} + (\partial \mathbf{B}/\partial t)_{conv}].\mathbf{n}\,ds = 0.$$

One proceeds similarly with the last Maxwell equation.

III. Three identities.

i) $e_{ijh}\, x^i_{,I}\, x^j_{,J}\, x^h_{,H}\, J = e_{IJH}$;

ii) $e_{IJH}\, X^I_{,i}\, X^J_{,j}\, X^I_{,i}\, J^{-1} = e_{ijh}$,

iii) $(rot\,\mathbf{a}) = J^{-1}\mathbf{F}\, rot_R\,(\mathbf{F}^T\mathbf{a})$

Proof of the first two identities:

$$\int_V dv = \int_V d\mathbf{x}\wedge d\mathbf{y}.d\mathbf{z} = \int_V (e_{ijh}\, dx^i dy^j dz^h) =$$

$$\int_V (e_{ijh}\, x^i_{,I}\, y^j_{,J}\, z^h_{,H}\; dX^I dY^J dZ^H) = \int_V J dV = \int_V (e_{IJH}\, d\, X^I X^J dX^H).$$

4 Variational approach to electromagnetic fields.

4.1 Preliminary remarks

We will first propose a simple example of a variational procedure in electrostatics.

Consider the electrostatic energy $\boldsymbol{E}$ in a vacuum, due to the presence of the electric field $\mathbf{E} = -(\nabla\Phi)$, and the following variation:

$$\delta_X \boldsymbol{E} = \delta_X \int_V (\varepsilon_0/2)\,(\nabla\Phi)^2\, dV, \tag{4.1}$$

where the non fixed bounded domain V is also varied. V will eventually represent the whole physical space, in the limit of its diameter tending to infinity. The operator δ_X does not commute with the volume integral of the domain V. The subscript X recalls that the variation is performed at $\mathbf{X}$ fixed, $\mathbf{X}$ representing here a fictitious material point. Φ is here understood as $\Phi(\mathbf{x}(\mathbf{X}),\mathbf{X})$. Therefore, $\delta_X \boldsymbol{E}$ also reads:

$$\delta_X \boldsymbol{E} = \delta_X (\varepsilon_0/2) \int_V \{\delta_{\mathrm{x}} (\nabla\Phi)^2\, dV + div\,[\,(\nabla\Phi)^2/2)\,\delta_X \mathbf{x}]\} dV\,, \tag{4.2}$$

where the new operator δ_{x}, which appears in the first term of the r.h.s. of (4.2), commutes with both, the volume integral and the gradient operator ∇. By developing the variation, the following result is achieved

$$\delta_X \boldsymbol{E} = \varepsilon_0 \int_V \{div\,[(\nabla\Phi)\delta_{\mathrm{x}}\Phi + (\nabla\Phi)^2/2)\,\delta_X \mathbf{x}] - \Delta\Phi\delta_X\Phi\}\, dv \tag{4.3}$$

or equivalently

$$\delta_X \boldsymbol{E} = \varepsilon_0 \int_{\partial V} [(\nabla\Phi)\delta_{\mathrm{x}}\Phi + (\nabla\Phi)^2/2)\,\delta_X \mathbf{x}].\mathbf{n}\, ds - \varepsilon_0 \int_V \Delta\Phi\delta_X\Phi\, dv. \tag{4.3a}$$

At the electrical *equilibrium* and in the absence of charges, the following condition yields:

$$\Delta\Phi = 0. \tag{4.4}$$

Therefore, under this condition, the variation reduces to:

$$\delta_X \boldsymbol{E} = \varepsilon_0 \int_{\partial V} [(\nabla\Phi\,)\delta_{\mathrm{x}}\Phi + ((\nabla\Phi)^2/2)\,\delta_X \mathbf{x}].\mathbf{n}\, ds. \tag{4.5}$$

Assume now that the following condition also yields:

$$\delta_{\mathrm{x}}\,\Phi = -(\nabla\Phi\,).\,\delta_X \mathbf{x}, \tag{4.6}$$

which is equivalent to write that, on the boundary of the 'fixed points' domain, the condition

$$\delta_X \Phi = 0,$$

holds true, having taken into account the equality:

$$\delta_X \Phi = \delta_x \Phi + (\nabla\Phi). \delta_X \mathbf{X}. \tag{4.7}$$

The expression (4.5), under the condition (4.6), reads:

$$\delta_X \boldsymbol{E} = \varepsilon_0 \int_{\partial V} [-(\nabla\Phi) \otimes (\nabla\Phi) + (\nabla\Phi)^2/2)\, \mathbf{I}\,]\mathbf{n}.\delta_X \mathbf{x}\; ds. \tag{4.8}$$

The integrand of the surface integral (4.8) is the Maxwell stress $-\mathbf{t_M}$, which, combined with unit normal **n**, multiplies the elementary displacement δ_X **x** of a boundary point. Therefore, the expression (4.8) can be interpreted as the work done by the traction $-\mathbf{t_M}.\mathbf{n}$ for an elementary displacement δ_X**x**.

In this respect, *the Maxwell stress*, which is obtained by the expounded variational procedure, *is a Cauchy-like stress tensor*, as it is naturally related to the traction. This procedure is a classical variational procedure, which is adopted in mechanics for a 'moving region'. One performs the variation by fixing the material point, although in the present case these fixed points are fictitious, as there is not matter. The condition (4.6) expresses the fact that the variation of electric potential Φ is only due to its spatial gradient.

Some additional comments are worthy to be made. The expounded variational procedure, which recalls the *Noether procedure*, generally applies (there are not alternatives in most cases) when the energy density, which possibly depends on the gradient of some variable, is only defined *per unit volume of the 'moving' region.* In the language of continuum mechanics we would say that we are dealing with the energy density per unit volume of the current configuration. In solids, one introduces a reference configuration and possibly transforms the related energy density from the current to the reference configuration. In that case, δ_X- variation is the natural one. In liquids, the energy density is introduced per unit volume of the current configuration; reference configurations have a rather poor physical meaning. However, in the case of liquid crystals the energy depends on the gradient of the director fields (which are in turn related to the disclination lines). Such a dependence is reminder of a sort of elasticity and, thus, the notion of reference configuration (where, for instance, all the directors are aligned along one direction) acquires a physical meaning. The case of electromagnetic fields may be more difficult. However, in the presence of energy per unit volume of the current configuration, the aforementioned δ_X-variation eventually addresses to a Cauchy-like stress in the form of an energy stress. It is Cauchy-like as, although possibly non-symmetric, is naturally related to traction.

The discussion of this point provides a contribution to the controversy on whether the Ericksen tensor should be a Cauchy-like stress or a stress of different nature [15].

Remark.
The reader can check that the following equalities:

(1) $$\delta_X \int_V \mathrm{w}\,dv = \int_{\partial V} (\mathrm{w}\,\mathbf{I} - \mathbf{F}^{-1T}(\partial \mathrm{w}/\partial \mathbf{F}^{-1})\,\mathbf{n}.\delta_X \mathbf{x}\,ds\,.$$

The equality (1) holds true along with the equilibrium solution. Note that the current boundary ∂V is not fixed. By contrast, if the reference boundary $\partial \mathrm{V}$ is not fixed, the following equality

(2) $$\delta_x \int_{\mathrm{V}} \mathrm{W}\,dv = \int_{\partial \mathrm{V}} [\mathrm{W}\,\mathbf{I} - \mathbf{F}^{T}(\partial \mathrm{W}/\partial \mathbf{F})]\mathbf{n}.\delta_x \mathbf{X}\,dS.$$

holds true along the solution of the Eshelby equation.[see eq $(2.14)_1$].

4.2 The Lagrangian for electromagnetic fields in a vacuum

A Lagrangian density for the electromagnetic field in a vacuum is usually introduced in the literature. The interest for such a Lagrangian is that Maxwell equations stem from it as Lagrange equations in the field variables, by invoking the Hamilton's principle. This Lagrangian volume density reads:

$$L^f = \tfrac{1}{2}\,[\varepsilon_0 E^2 - (\mu_0)^{-1} B^2]. \tag{4.9}$$

In terms of the fields Φ and $\mathbf{A}$, such as introduced in Section **3.3**, this Lagrangian also reads:

$$L^f = \tfrac{1}{2}\,[\varepsilon_0 (-\nabla\Phi - \partial \mathbf{A}/\partial t)^2 - (\mu_0)^{-1} (rot\,\mathbf{A})^2\,]. \tag{4.10}$$

The Maxwell equations $(3.1)_2$ and $(3.1)_3$ are identically satisfied by the fields $\Phi(\mathbf{x},t)$ and $\mathbf{A}(\mathbf{x},t)$. The remaining equations $(3.1)_1$ and $(3.1)_4$ stem from the following Lagrange equations, respectively:

$$\partial/\partial t\;[\partial L^f/\partial(\Phi_{,t})] - \partial L^f/\partial\Phi + div\,[\partial L^f/\partial(\nabla\Phi)] = 0 \tag{4.11}$$

$$\partial/\partial t\,[\partial L^f/\partial(\mathbf{A}_{,t})] - \partial L^f/\partial \mathbf{A} + div\,[\partial L^f/\partial(\nabla \mathbf{A})] = \mathbf{0}. \tag{4.12}$$

where $\Phi_{,t} \equiv \partial\Phi/\partial t$ and $\mathbf{A}_{,t} \equiv \partial \mathbf{A}/\partial t$.

As the Lagrangian does not depend explicitly on ϕ and $\phi_{,t}$, the equation (4.11) reduces to

$$div\,[\partial L^f/\partial(\nabla\Phi)] = 0\,, \tag{4.11a}$$

which entails $(3.1)_1$. The computation is straightforward. Equation $(3.1)_4$ stems from equation (4.11), having noted that

$$\partial L^f / \partial (\mathbf{A}_{,t}) = -\varepsilon_0 \mathbf{E} \tag{4.13}$$

and that

$$div\, [\partial L^f / \partial(\nabla \mathbf{A})] = (\mu_0)^{-1} rot\,(rot\,\mathbf{A}) \equiv$$

$$\equiv (\mu_0)^{-1} rot\,\mathbf{B}. \tag{4.14}$$

Also note that

$$[\partial L^f / \partial(\nabla \mathbf{A})] = (\mu_0)^{-1} (rot\,\mathbf{A}). \tag{4.14a}$$

The latter formula can be achieved in components as follows:

$$[\partial L^f / \partial(\nabla A_{i,j})]_{,j} = [e_{lhk}\, e_{lmn}\, A_{k,h}\, (\partial A_{n,m} / \partial A_{i,j})]_{,j}$$

$$= (e_{lhk}\, e_{lmn}\, A_{k,h} \delta_{ni} \delta_{mj})_{,j} =$$

$$= e_{lhk}\, e_{lji}\, A_{k,hj} = (rot\, rot\, \mathbf{A})_{i}. \equiv (rot\,\mathbf{B})_{i}. \tag{4.15}$$

In a moving frame

Basing on the equations (3.19), a scalar potential in the moving frame could not be introduced in a natural way. In fact, the time convected derivative in the equation $(3.19)_3$ does not commute with the spatial *rot* operator. Nonetheless, one can still introduce the vector potential **A** and rewrite the equation $(3.19)_3$ in terms of Φ and of **A** as follows:

$$rot\, \mathcal{E} + (\partial \mathbf{B} / \partial t)_{conv} \equiv rot\, [\mathcal{E} + (\partial \mathbf{A} / \partial t) - \mathbf{v} \wedge (rot\,\mathbf{A})] = 0, \tag{4.16}$$

whence

$$\mathcal{E} = -\nabla\Phi - (\partial \mathbf{A} / \partial t) + \mathbf{v} \wedge (rot\,\mathbf{A}). \tag{4..17}$$

This expression is consistent with equation $(3.12)_2$. If one assumes that $\mathbf{A} = \mathbf{A}(\mathbf{x}(t),t)$, $\mathcal{E}$ equivalently reads:

$$\mathcal{E} = -\nabla(\Phi - \mathbf{A}.\mathbf{v}) - \mathbf{A}^{\bullet}, \tag{4.18}$$

by taking into account the following identity:

$$\mathbf{v} \wedge (rot\,\mathbf{A}) = \nabla(\mathbf{A}.\mathbf{v}) - (\nabla\mathbf{A})\mathbf{v} = \nabla(\mathbf{A}.\mathbf{v}) + \mathbf{A}^{\bullet} - (\partial \mathbf{A} / \partial t)_{\mathbf{x}}. \tag{4.19}$$

$\mathbf{A}^{\bullet}$ denotes the total time derivative; $\mathbf{v} \equiv \mathbf{x}^{\bullet}$

We next discuss the introduction of a Lagrangian for a charged moving point. This Lagrangian will represent a reference-point for the treatment of the interaction-Lagrangian for a continuum, as we will see in the subsequent Section **4.4**.

The field (4.18) is the electric field that acts on a charge q, which is located at **x** and moves with velocity **v**. The expression (4.18) is currently adopted in writing the Lorentz force, which reads:

$$\mathbf{F} = -\, q\,[\nabla(\Phi - \mathbf{A}.\mathbf{v}) + \mathbf{A}^{\bullet}], \tag{4.20}$$

The related Lagrangian for the moving charge of mass m_q can be written as:

$$L(\mathbf{x}, \mathbf{x}^{\bullet}, t) = \tfrac{1}{2}\, m_q\, (\mathbf{x}^{\bullet})^2 - q\,[\Phi(\mathbf{x}(t),t)) - \mathbf{A}(\mathbf{x}(t),t)).\, \mathbf{x}^{\bullet}], \tag{4.21}$$

having assumed that the scalar potential φ is here $\varphi = \Phi(\mathbf{x}(t),t)) - \mathbf{A}(\mathbf{x}(t),t)).\, \mathbf{x}^{\bullet}$. The corresponding Lagrange equation reads:

$$d/dt\,[\partial L/\partial(\mathbf{x}^{\bullet}) - (\partial L/\partial \mathbf{x}) = 0. \tag{4.22}$$

The latter, in turn, provides the following equation:

$$d/dt\,[\, m_q \mathbf{x}^{\bullet} + q\,\mathbf{A}] = -\, q\,\nabla(\Phi - \mathbf{A}.\mathbf{v}), \tag{4.23}$$

which represents the balance equation for the canonical momentum. Should the r.h.s. of equation (4.23) vanish, the *canonical momentum* would be conserved. It is worth noticing that the *canonical momentum* associated with the Lagrangian (4.20) is

$$(\partial L/\partial\, \mathbf{x}^{\bullet}) = m_q\, \mathbf{x}^{\bullet} + q\,\mathbf{A}. \tag{4.24}$$

A Lagrangian for pure electromagnetic fields in moving frames cannot be established yet, basing on (4.17) and (4.18). In fact, the fields of interest $\mathcal{E}$ and **B** pertain to different context. From the mathematical standpoint $\mathcal{E}$ has a twofold dependence: on time t explicitly, and through **x**(t). At the contrary, the field **B**, such as introduced in (3.14) and in (4.15), depends on time t only explicitly. Accordingly, **A** should also depend on **x** and on time t only explicitly. By introducing **A**(**x**(t),t), it would not be clear whether the Lagrangian should depend on $\mathbf{A}^{\bullet}$ or on $\mathbf{A}_{,t}$. Nonetheless, one can formally rewrite the Lagrangian (4.8) in terms of $\mathcal{E}$ and $\mathbf{B}(\mathbf{x}, t)$ as follows:

$$L^f = \tfrac{1}{2}\,[\varepsilon_0\,(\mathcal{E} - \mathbf{v} \wedge \mathbf{B})^2 - (\mu_0)^{-1}\, B^2], \tag{4.25}$$

and account for the dependence of L^f on $\mathbf{x}^{\bullet}$. In this case, the derivative of L^f with respect to $\mathbf{x}^{\bullet}$ would be the following:

$$(\partial L^{f} / \partial \mathbf{x}^{\bullet}) = \varepsilon_0 (\mathbf{E} \wedge \mathbf{B}). \qquad (4.26)$$

This result is consistent with the formula (3.36) as both represent the momentum of the electromagnetic field in a vacuum, although the re-formulation (4.25) of the Lagrangian stands on a rather weak basis. One could also introduce the magnetic induction in the moving frame $\mathcal{B} = [\mathbf{B} - (1/c^2)(\mathbf{v} \wedge \mathbf{E})]$ (see formula (3.23)) and rewrite the Lagrangian in terms of $\mathcal{E}$ and $\mathcal{B}$ respectively, [33]. Such a Lagrangian, however, would lead to a *vanishing momentum* for the pure electromagnetic fields, as one can easily check.

We will see that the expression (4.26) for the momentum is validated by the general approach which is proposed in the subsequent Sections and which relies on firmer bases. In the following, the problem of a variational approach for an electromagnetic deformable body, which is immersed in the physical space, will be discussed.

4.3 Material form of the field-Lagrangian

We will be concerned first with the *pure* electromagnetic fields, as in the previous Section. The Lagrangian density (4.9) can be rewritten per unit volume of a different configuration. The latter will be identified with the reference configuration a solid material when the related Lagrangian will be introduced. By appealing to the invariance of the global Lagrangian, the Lagrangian density per unit volume of the new configuration will be $L^F = J\, L^f$. The fields **E** and **B** also transform, in accordance with equations (3.29) and (3.30). They have to be understood now as depending on the new domain variables **X**, on time t and on the (fictitious, for the moment) deformation and/or motion. Basing on this transformation of **E** and **B**, one introduces the *material form* of the field-Lagrangian L^F, as follows:

$$L^F = \tfrac{1}{2}\ [J\, \varepsilon_0 \boldsymbol{\mathcal{E}}^{*}.\mathbf{C}^{-1} \boldsymbol{\mathcal{E}}^{*} - (\mu_0\, J)^{-1} \boldsymbol{\mathfrak{B}}.\mathbf{C} \boldsymbol{\mathfrak{B}}], \qquad (4.27)$$

by taking into account the equations (3.29) and (3.30).

With the help of the equations (3.34) and (3.35), L^F can be also expressed in terms the field variables ϕ and $\mathfrak{A}$:

$$L^F = \tfrac{1}{2} \{J\, \varepsilon_0 [-\nabla_R \phi - \mathfrak{A}^{\bullet} + \mathbf{V} \wedge rot_R\, \mathfrak{A}].\, \mathbf{C}^{-1} [-\nabla_R \phi - \mathfrak{A}^{\bullet} + \\ + \mathbf{V} \wedge rot_R\, \mathfrak{A}] - (\mu_0\, J)^{-1} (rot_R\, \mathfrak{A}).\, \mathbf{C}\, (rot_R\, \mathfrak{A})\}. \qquad (4.28)$$

Eventually, the Lagrangian depends on the following fields:

$$L^F = L^F(\phi, \phi^{\bullet}, \nabla_R \phi, \mathfrak{A}, \mathfrak{A}^{\bullet}, \nabla_R \mathfrak{A}, \mathbf{x}, \mathbf{x}^{\bullet}, \mathbf{F}), \qquad (4.29)$$

where the dependence on $\mathbf{x}^{\bullet}$ is through **V**, as. $\mathbf{x}^{\bullet} = -\mathbf{F}\mathbf{V}$. The related Lagrange equations are

$$d/dt\, [\partial L^F / \partial(\phi^{\bullet})] - \partial L^F / \partial \phi + div_R [\partial L^F / \partial(\nabla_R \phi)] = 0$$

$$d/dt\,[\partial L^F/\partial(\mathfrak{A}^\bullet)] - \partial L^F/\partial\mathfrak{A} + div_R\,[\partial L^F/\partial(\nabla_R\mathfrak{A})] = 0$$

$$d/dt\,[\partial L^F/\partial(\mathbf{x}^\bullet)] - \partial L^F/\partial\mathbf{x} + div_R\,[\partial L^F/\partial\mathbf{F}] = 0 \qquad (4.30)$$

We will show that the first two equations correspond to the Maxwell equations $(3.28)_1$ and $(3.28)_4$, provided that suitable identifications are introduced for the fields $\mathfrak{D}$ and $\mathfrak{H}$ in terms of $\mathfrak{E}$ and $\mathfrak{B}$.

First we note that the material Lagrangian preserves the full invariance of only two of the four Maxwell equations expressed in material form, namely of the equations $(3.28)_2$ and $(3.28)_3$. The remaining two equations, $(4.30)_{1,2}$, which are expressed in the Lorentz form (i.e. either in terms of $\mathfrak{E}$ and $\mathfrak{B}$ or in terms of ϕ and $\mathfrak{A}$) give unexpected results if the fields $\mathfrak{D}$ and $\mathfrak{H}$ are compared with the classical expressions (3.2).

In fact, having noted that L^F does not explicitly depend either on ϕ or on $\phi^\bullet$, the first Lagrange equation $(4.30)_1$ results into the following one:

$$div_R\,[\partial L^F/\partial(\nabla_R\phi)] = -\,div_R\,[\varepsilon_0\,J\,\mathbf{C}^{-1}(\mathfrak{E} + \mathbf{V}\wedge\mathfrak{B})] = 0, \qquad (4.31)$$

as

$$[\partial L^F/\partial(\nabla_R\phi)] = -\,\varepsilon_0\,J\,\mathbf{C}^{-1}(\mathfrak{E} + \mathbf{V}\wedge\mathfrak{B}). \qquad (4.31a)$$

A comparison of equation $(4.30)_1$ with the equations (4.11) and $(3.28)_1$ suggests the following identification for the *material form* of the electric displacement $\mathfrak{D}_o$ in a vacuum:

$$\varepsilon_0\,J\,\mathbf{C}^{-1}(\mathfrak{E} + \mathbf{V}\wedge\mathfrak{B}) \equiv \mathfrak{D}_o. \qquad (4.32)$$

Note that $\mathfrak{E} + \mathbf{V}\wedge\mathfrak{B} \equiv \mathfrak{E}^*$; [see formulas (3.30)].

With reference to the formulas (3.30), we also assume that the electric displacement in a material reads:

$$\mathfrak{D} = \mathfrak{D}_o + \mathcal{P}. \qquad (4.33)$$

It is worth to remarking that, by adhering formally to Lorentz's view, one would have been tempted to propose (though incorrectly) the following identification $\mathfrak{D}_o = \varepsilon_0\,(\mathfrak{E} + \mathbf{V}\wedge\mathfrak{B})$. The second equation, $(4.30)_2$, leads to the following equation:

$$rot_R\,\{(\mu_0)^{-1}\,J^{-1}\,\mathbf{C}\mathfrak{B} + \mathbf{V}\wedge\varepsilon_0\,J\,\mathbf{C}^{-1}(\mathfrak{E} + \mathbf{V}\wedge\mathfrak{B})=$$

$$= d/dt\,|_X\,[\varepsilon_0\,J\,\mathbf{C}^{-1}(\mathfrak{E} + \mathbf{V}\wedge\mathfrak{B})]. \qquad (4.34)$$

This result is achieved by performing explicitly the computations and by taking into account that L^F does not depend explicitly on $\mathfrak{A}$ and that

$$\partial L^F/\partial(\mathfrak{A}^{\bullet}) = -\ \varepsilon_0 J\, C^{-1}(\mathfrak{E} + V \wedge \mathfrak{B}) \equiv -\mathfrak{D}_o. \tag{4.35}$$

The quantity $[\partial L^F/\partial(\nabla_R \mathfrak{A})]$ can be evaluated as in Section **4.2**, (formulas (4.14) and (4.15)), so that the following relationship yields:

$$div_R\, [\partial L^F/\partial(\nabla_R \mathfrak{A})] =$$

$$= rot_R\, \{(\mu_0)^{-1} J^{-1}\, \mathbf{C}\mathfrak{B} + \mathbf{V} \wedge \varepsilon_0 J\, \mathbf{C}^{-1}(\mathfrak{E} + \mathbf{V} \wedge \mathfrak{B}). \tag{4.36}$$

In order to identify the equation (4.34) with the equation $(3.28)_4$, (in the absence of free charges and of free currents), one is suggested, once more, that the identification (4.32) holds true and that, along with it, the following identification also yields:

$$\mathfrak{H}_o = (\mu_0 J)^{-1}\, \mathbf{C}\mathfrak{B} + \mathbf{V} \wedge \mathfrak{D}_o \equiv (\mu_0 J)^{-1}\, \mathbf{C}\mathfrak{B}^*, \tag{4.37}$$

as

$$[\partial L^F/\partial(\nabla_R \mathfrak{A})] = rot_R\, \mathfrak{H}_o\,. \tag{4.37a}$$

Notice that the expression (4.37) is consistent with the equations (3.31), as $\mathfrak{H} = \mathfrak{H}_o - J^{-1}\mathfrak{M}^*$.

The third equation, equation $(4.30)_3$, leads to the balance of the momentum for the electromagnetic field.

$$\partial L^F/\partial \mathbf{x}^{\bullet} = -\ \mathbf{F}^{-1T}(\partial L^F/\partial \mathbf{V}) = \mathbf{F}^{-1T}\,[(\varepsilon_0 J\, \mathbf{C}^{-1}(\mathfrak{E} + \mathbf{V} \wedge \mathfrak{B}) \wedge \mathfrak{B}] =$$

$$= \mathbf{F}^{-1T}(\mathfrak{D}_0 \wedge \mathfrak{B}) \tag{4.38}$$

Expand the expression (4.38) in components and find the following result:

$$(\partial L^F/\partial \mathbf{x}^\bullet)_i = X^I_{,i}\, e_{IJH}\, (\mathcal{D}_0)^J (\mathcal{B})^H =$$

$$= e_{ijh}\, x^i_{,I}\, x^j_{,J}\, x^h_{,H}\, J^{-1}\, X^I_{,i}\, (\mathcal{D}_0)^J (\mathcal{B})^H =$$

$$= e_{ijh}\, x^j_{,J}\, x^h_{,H}\, J^{-1}\, (\mathcal{D}_0)^J (\mathcal{B})^H =$$

$$= e_{ijh}\, J\, (\mathbf{D_0})^J (\mathbf{B})^H \equiv J(\mathbf{D_0} \wedge \mathbf{B})_i = J(\mathbf{E} \wedge \mathbf{H_0})_i / c^2 \qquad (4.39)$$

where

$$\mathbf{D_0} = J^{-1}\mathbf{F}\mathcal{D}_0 = \varepsilon_0\, \mathbf{E} \text{ and } \mathbf{H_0} = (\mu_0)^{-1}\, \mathbf{B}.$$

If we express the momentum per unit volume of the current configuration, formula (4.39) reduces to formula (4.26), and reads:

$$J^{-1}(\partial L^F/\partial \mathbf{x}^\bullet) = \mathbf{D_0} \wedge \mathbf{B}. \qquad (4.40)$$

Note that the expression $\mathbf{D_0} \wedge \mathbf{B}$ represents the Minkowski form of the momentum in a vacuum. By contrast, the expression $[(\mathbf{E} \wedge \mathbf{H_0})/c^2]$ represents the Abraham form of the momentum in a vacuum. They do no differ from one another, in a vacuum. In a linear isotropic non-magnetised dielectric, the *Minkowski momentum* reads: $\mathbf{D} \wedge \mathbf{B}$, where $\mathbf{D} = \varepsilon_0 \varepsilon \mathbf{E}$, whereas the *Abraham momentum* remains unchanged with respect to that in a vacuum. [7-9].

For future use, also note that:

$$(\partial L^F/\partial \mathbf{V}) \equiv -\mathbf{F}^T(\partial L^F/\partial \mathbf{x}^\bullet) = -\mathcal{D}_0 \wedge \mathcal{B} = -J(\mathbf{D_0} \wedge \mathbf{B}). \qquad (4.41)$$

Evaluate now the quantity $(\partial L^F/\partial \mathbf{F})$ in the equation $(4.30)_3$. This quantity is interpreted as the electromagnetic stress tensor in the *Piola-Kirchhoff form*. It explicitly reads:

$$(\partial L^F/\partial \mathbf{F}) = -\mathbf{F}^{-1T}\{\mathcal{E}^* \otimes \mathcal{D}_0 + [(\mu_0)^{-1} J^{-1}\, \mathbf{C}\mathcal{B} \otimes \mathcal{B}]\} +$$

$$+ \tfrac{1}{2}\mathbf{F}^{-1T}\{\mathcal{E}^*.\mathcal{D}_0 + [(\mu_0)^{-1} J^{-1}\, \mathbf{C}\mathcal{B}].\mathcal{B}\} +$$

$$+ \mathbf{F}^{-1T}\{(\mathcal{D}_0 \wedge \mathcal{B}) \otimes \mathbf{V}\}, \qquad (4.42)$$

or, equivalently,

$$-(\partial L^F/\partial \mathbf{F}) = \mathbf{E} \otimes \mathcal{D}_0 + [(\mu_0)^{-1}\mathbf{B} \otimes \mathcal{B}] +$$

$$-\tfrac{1}{2} J\, [\varepsilon_0 E^2 + (1/\mu_0)\, B^2]\, \mathbf{F}^{-1T} - J\, [(\mathbf{D_0} \wedge \mathbf{B}) \otimes \mathbf{V}]. \qquad (4.43)$$

The corresponding *Cauchy form* is

$$\boldsymbol{t}^{F} \equiv - J^{-1}(\partial L^{F}/\partial \mathbf{F})\,\mathbf{F}^{T} = [\varepsilon_0 \mathbf{E} \otimes \mathbf{E} + (\mu_0)^{-1}\mathbf{B} \otimes \mathbf{B} - \tfrac{1}{2}[\varepsilon_0 E^2 + (1/\mu_0) B^2]\,\mathbf{I} - [\varepsilon_0(\mathbf{E} \wedge \mathbf{B}) \otimes \mathbf{v}] \equiv$$

$$\equiv [\mathbf{t}_M - \mathbf{p}^{em} \otimes \mathbf{v}], \tag{4.44}$$

where $\mathbf{p}^{em} \equiv \varepsilon_0 (\mathbf{E} \wedge \mathbf{B})$, in accordance with the formulas (3.40) and (4.26).

5 The Lagrangian for an electromagnetic body

5.1 Dielectrics

The general attitude in physics is to assume that the total Lagrangian for an electromagnetic material is given by the superposition of the field-Lagrangian L^F, the interaction-Lagrangian L^{INT}, which accounts for the interactions of matter and pure fields, and of the matter-Lagrangian L^M. Hereafter we disregard the presence of electric free charges and currents. L^{INT} and L^M are assumed to depend, among the other variable, also on the polarisation. This having stated, the Lagrangian for an electromagnetic body per unit volume of the current configuration reads:

$$\mathcal{L} = L^f + L^{int} + L^M. \tag{5.1}$$

In the following we will introduce a specific form for the interaction Lagrangian L^{int}, basing on the arguments developed in Section **3.** First we note that, by recalling formula (3.16), the contribution to the electric current in a dielectrics can be written as

$$\mathbf{j}_P \equiv - (div\,\mathbf{P})\,\mathbf{v} + (\partial \mathbf{P}/\partial t)_{conv}, \tag{5.2}$$

where $-(div\,\mathbf{P}) \equiv \rho_P$ is the polarisation charge and $(\partial \mathbf{P}/\partial t)_{conv}$ is explicitly given by the formula (3.17).

Next, we suggest that the Lagrangian that accounts for the interaction effects can be based on the Lagrangian of a moving point, as follows. Replace in the Lagrangian (4.21) the point-wise charge q with the charge density ρ_P and the 'current' qv by $\mathbf{j}_P$. Assume that

$$L^{int} = \mathbf{j}_P \,.\, \mathbf{A} + (div\,\mathbf{P})\,\Phi \tag{5.3}$$

or, equivalently, that

$$L^{int} = \mathbf{A}.[(\partial \mathbf{P}/\partial t) + rot\,(\mathbf{P} \wedge \mathbf{v})] + (div\,\mathbf{P})\,\Phi, \tag{5.4}$$

by appealing to the equalities (5.2) and (3.17).

By expanding the r.h.s of the expression (5.4) and after a rearrangement of the terms, one writes:

$$L^{int} = (\partial/\partial t)(\mathbf{A}.\mathbf{P}) + div\,(\mathbf{P}\,\Phi) + div\,[(\mathbf{P}\wedge\mathbf{v})\wedge\mathbf{A}] + \\ -\mathbf{P}.(\partial\mathbf{A}/\partial t) - \mathbf{P}.\nabla(\Phi) + \mathbf{P}.(\mathbf{v}\wedge rot\,\mathbf{A}). \tag{5.5}$$

Note that the total time derivative $(\partial/\partial t)(\mathbf{A}.\mathbf{P})$ and the total divergence term $div[(\mathbf{P}\wedge\mathbf{v})\wedge\mathbf{A} + \mathbf{P}\,\Phi]$ can be both omitted in the Lagrangian, as they do not affect the Lagrange equations. Hence, L^{int} can be re-written in the following simplified form:

$$L^{i} = \mathbf{P}.[(-\partial\mathbf{A}/\partial t) - \nabla(\Phi) + (\mathbf{v}\wedge rot\,\mathbf{A})]. \tag{5.6}$$

It is worth remarking that the Lagrangian density L^{int}, as written in the form (5.4), depends explicitly on Φ and on $\mathbf{A}$, whereas L^{i}, like L^{f}, depends only on the first order derivatives of Φ and on $\mathbf{A}$.

One recognises that the expression in the square brackets, in the formula (5.6), represents the electric field $\mathcal{E}$ such as introduced in (3.12) and (3.8). Accordingly, one writes the Lagrangian densities as follows:

$$L^{i} = \mathbf{P}.\,\mathcal{E} \tag{5.7}$$

or, in the reference configuration:

$$L^{I} = J\,L^{i} = J\mathbf{P}.\mathcal{E} = \Pi.\mathcal{E} \tag{5.8}$$

where $\Pi \equiv J\mathbf{P}$ represents the polarisation per unit mass.

In the reference configuration and in the full material form one can also write:

$$L^{I} = J\,(J^{-1}\mathbf{F}\,\mathcal{P}).\mathcal{E} = \mathcal{P}.\mathfrak{E}, \tag{5.8a}$$

with reference to the formulas (3.29) and (3.30).

5.2 In the presence of magnetisation

Magnetisation contributes to the interaction Lagrangian through the term ($rot\ \mathcal{M}$). In fact, this term is assimilated to an electric current and the interaction Lagrangian modifies as follows:

$$L^{int} = (\partial/\partial t)(\mathbf{A}.\mathbf{P}) + div\,[(\mathbf{P}\,\Phi) + (\mathbf{P}\wedge\mathbf{v})\wedge\mathbf{A} + \mathcal{M}\wedge\mathbf{A}] +$$
$$+\,\mathbf{P}.\,[-(\partial\mathbf{A}/\partial t) - \nabla(\Phi) + (\mathbf{v}\wedge rot\,\mathbf{A})] + \mathcal{M}.\,rot\,\mathbf{A}. \tag{5.9}$$

Once again, one can disregard the total time derivative and the total divergence terms, as they do not alter the Lagrange equations. It is not difficult to show that the interaction Lagrangian densities, in this case, read:

$$L^{i} = \mathbf{P}.\,\mathcal{E} + \mathcal{M}\mathbf{B} = J^{-1}(\mathcal{P}.\mathfrak{E} + \mathfrak{M}^{*}.\mathfrak{B}) \tag{5.10}$$

or, per unit volume of the reference configuration,

$$L^{I} = \mathcal{P}.\mathfrak{E} + \mathfrak{M}^{*}.\mathfrak{B} \equiv \Pi.\mathcal{E} + J\,\mathcal{M}\mathbf{B}, \tag{5.10a}$$

having taken into account the relationships (3.30).

In the following, L^{I} has to be understood as $L^{I}(\nabla_{R}\phi,\ \mathfrak{A}^{\bullet}, \nabla_{R}\mathfrak{A},\ \mathcal{P},\ \mathfrak{M}^{*})$.

One could have started by considering the magnetisation in the *material* frame ($\mathbf{F}^{T}\ \mathbf{M}$). In this case, its contribution to L^{int} would have been the following:

$$J^{-1}\,(rot_{R}\,\mathfrak{M}^{*}).\mathfrak{A}, \tag{5.11}$$

It is not difficult to check that one would have been led to the same Lagrangian density given by the formula (5.10a). In this respect, one has to take into account the following identity :
$J^{-1}\,(rot_{R}\,\mathfrak{M}^{*}).\mathfrak{A} \equiv (rot\ \mathcal{M}).\ \mathbf{A}$, the identity in Appendix **3.III.**iii), the relationships (3.36) and the following one: $\mathfrak{M}^{*} \equiv \mathbf{F}^{T}\,\mathcal{M}$.

Next we consider the matter-Lagrangian L^{M}, namely the contribution of the bulk. L^{M} is assumed to depend on the polarisation Π and on the magnetisation ($J^{-1}\,\mathcal{M}$), both per unit mass, as suggested by the expression (5.10a). This dependence is through the fields $\mathcal{P}$ and $\mathfrak{M}^{*}$, respectively, in the form written hereafter. The dependence on the spatial derivatives of these fields is neglected, as we intend to disregard here dispersion phenomena, for sake of simplicity. The dependence on their time derivatives is neglected as well. We also intend to restrict to non-saturated magnetism (no hysteresis). In this case, most of the arguments that have been expounded for polarisation can be repeated for magnetisation, although the physical nature of magnetisation is different and would need a separate treatment.

L^M is expressed here per unit volume of the reference configuration and reads:

$$L^M = \tfrac{1}{2}\rho_o v^2 - W(\mathbf{F}, \Pi, J^{-1}\mathcal{M}, \mathbf{X}) =$$
$$= \tfrac{1}{2}\rho_o v^2 - W(\mathbf{F}, \mathbf{F}\mathcal{P}, (J^{-1}\mathbf{F}^{-1T}\mathfrak{M}^*), \mathbf{X}), \tag{5.12}$$

We recall that the stored energy can be also expressed per unit mass by writing $W = \rho_o\Sigma$. With reference to the field equations $(4.30)_{1,2}$, in which L^F is replaced by the total Lagrangian $L \equiv L^F + L^I + L^M$, we note that L^M does not contribute to these equations, whereas L^I does contribute. More specifically, by replacing L^F with L, the equation (4.31) modifies as follows:

$$div_R\,[\partial L/\partial(\nabla_R\phi)] = -\,div_R\,[\varepsilon_0 J\,\mathbf{C}^{-1}(\mathcal{E} + \mathbf{V}\wedge\mathfrak{B}) + \mathcal{P}] = 0. \tag{5.13}$$

This result suggests, once again, the following identification: $\varepsilon_0 J\,\mathbf{C}^{-1}(\mathcal{E} + \mathbf{V}\wedge\mathfrak{B}) + \mathcal{P} \equiv \mathfrak{D}$, in accordance with the equation $(3.28)_1$, and with the relationships (3.30), (4.32) and (4.33). The equation (4.34) also modifies, consistently with the equation $(3.28)_4$. In fact, formula (4.35) results into the following one:,

$$\partial L/\partial(\mathfrak{A}^\bullet) = -\,\mathfrak{D}. \tag{5.14}$$

Accordingly, the formula (4.37) results into the following one:

$$\mathfrak{H} = (\mu_0 J)^{-1}\mathbf{C}\mathfrak{B} + \mathbf{V}\wedge\mathfrak{D} - \mathfrak{M} \equiv (\mu_0 J)^{-1}\mathbf{C}\mathfrak{B}^* - \mathfrak{M}^*. \tag{5.15}$$

5.3 The stress tensor

Next we will be concerned with the contribution of L^I and of L^M to the total stress tensor which has to be added to the contribution due to the pure fields given in Section **4.3**, formulas (4.39) and subsequent ones. First we note that L^I *does not contribute to the stress* (as it does not depend on **F**) and that L^M depends on **F** through W. The dependence of W on **F** is, in turn, twofold: an explicit one and through the $\mathcal{P}$ and $\mathfrak{M}^*$. Hence, differentiation of L^M with respect to **F** gives the following result:

$$\partial L^M/\partial\mathbf{F} = -\,(\partial W/\partial\mathbf{F})_{\mathcal{P},\mathfrak{M}^*} - (\partial W/\partial(\mathbf{F}\mathcal{P})).(\partial(\mathbf{F}\mathcal{P})/\partial\mathbf{F}) +$$
$$-\,(\partial W/\partial(J\mathbf{F}^{-1T}\mathfrak{M}^*)).(\partial(J\mathbf{F}^{-1T}\mathfrak{M}^*)/\partial\mathbf{F}) =$$
$$= -\,(\partial W/\partial\mathbf{F})_{\mathcal{P},\mathfrak{M}^*} - (\partial W/\partial\Pi)\otimes\mathcal{P} +$$
$$+\,J\mathcal{M}\otimes\mathbf{F}^{-1}(\partial W/\partial(J\mathbf{F}^{-1T}\mathfrak{M}^*)) +$$
$$-\,[(\partial W/\partial(J\mathbf{F}^{-1T}\mathfrak{M}^*)).(J\mathbf{F}^{-1T}\mathfrak{M}^*)]\,\mathbf{F}^{-1T}. \tag{5.16}$$

Formula (5.16) represents the contribution of L^M and of L^I to the total stress, in the Piola-Kirchhoff form. In the Cauchy form, these contributions read:

$$J^{-1}(\partial L^{I+M}/\partial \mathbf{F})\,\mathbf{F}^T = J^{-1}(\partial L^M/\partial \mathbf{F})\,\mathbf{F}^T = -J^{-1}(\partial W/\partial \mathbf{F})\,\mathbf{F}^T +$$
$$-J^{-1}[(\partial W/\partial \Pi)] \otimes \mathcal{P}\,\mathbf{F}^T - \mathcal{M} \otimes \mathbf{F}^{-1}[\partial W/\partial(J\mathbf{F}^{-1T}\mathfrak{M}^*)]\,\mathbf{F}^T +$$
$$-\ [(\partial W/\partial(J\mathbf{F}^{-1T}\mathfrak{M}^*)).(\mathbf{F}^{-1T}\mathfrak{M}^*)]\,\mathbf{F}^{-1T}\ . \tag{5.17}$$

The subscripts $_{\mathcal{P},\mathfrak{M}^*}$ have been dropped in the formulas (5.17) and will be omitted hereafter. The expression (5.17) combines with the additional Lagrange equations in $\mathcal{P}$ and in $\mathfrak{M}^*$. As the Lagrangian does not depend on the derivatives of $\mathcal{P}$ and of $\mathfrak{M}^*$, the related Lagrange equations reduce, respectively, to the following ones:

$$\partial L/\partial \mathcal{P} = 0. \tag{5.18}$$

$$\partial L/\partial \mathfrak{M}^* = 0. \tag{5.19}$$

These equations explicitly read:

$$\mathfrak{E} - (\partial W/\partial \mathcal{P}) \equiv \mathfrak{E} - (\partial W/\partial \Pi).(\partial(\mathbf{F}\mathcal{P})/\partial \mathcal{P}) \equiv$$
$$\equiv \mathfrak{E} - \mathbf{F}^T(\partial W/\partial \Pi) = 0. \tag{5.20}$$

$$\mathfrak{B} - (\partial W/\partial(J\mathbf{F}^{-1T}\mathfrak{M}^*)).(\partial(J\mathbf{F}^{-1T}\mathfrak{M}^*)/\partial \mathfrak{M}^*) \equiv$$
$$\equiv \mathfrak{B} - J\mathbf{F}^{-1}(\partial W/\partial(J\mathcal{M})) = 0. \tag{5.21}$$

The expressions (5.20) and (5.21) also read, respectively:

$$\partial W/\partial \Pi - \mathbf{F}^{-1T}\mathfrak{E} \equiv \partial W/\partial \Pi - \mathcal{E} = 0 \tag{5.22}$$

and

$$\partial W/\partial(J\mathcal{M}) - \mathbf{B} \equiv 0. \tag{5.23}$$

with reference to the formulas (3.29) and (3.30). As a result, the equations (5.16) and (5.17) simplify, respectively, as follows:

$$\partial L^M/\partial \mathbf{F} = -(\partial W/\partial \mathbf{F}) + \\ - \mathcal{E}\otimes\mathcal{P} + \mathcal{M}\otimes\mathfrak{B} - (\mathfrak{M}^*.\mathfrak{B})\,\mathbf{F}^{-1T}, \tag{5.24}$$

$$J^{-1}(\partial L^M/\partial \mathbf{F})\,\mathbf{F}^T = -J^{-1}(\partial W/\partial \mathbf{F})\,\mathbf{F}^T + \\ - J^{-1}\,\mathcal{E}\otimes\Pi + \mathcal{M}\otimes\mathbf{B} - (\mathcal{M}\mathbf{B})\,\mathbf{I} = \\ = -J^{-1}(\partial W/\partial \mathbf{F})\,\mathbf{F}^T - \mathcal{E}\otimes\mathbf{P} + \mathcal{M}\otimes\mathbf{B} - (\mathcal{M}\mathbf{B})\,\mathbf{I}. \tag{5.25}$$

These results are in accordance with those expounded in Section **3.7**, to which the reader is referred for comparison.

The relationships $(\partial W/\partial\mathcal{P}) = \mathfrak{E}$ and $\partial W/\partial\mathfrak{M}^* = \mathfrak{B}$, which stem from (5.18) and (5.19), are usually understood as identities or as constitutive equations, in textbooks and in some literature. Here, they are rather conceived as equilibrium equations. Should W depend also on the time and spatial derivatives of $\mathcal{P}$ and $\mathfrak{M}^*$, the related Lagrange equations would be partial differential equations. In this case, the aforementioned simple relationships would be lost and possible more complex relationships could follow from the solution of the problem.
In addition, in the presence of more complex functional dependence of the Lagrangian on $\mathcal{P}$ and $\mathfrak{M}^*$, the *symmetry of the total Cauchy-like* stress tensor [see formula (3.46)] may be also lost. This can be the case of ferromagnetism. This can be the case, for instance, of the response of a crystal at high-frequency regime, in the so-called optical mode, (in the terminology of solid-state physics). However, the Lagrangian density L is certainly consistent with the low-frequency regime, (the so-called acoustical mode).

5.4 The momentum

The equation $(4.30)_3$ has been interpreted as the equation of balance of momentum for the pure field. In this view, by replacing L^F by L in the equation $(4.30)_3$, we can associate the electromechanical momentum to the *canonical momentum* $(\partial L/\partial \mathbf{x}^{\bullet})$. L depends on the additional variables $\mathcal{P}$ and $\mathfrak{M}^*$ (with respect to L^F). As the fields $\mathcal{P}$ and $\mathfrak{M}^*$ do not depend on $\mathbf{v}$, *the interaction-Lagrangian does not contribute to the momentum.* The matter-Lagrangian contributes only through the term $\rho_o\,\mathbf{v}$. Thus, taking into account the expressions (4.38) and (4.39), one writes:

$$(\partial L/\partial \mathbf{v}) = \rho_o\,\mathbf{v} - \mathbf{F}^{-1T}(\partial L/\partial \mathbf{V}) = \rho_o\,\mathbf{v} + \mathbf{F}^{-1T}(\mathcal{D}_0\wedge\mathfrak{B}) = \\ = \rho_o\,\mathbf{v} + J(\mathbf{D}_0\wedge\mathbf{B}), \tag{5.26}$$

or, per unit volume of the current configuration,

$$J^{-1}(\partial L/\partial \mathbf{v}) = \rho\,\mathbf{v} + (\mathbf{D}_0\wedge\mathbf{B}). \tag{5.27}$$

where $\rho = \rho_o J^{-1}$.

Thus, the total momentum is given by the superposition of the mechanical momentum plus the momentum of the pure field, such as given by the formula (4.39).

5.5 The pseudomomentum

Along with the canonical momentum $(\partial L/\partial \mathbf{v})$, it is worthy to considering the remaining two canonical momenta $[\partial L/\partial(\phi^{\bullet})]$ and $[\partial L/\partial(\mathfrak{A}^{\bullet})]$. The first one identically vanishes, whereas the second one is given by the equation (5.14). We will see below that all canonical momenta contribute to the *pseudomomentum*.

The pseudomomentum is the quantity that is conserved in a homogeneous body, for the *infinitesimal displacement* $\delta_x \mathbf{X}$ (see Section **2.**) of the Lagrangian, [7-9]. By contrast, the *physical* momentum is quantity conserved in the physical space for the *infinitesimal displacement* $\delta_X\, \mathbf{x}$ of the Lagrangian L, $\delta_X\, \mathbf{x}$ representing the classical displacement of a material point.

The invariance of the Lagrangian, with respect to the infinitesimal translation $\delta_x \mathbf{X}$, has to be performed by introducing a new Lagrangian density $\mathscr{L}$ per unit volume of the current configuration. This Lagrangian is equal in value to $\mathfrak{L}$, which was introduced by formula (5.1), but depends on different variables. More specifically, these are: ϕ, $\phi_{,t}$, $\nabla\phi$, $\mathfrak{A}$, $\mathfrak{A}_{,t}$, $\nabla\mathfrak{A}$, $\mathbf{X}$, $\mathbf{V}$, $\mathbf{F}^{-1}$, $\mathcal{P}$ and $\mathfrak{M}^*$. The mentioned invariance shows in evidence the following quantity:

$$\boldsymbol{p}_o = -\,\mathbf{F}^{\mathrm{T}}(\partial L/\partial \mathbf{v}) - (\nabla_R\, \mathfrak{A})^{\mathrm{T}}(\partial L/\partial \mathfrak{A}^{\bullet}), \tag{5.28}$$

where $L = J\mathfrak{L}$, is the Lagrangian introduced in the previous sections.

By taking into account the formulas (5.14) and (5.27), the expression (5.28) can be developed as follows:

$$\begin{aligned}
\boldsymbol{p}_o &= -\,\mathbf{F}^{\mathrm{T}}(\partial L/\partial \mathbf{v}) - [(\nabla_R\, \mathfrak{A}) - (\nabla_R\, \mathfrak{A})^{\mathrm{T}}](\partial L/\partial \mathfrak{A}^{\bullet}) + (\nabla_R\, \mathfrak{A})(\partial L/\partial \mathfrak{A}^{\bullet}) = \\
&= -\,\mathbf{F}^{\mathrm{T}}\rho_o\mathbf{v} - \mathfrak{D}_0 \wedge \mathfrak{B} - (rot_R\, \mathfrak{A})\mathfrak{D} + (\nabla_R\, \mathfrak{A})\,\mathfrak{D} = \\
&= \rho_o\mathbf{CV} - \mathfrak{D}_0 \wedge \mathfrak{B} - \mathfrak{B} \wedge \mathfrak{D} + (\nabla_R\, \mathfrak{A})\,\mathfrak{D} = \\
&= \rho_o\mathbf{CV} + (\mathfrak{D} - \mathfrak{D}_0) \wedge \mathfrak{B} + (\nabla_R\, \mathfrak{A})\,\mathfrak{D}.
\end{aligned} \tag{5.29}$$

If one disregards the unimportant term $[(\nabla_R\, \mathfrak{A})\,\mathfrak{D}]$ (which can be made to vanish), one can write the *pseudomomentum density*, per unit volume of the reference configuration, as follows:

$$\boldsymbol{p}_R = \rho_o\mathbf{CV} + \mathcal{P} \wedge \mathfrak{B}, \tag{5.30}$$

taking into account that $\mathcal{D} - \mathcal{D}_0 \equiv \mathcal{P}$, (formula (4.33)).

The expression (5.30) can be also written in the following equivalent form:

$$\boldsymbol{p}_R = \rho_o \mathbf{CV} + \mathbf{JP} \wedge \mathbf{B}, \tag{5.31}$$

by taking into account the identity (3.32). This expression, in turn, leads to the pseudomomentum density per unit volume of the current configuration, which reads:

$$\boldsymbol{p} = \rho \mathbf{CV} + \mathbf{P} \wedge \mathbf{B}. \tag{5.32}$$

5.6 The [Eshelby-like] material energy stress

The natural counterpart of the pseudomomentum $\boldsymbol{p}_R$ is the *material energy stress*. This is an Eshelby-like stress and reads:

$$\mathbf{b} = -\ \mathrm{L}\,\mathbf{I} + \mathbf{F}^{\mathrm{T}}\,(\partial \mathrm{L}/\partial \mathbf{F})_{\,\mathbf{expl}} + (\nabla_{\mathrm{R}}\phi) \otimes [\partial\,\mathrm{L}\,/\partial(\nabla_{\mathrm{R}}\phi)] +$$
$$+ (\nabla_{\mathrm{R}}\ \mathfrak{A})^{\mathrm{T}}\,[\partial\,\mathrm{L}/\partial(\nabla_{\mathrm{R}}\mathfrak{A}\,)]. \tag{5.33}$$

It is worth recalling that

$$\partial\,\mathrm{L}/\partial(\nabla_{\mathrm{R}}\mathfrak{A}) = \mathbf{V} \wedge \mathcal{D}_0 + (\mu_0\,\mathrm{J})^{-1}\,\mathbf{C}\mathfrak{B}\ - \mathfrak{M}^{*} \equiv \mathfrak{H} \tag{5.34}$$

The quantities $\boldsymbol{p}_R$ and $\mathbf{b}$ are related one another by the following balance law:

$$\partial\,/\partial\,\mathrm{t}\,|_{\mathrm{X}}\ \ \boldsymbol{p}_R - \mathit{div}_R\,\mathbf{b} = (\partial \mathrm{L}/\partial \mathbf{X})_{\mathbf{expl}}\,. \tag{5.35}$$

From the physical standpoint, we can interpret the term $(\partial L/\partial X)_{expl}$, once again, as a force density due to an inhomogeneity. From the mathematical standpoint, $(\partial L/\partial X)_{expl}$ has to be understood in the distributional sense.

The equation (5.35) stems from a variational principle such as expounded in Section **2** [see formula (2.3)] and in Section **4** [**Remark** (2), at the end of Section **4.1**]. In the present case, such a variational principle is extended to dynamics as a Hamilton-like principle. The invariance of the global Lagrangian $\int_V \mathcal{L}\,\mathrm{dv}$ with respect to the elementary translation $\delta_x \mathbf{X}$, along the solutions of the equation (5.35), leads to the conservation of the total pseudomomentum

$$\partial\,/\partial\,\mathrm{t}\,|_{\mathrm{X}}\ \ \int_V \boldsymbol{p} = 0 \tag{5.36}$$

for a *homogeneous unbounded body*. If the material is not homogeneous, one can evaluate the material force through the equation (5.35), once $\boldsymbol{p}_R$ and **b** are known.

Herein after, we restrict to the Lagrangian L^{I+M} and evaluate, for future applications, the corresponding material tensor **b***as follows:

$$\mathbf{b}^* = -\ L^{I+M}\ \mathbf{I}+F^{T}\,(\partial L^{M}/\partial\mathbf{F})_{expl} + (\nabla_R\phi)\otimes[\partial L^{I}/\partial(\nabla_R\phi)] +$$

$$+ (\nabla_R\ \mathfrak{A})^{T}\,[\partial L^{I}/\partial(\nabla_R\mathfrak{A})], \tag{5.37}$$

after noting that L^{I} does not depend on **F** and that L^{M} does not depend on $(\nabla_R\phi)$ neither on $(\nabla_R\ \mathfrak{A})$. The expression (5.37) explicitly reads:

$$\mathbf{b}^* = -\,[(\mathcal{P}.\mathfrak{E} + 2\mathfrak{M}^*.\mathfrak{B} + (\tfrac{1}{2}\,\rho_o\,v^2 - W)\,]\mathbf{I} - \mathbf{F}^{T}(\partial W/\partial\mathbf{F})_{expl} +$$

$$-\ \mathfrak{E}\otimes\mathcal{P} + \mathfrak{M}^*\otimes\mathfrak{B} + (\nabla_R\ \mathfrak{A})\,[\partial L^{I}/\partial(\nabla_R\mathfrak{A})] - \mathfrak{A}^{\bullet}\otimes[\partial L^{I}/\partial(\nabla_R\phi)], \tag{5.38}$$

having taken into account that:

$$(\nabla_R\phi) = -\,(\mathfrak{E} + \mathfrak{A}^{\bullet})$$

$$\partial L^{I}/\partial(\nabla_R\phi) = -\ \mathcal{P}$$

$$\partial L^{I}/\partial(\nabla_R\mathfrak{A}) = -\ \mathfrak{M}^*.$$

$$[\,(\nabla_R\ \mathfrak{A}) - (\nabla_R\ \mathfrak{A})^{T}]\,[\partial L^{I}/\partial(\nabla_R\mathfrak{A})] = \mathfrak{M}^*\otimes\mathfrak{B} - (\mathfrak{M}^*.\mathfrak{B})]\mathbf{I}\ . \tag{5.39}$$

By re-arranging the terms and disregarding the unimportant terms $(\nabla_R\ \mathfrak{A})\,[\partial L^{I}/\partial(\nabla_R\mathfrak{A})] - \mathfrak{A}^{\bullet}\otimes[\partial L^{I}/\partial(\nabla_R\phi)]$ which can be shown to contribute nothing when integrated over the whole body, the expression (5.38) also reads:

$$\mathbf{b}^{**} = [(W - \mathcal{P}.\mathfrak{E} - \mathfrak{M}^*.\mathfrak{B}) - \tfrac{1}{2}\,\rho_o\,v^2 - (\mathfrak{M}^*.\mathfrak{B})]\mathbf{I} +$$

$$-\,\mathbf{F}^{T}\,(\partial W/\partial\mathbf{F}) - \mathfrak{E}\otimes\mathcal{P} + \mathfrak{M}^*\otimes\mathfrak{B}, \tag{5.40}$$

where the expression in the round brackets in the r.h.s. suggests the possibility of introducing a stored energy $W(\mathfrak{E}, \mathfrak{B}, \ldots)$, through a Legendre transformation. All the results of Sections **5.5** and **5.6** have been presented in terms of the Lagrangian L. It is worth remarking that they can be equally recovered by requiring the invariance of the global Lagrangian $\int_V \mathcal{L}\,dv$ with respect to the elementary translation $\delta_x\mathbf{X}$.

Should the inhomogeneity 'move' with respect to the material, one could also evaluate the *power expended by the material force* if the 're-location' in time (idest the velocity) can be associated with the inhomogeneity.

Material boundaries, surfaces across which phase-transitions occur and cracks can be viewed as inhomogeneities. Hence, one envisages interesting applications of the theory to the behaviour of defective electromagnetic materials. Applications to the various fields of material science can be found in the literature, a partial list of which is reported in the list of références below. This list is made of two parts, numbered references referred to in the text and an additional list in the Oxford style for further research.

REFERENCES

[1] Eshelby J.D., (1951), Force on an Elastic Singularity, *Phil. Tran. Roy. Soc. Lond.*, A244, 87-112.
[2] Eshelby J.D., (1956), Continuum Theory of lattice Defects, in : *Progress in Solid State Physics*, Eds. F. Seitz and D.Turnbull, Vol.3, p.79, Academic Press, New York.
[3] Eshelby J.D., (1975), Elastic Energy-momentum Tensor, J. of Elasticity, 5, 321-335.
[4] Eshelby J.D., (1970), Energy Relations and the Energy-momentum Tensor in Continuum Mechanics, in : *Inelastic Behavior of Solids*, Eds. M.F. Kanninen, W.F. Adler, A.R. Rosenfeld, and R.I. Jaffe, pp.77-114, McGraw Hill, New York.
[5] Eshelby J.D., (1982), Aspects of Dislocation Theory, in : *Mechanics of Solids* (The Rodney Hill 60th Anniversary Volume), eds. H.G. Hopkins and M. Sewell, pp.185-255, Pergamon Press, Oxford.
[6] Nelson D.F., (1979), *Electric,Optic and Acoustic Interactions in Dielectrics*, John Wiley, New York.
[7] Nelson D.F., (1991), Momentum, Pseudomomentum and Wave Momentum : Toward Resolving the Minkowski-Abraham Controversy, *Phys. Rev.*, A44, 3905-3916.
[8] Peierls R., (1979), *Surprises in Theoretical Physics,* Princeton Univ.Press.
[9] Peierls R., (1985), Momentum and Pseudomomentum of Light and Sound, i n : *Highlights of Condensed-Matter Physics*, Ed. M. Tosi, Corso LXXXIX, pp.237-255, Soc.Ital. Fisica, Bologna.
[10] Eringen A.C. and Maugin G.A., (1990), *Electrodynamics of Continua* , Two volumes, Springer-Verlag, New York.
[11] Maugin G.A., (1988), *Continuum Mechanics of Electromagnetic Solids*, (Vol.33 of Series in Applied Mathematics and Mechanics), North-Holland, Amsterdam.
[12] Eshelby J.D., (1980), The Force on a Disclination in a Liquid Crystal, *Phil. Mag.,* A42, 354-367.
[13] Ericksen J.L., (1995), Remarks concerning Forces on Line Defects, *Zeit. Angew .Math.Phys.*, (Special issue : Theoretical, Experimental and Numerical Contributions to the Mechanics of Fluids and Solids, dedicated to P.M.Naghdi), 46S, 247-2271.
[14] Kröner E., (1993), Configurational and Material Forces in the Theory of Defects in Ordered Structures, *CMDS 7 Proceedings*, K.H. Anthony and H.J. Wagner Editors, Material Science Forum, 123-125, 447-454.
[15] Maugin G.A. and Trimarco C., (1995b), On Material and Physical Forces in Liquid

Crystals, *Int. J. Engng. Sci.*, 33, 1663-1678.
[16] Ogden R.W., (1984), *Nonlinear Elastic Deformations*, Ellis Horwood, Chichester, U.K (Dover reprint, New York, 1997).
[17] Reissner E., (1953), Variational Theorem for Finite Elastic Deformations, *J. Math. and Phys.* (MIT), 32, 129-135.
[18] Maugin G.A. and Trimarco C., (1993), Note on Mixed Variational Principle in Finite Elasticity, *Rend. Mat. Accad. Lincei*, IX-III, 69-74.
[19] Knops R., Trimarco C. and Williams H.T., Uniqueness and Complementary Energy in Finite Elasticity. *(Forthcoming).*
[20] Hanyga A., (1985), *Mathematical Theory of Nonlinear Elasticity*, Ellis Horwood,,Chichester, U.K.
[21] Golebiewska-Herrmann A., (1983), Lagrangian Formulation of Continuum Mechanics, *Physica,* 118A, 300-314.
[22] Pack Y.E. and Herrmann G., (1986a), Conservation Laws and the Material Momentum Tensor for the Elastic Dielectric, *Int. J. Engng. Sci.*, 24, 1365-1374.
[23] Maxwell J. C., (1891), *A Treatise on Electricity and Magnetism*, volumes I and II, Oxford Classic Text in the physical Sciences, Clarendon Press, Oxford, ,(1998, reprint)
[24] Becker R., *Electromagnetic Fields and Interactions*, Dover Publ., NewYork, (1982, reprint).
[25] Stratton J. A. (1941), *Electromagnetic Theory,* McGraw-Hill, New York.
[26] Jackson J.D., (1962), *Classical Electrodynamics*, J. Wiley & Sons,N.Y.
[27] Tamm I. E., (1979), *Fundamentals of the Theory of Electricity*, translated from the 1976 Russian edition, MIR Publishers.
[28] Landau L.D. and Lifschitz E.M., (1960), *Electrodynamics of Continuous Media*, v. 8 of Course of Theoretical Physics, Pergamon Press, Oxford.
[29] Penfield P. and Haus H. A., (1967), *Electrodynamics of Moving Media*, M.I.T. Press, Cambridge, Massachusetts.
[30] Pao Y. H., (1978), Electromagnetic Forces in Deformable Continua, *Mechanics Today*,vol 4, S. Nemat-Nasser Editor, Pergamon Press, New York.
[31] Trimarco C.: (1994). How Multipole Electric Moments Enter into Macroscopic Maxwell Equations, *Il Nuovo Cimento* B, 109, 533 -540.
[32] Toupin R.A., (1956), The Elastic Dielectrics, *J. Rational Mech and Anal.*, 5, (849-915).
[33] Schoeller H. and Thellung A., (1992) Lagrangian Formalism and Conservation Law for Electrodynamics in Nonlinear Elastic Dielectrics, *Annals of Physics*, 220-1.

Additional references for further studies

Chadwick P., (1975), Applications of an Energy-momentum Tensor in Nonlinear Elastostatics, *J. Elasticity*, 5, 250-258.
Cherepanov G.P., (1989), Remark on the Dynamic Invariant or Path-independent Integral, *Int .J Solids Structures*, 25, 1267-9.
Dascalu C. and Maugin G.A., (1994), Energy-release Rates and Path-independent Integrals in Electroelastic Crack Propagation, *Int.J.Engng.Sci.,* 32, 755-765.
Edelen D.G.B., (1981), Aspects of Variational Arguments in the Theory of Elasticity : Facts and Folklore, *Int. J. Solids Structures*, 17, 729-740.
Ericksen J.L., (1977), Special topics in Elastostatics, in *: Advances in Applied Mechanics*, Ed.

C-S.Yih, Vol.17, pp.189-244, Academic Press, New York.
Ericksen J.L., (1991), *Introduction to the Thermomechanics of Solids*, Chapman and Hall, London.
Ericksen J.L., (1997), Equilibrium Theory for X-ray Observations of Crystals, *Arch. Rat. Mech. Anal.*, 139, 181-200.
Fomethe A. and Maugin G.A., (1996), Material Forces in Thermoelastic Ferromagnets *Cont. Mech. and Thermodynamics*, 8, 275-292.
Golebiewska-Herrmann A., (1981), On Conservation Laws of Continuum Mechanics, *Int. J. Solids Structures*, 17, 1-9.
Green A.E., (1973), On Some General Formulae in Finite Elastostatics, *Arch.Rat.Mech.Anal.*, 50, 73-80.
Grinfeld M.A. (1991), *Thermodynamic Methods in the Theory of Heterogeneous* Systems, ISIMM Series, Longman, Harlow, Essex.
Gurtin M.E. , (2000), *Configurational Forces as Basic Concepts of Continuum Physics*, Springer-Verlag, New York.
Herrmann G., (1980), Some Applications of Invariant Variational Principles in Mechanics of Solids, in : *Variational Methods in the Mechanics of Solids* (IUTAM Symp.,Evanston, 1978), pp.145-150, Pergamon Press, Oxford.
Hill R., (1986), Energy-momentum Tensor in Elastostatics : Some reflections on the General Theory, J. Mech. Phys. Solids, 34, 305-317.
Knowles J.K., and Sternberg E., (1972), Class of Conservation Laws in Linearized and Finite Elastostatics, *Arch. Rat. Mech. Anal.*, 44, 187-211.
Maugin G.A., (1993), *Material Inhomogeneities in Elasticity*, Chapman and Hall, London (Volume 3 in Series « Applied Mathematics and Mathematical Computation).
Maugin G.A., (1997), Momentum and Pseudomomentum in Matter, *GAMM-Mitteilungen*, Heft 1, 37-51.
Maugin G.A., (1999), *Nonlinear Waves in Elastic Crystals*, Oxford Texts in Applied Mathematics, Oxford University Press, U.K.
Maugin G.A. and Epstein M., (1991), The Electroelastic Energy-momentum Tensor, *Proc. Roy. Soc. Lond.*, A433, 299-312.
Maugin G.A. and Trimarco C., (1991), Pseudo-momentum and Material Forces in Electromagnetic Solids, *Int. J.Appl. Electromagn. Mat.*, 2, 207-216.
Maugin G.A. and Trimarco C., (1991), Pseudo-quantité de mouvement et milieux élastiques inhomogènes, *C.R.Acad.Sci.Paris*, II-313, 851-856.
Maugin G.A. and Trimarco C., (1992), Pseudo-momentum and Material Forces in Nonlinear Elasticity : Variational Formulations and Application to Brittle Fracture, *Acta Mechanica*, 94, 1-28.
Maugin G.A. and Trimarco C., (1993), Material Conservation Laws in Continuum Mechanics and the Electrodynamics of Continua, in : *Advances in Modern Continuum Mechanics*, Ed. G. Ferrarese, pp.131-149, Pitagora, Bologna.
Maugin G.A. and Trimarco c., (1995), The Dynamics of Configurational Forces at Phase-Transition Fronts (70th Anniversary of J.L.Ericksen), *Meccanica*, 30, 605-619.
Maugin G.A. and Trimarco C., (1995), Dissipation of Configurational Forces in Defective Elastic Solids, *Arch.Mech.* (Poland), 47, 81-99.
Maugin G.A. and Trimarco C., (1995), Configurational Forces and Coherent Phase- transition Fronts in Thermoelastic Solids (IUTAM Symp., Nottingham, 1994), in : *Anisotropy*,

Inhomogeneity and Nonlinearity in Solid Mechanics (A.J.M. Spencer Anniversary Volume), Eds. A. H. England and D. F. Parler, pp.345-350, Kluwer, Amsterdam.

Maugin G.A. and Trimarco C ., (1997), Driving Force on Phase-transition Fronts in Thermoelectroelastic Crystals, *Mathematics and Mechanics of Solids*, 2, 199-214

Norris, A.N., The energy of a growing elastic surface. *Int. J Solids Structures*, 36, 5237-5252.

Nowacki J. P. , Trimarco C. (1990): Note on thermal inclusion in elastic dielectric material. *Atti Sem. Mat. Fis. Univ. Modena*, XXXVIII, 371-378.

Sabir M. and Maugin G.A., (1996), On the Fracture of Paramagnets and Soft Ferromagnets, *Int.J .Non-linear Mechanics*, 31, 425-440.

Trimarco C . (1989): On the electrostatics of a rigid homogeneous isotropic and dielectric. *Int. J. Engng. Sci.*, 27, 1569-1579.

Trimarco C . (1992): The electric capacitance of a rigid dielectric structure. *Int. J. of Solids and Structures*, 29, 1647-1655.

Trimarco C. (1994): The Toupin-Mindlin Theory of Dielectrics in The Light of View of Mossotti's Idea. *Bull. of Polish Academy of Sciences*, 42, n. 3, 429-438.

Trimarco C. (1999): Microscopic Variables and Macroscopic Quantities. In : *Geometry, Continua and Microstructures,* Série Mathématique: Travaux en cours, Ed. G.A. Maugin, pp.121-129, Herrmann, Paris.

Trimarco C. (1999): Hamiltonian versus Lagrangian Forms. In : *Monographs and Surveys in Pure and Applied Maths series*, 103-108, Chapman and Hall/CRC, ed.G.Iooss, O.Guès, A. Nouri.

Truesdell C.A., and Noll W., (1965), Nonlinear Field Theories of Mechanics, in : *Handbuch der Physik*, Bd.III/3, ed. S. Flügge, Springer-Verlag, Berlin.

Truesdell C.A., and Toupin R.A., (1960), The Classical Theory of Fields, in : *Handbuch der Physik*, Bd.III/1, ed. S. Flügge, Springer-Verlag, Berlin.

Configurational Mechanics Applied to Strength - of - Materials

R. Kienzler[1], and G. Herrmann[2]

[1]University of Bremen, Bremen, Germany
[2]Stanford University, Stanford, CA, USA

Abstract: Defects, especially cracks, are treated on the basis of engineering-type theories of bars, beams, shafts, discs, plates and shells. The present contribution is concerned with conservation and balance laws in one- and two-dimensional theories of Strength - of - Materials and provides a method to calculate stress-intensity factors for structural members with cracks.

Introduction

Theories which fall under the general phrase "Strength - of - Materials" are to be understood as an approximate mathematical description of the static and dynamic response to various loadings of elastically deformable bodies whose one dimension is either considerably larger or considerably smaller than the other two. In the former case, if such a body, considered straight, is subjected to loads along its axis, it is called a bar in tension - compression, if it is subjected to twisting moments, it is called a shaft in torsion, and if it is subjected to bending moments and/or transverse loads it is called a beam in bending. Curved bodies, i. e., arches and frames, may also be considered. In the latter case, the body may be characterized by its middle surface. If the middle surface is plane and the body is subjected to in - plane forces, it is called a disc, if it is subjected to bending moments and/or transverse loads it is called a plate, and if the middle surface is curved it is called a shell.

One of the important tasks facing an engineer in practice is the assessment of such structural components against failure, particularly against fracture. That is, calculation of crack-extension forces or energy-release rates which via Irwin's relation determine the stress-intensity factors. Quite often, extensive analytical and/or numerical work may be involved to achieve the desired goal.

It is a remarkable circumstance that defects in elastic solids were considered theoretically *first* within the context of the theory of elasticity. Some forty years *later* it was found that some elements of fracture mechanics might be developed on the basis of the much older and simpler theories of strength-of-materials. In this way, failure of bars and beams due to fracture might easily be treated in an undergraduate course for students in engineering.

In Section 1, conservation and balance laws for bars, shafts and beams are established as a derivative of the product of the coordinate x raised to the k-th power and the potential energy density. Energy-release rates and stress-intensity factors are calculated for several examples. Section 2 is concerned with plates an shells (discs, i. e., problems of plane elasticity, are treated in other contributions of this course, e. g., in the Lecture Note of Herrmann and Kienzler entitled "Conservation Laws and Their Application in Configurtional Mechanics"). Conservation laws are established by the application of invariant differential operators. Section 3 concludes this contribution with some final remarks.

1 Bars, Shafts and Beams

1.1 Governing Equations

Bars in tension-compression. We consider a uniform elastic bar, of unspecified length ℓ and cross-section A which may be subjected to end loads N^l, N^r and distributed applied axial loads n, measured per unit length of the bar, cf. Figure 1.1.

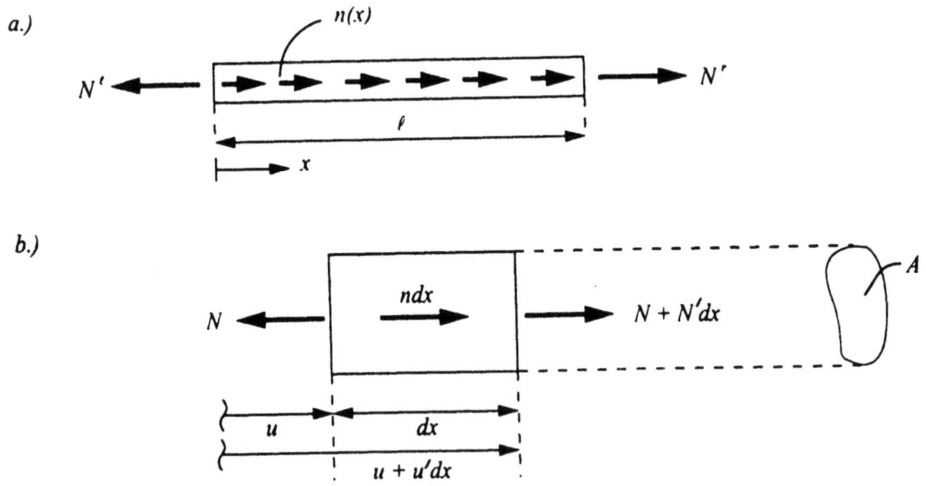

Fig.1.1 a. Bar under tension-compression
b. Deformed infinitesimal element (cross-section A) of a bar

Equilibrium of forces in the x -direction requires

$$N' = -n , \tag{1.1}$$

where N is the resultant internal axial force and primes indicate differentiation with respect to the axial coordinate $x_1 = x$.

Let u be the displacement of any cross-section, then u' is the strain ε

$$u' = \varepsilon . \tag{1.2}$$

Hooke's law is expressed as

$$N = EA\varepsilon , \tag{1.3}$$

where E is Young's modulus of the material. The product EA is called the axial stiffness. The bar may be inhomogeneous because either $E(x)$ or $A(x)$ or both may vary along the axis.

The strain energy W and the potential of the axis loads V , per unit length of the bar, are given, respectively, by

$$W = \frac{1}{2} EAu'^2 = \frac{1}{2}\frac{N^2}{EA} , \tag{1.4}$$

$$V = -nu . \tag{1.5}$$

Shafts in torsion. The basic equations of torsion of circular shafts are given by equations which are analogous to those of bars. As in Figure 1.2, the shaft my be subjected to end torques T^l and T^r , as well to a torque per unit of length t , applied along the shaft. Equilibrium of an element of the shaft requires

$$T' = -t , \tag{1.6}$$

where T is the resultant of distributed shear in each cross-section and may be called internal torque.

The kinematic relation may be expresses as

$$\gamma = R\, \vartheta', \tag{1.7}$$

where ϑ is the rotation of the cross-section, γ is the angle or twist per unit of length or the shear strain and R is the radius of the cross-section. The elastic law can then be written as

$$T = GI_p\,\gamma, \tag{1.8}$$

where G is the shear modulus and I_p the polar moment of inertia of the cross-section. Again, the shaft may be inhomogeneous, because $G(x)$ or $I_p(x)$ or both may vary along the axis of the shaft. The strain energy W and the potential V of the external loads, per unit of length, may be written similarly as for bars, as

$$W = \frac{1}{2} GI_p\ \vartheta'^2 = \frac{1}{2}\frac{T^2}{GI_p}, \tag{1.9}$$

$$V = -t\,\vartheta. \tag{1.10}$$

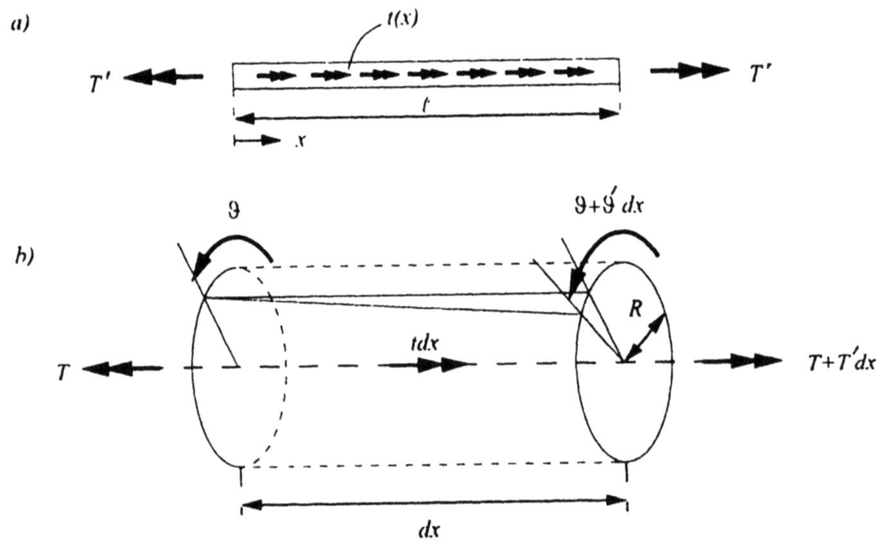

Fig. 1.2 a. Circular shaft under Torsion
b. Deformed infinitesimal element of a shaft

Beams in bending. The classical, or elementary, or Bernoulli-Euler theory of beams is different form the theories of bars and shafts. While still one-dimensional in the sense that all quantities depend on only one independent variable, namely the axial coordinate x, the deformed configuration described by the deflection $w(x)$, lies in a plane, because $w(x)$ is normal to x. The beam, see Figure 1.3 may be subjected to end moments M^l and M^r, and transverse forces Q^l and Q^r, and transverse distributed loads q measured per unit of length.

As is seen from Figure 1.3, equilibrium of forces and moments requires, respectively,

$$Q' = -q\,, \tag{1.11}$$

$$M' = Q\,, \tag{1.12}$$

where Q is the internal shear force and M the internal bending moment.

The kinematical constraint requiring plane cross-sections to remain plane and normal to the central axis implies

$$\psi = -w'\,, \tag{1.13}$$

where ψ is the angle of rotation of the cross-section. The elastic relation expresses proportionality between the curvature ψ' and the bending moment M, the factor being the bending stiffness EI and I the area moment of inertia. Thus

$$M = EI\psi' \qquad \text{or} \qquad M = -EIw''\,. \tag{1.14}$$

It is a feature of the elementary beam theory that the shear force Q is not related through a form of the elastic law to some kinematic quantity, but is, rather, to be determined from equation (1.12) expressing equilibrium of moments.

The strain energy W per unit of length in elementary beam theory may be written as

$$W = \frac{1}{2}EI\psi'^2 = \frac{1}{2}Ew''^2 = \frac{1}{2}\frac{M^2}{EI} \tag{1.15}$$

and the potential of the externally applied loads per unit of length V as

$$V = -qw\,. \tag{1.16}$$

Thus the Lagrangian function $L = -(W + V) = L(x, w'')$ depends on derivatives of second order and describes a one-dimensional grade two material or, more precisely, a one-dimensional Cosserat continuum with constrained rotation (Eshelby, 1975).

1.2 Balance and Conservation Laws for Bars and Shafts

We propose to establish all balance and conservation laws of interest to us in a unified and comprehensive fashion by considering the derivative of the product of the coordinate x raised to the k-th power and the potential energy density $W + V$. Thus, we evaluate the derivative of the expression $x^k(W+V)$ and attempt to determine the value of k for which conservation laws exist. We obtain in a first step

$$\left[x^k(W+V)\right]' = x^k(W+V)' + kx^{k-1}(W+V)\,. \tag{1.17}$$

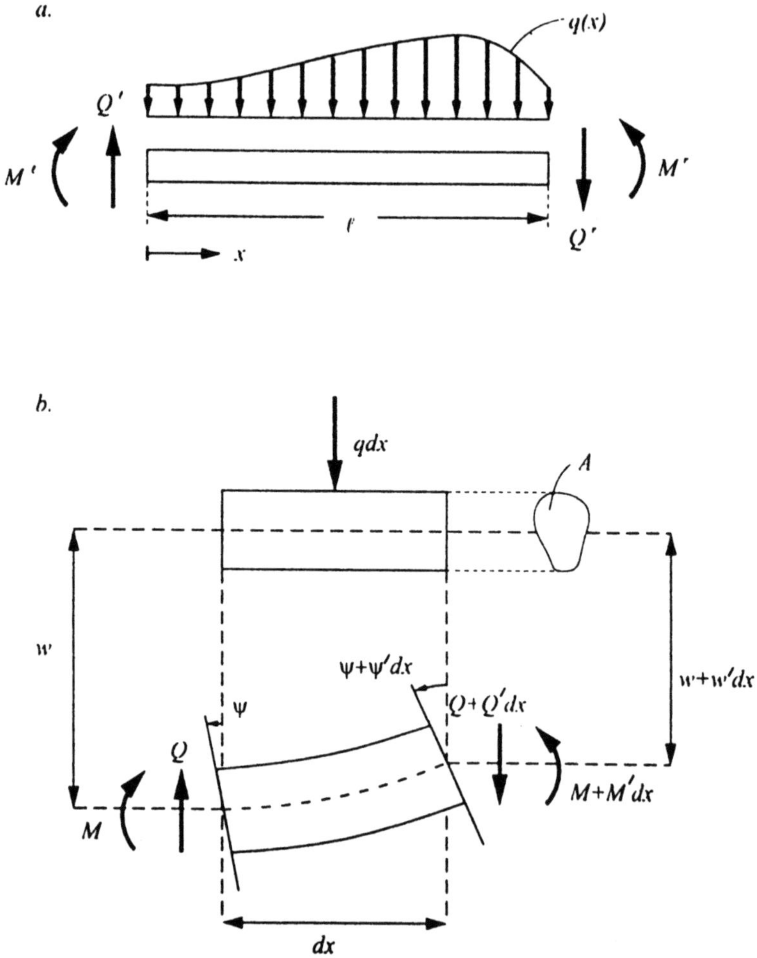

Fig. 1.3 a. Beam under bending
b. Infinitesimal beam element (cross-section A) in the undeformed and deformed configuration

We start with bars in tension-compression and assume that EA and n are not constant along the bar. The corresponding results for shafts in torsion will be given at the end of this Sub-section. Substitution of the expression for W and V given in equations (1.4) and (1.5), respectively, and of the equation of equilibrium (1.1), together with partial integration where appropriate, leads to the intermediate result

$$\left[x^k(W+V-Nu')\right]' = x^k\left(\frac{1}{2}(EA)'\,u'^2 - n'u\right) + kx^{k-1}(W+V-EAu'^2).$$

Further similar manipulations on the right-hand side of the above equation yield the final result

$$\left[x^k B - kx^{k-1}H + \frac{1}{2}k\,(k-1)\,x^{k-2}R\right]' =$$

$$= -x^k b - kx^{k-1}h - \frac{1}{2}k(k-1)x^{k-2}r - \frac{1}{4}k(k-1)(k-2)x^{k-3}EAu^2 \tag{1.18}$$

Here the abbreviations are introduced as follows

material force $\quad B = W + V - Nu'\,,$ (1.19a)

material virial $\quad H = -\frac{1}{2}EAu'u\,,$ (1.19b)

second-order virial $\quad R = -\frac{1}{2}EAu^2\,,$ (1.19c)

material loading $\quad b = -\frac{1}{2}(EA)'u'^2 + n'u\,,$ (1.19d)

material virial loading $\quad h = \frac{3}{2}nu\,,$ (1.19e)

second-order virial loading $\quad r = \frac{1}{2}(EA)'u^2\,.$ (1.19f)

For $k = 0$, i. e., when considering the gradient of the potential energy density $W + V$, it is seen that the balance law

$$B' = -b \tag{1.20}$$

is obtained, with embodies the balance of linear material momentum B, corresponding to the balance of physical momentum given by equation (1.1).

The source term b vanishes if the stiffness EA is uniform and the applied loading along the bar n is constant, and in this case the conservation law results

$$B' = 0\,; \qquad B = const\,. \tag{1.21}$$

For $k = 1$, i. e., when considering the (first-order) scalar moment (virial), the balance law is obtained

$$(xB - H)' = -xb - h. \tag{1.22}$$

The source term vanishes if EA is uniform and, in addition, the bar is load-free along its length, $n = 0$. The associated conservation law is then

$$(xB - H)' = 0; \qquad xB - H = const. \tag{1.23}$$

For $k = 2$, i. e., when considering the second-order scalar moment, we obtain the balance law

$$(x^2B - 2xH + R)' = -x^2b - 2xh - r. \tag{1.24}$$

The source term on the right vanishes under the same (no additional) conditions as in the previous case, i. e., $(EA)' = 0$ and $n = 0$, and the conservation law reads

$$(x^2B - 2xH + R)' = 0; \quad x^2B - 2xH + R = const. \tag{1.25}$$

It is seen from inspection of the last term on the right-hand side of equation (1.18) that it can not be made to vanish for $k \geq 3$. Thus no further *conservation* laws using the scheme proposed here seem to exist. The number of *balance* laws, however, is obviously infinite.

The three conservation laws (1.21), (1.23) and (1.25) agree with those given in the Lecture Notes of Herrmann and Kienzler, where the bar with constant axial stiffness under end loading $(EA = const., n = 0)$ was taken as an example to illustrate the application of Noether's formalism and the Neutral Action method.

There exist several other methods of obtaining the balance and conservation laws above which should be mentioned for completeness. One would consist in suitable interpretation of the physical and material balance laws of plane elastostatics (Günther, 1962), another by means of the virtual-work theorem (Kienzler and Herrmann, 1986 a), by a variational principle with varying boundaries (Kienzler, 1986) and finally, by duality considerations, interchanging dependent and independent variables in the Lagrangian (Rösel, 1986).

If one is interested in establishing conservation laws for nonuniform bars $(EA)' \neq 0$ under arbitrary axial loading $n \neq 0$, the Neutral Action method provides the required efficient tool. Details of the fairly extensive derivations are given in Kienzler and Herrmann (2000). Here, for completenes, we list some results. For inhomogeneous bars without distributed loading, e. g., $EA \neq const, n = 0$.

$$(EA\,B)' = 0\,, \tag{1.26a}$$

$$\Big(EA\int\frac{dx}{EA}B - H\Big)' = 0\,, \tag{1.26b}$$

$$\Big(2EA\int\frac{1}{EA}\Big(\int\frac{dx}{EA}\Big)'dx\,B - 2\int\frac{dx}{EA}H + \frac{R}{EA}\Big)' = 0, \tag{1.26c}$$

$$N' = 0\,, \tag{1.26d}$$

$$\Big(\int\frac{dx}{EA}N - \frac{V}{EA}\Big)' = 0\,. \tag{1.26e}$$

(The quantiy $V = EAu$ is the so-called virial and should not be confused with the potential of external forces for which the same symbol is used).

With the stiffness distribution $E(x)A(x)$ given explicitly, the integrals above can readily be evaluated and integration constants are of no consequence. For $EA = const.$, the above set of equations (1.26 a-c) obviously reduced to the set (1.21), (1.23) and (1.25). Equations (1.26d) and (1.26e) are physical conservation laws.

For a homogeneous bar, e. g., $EA = EA_0 = const.$, under constant external loading $n = n_0 = const.$ we find

$$B' = 0\,, \tag{1.27a}$$

$$\Big[Bx - H - \frac{1}{4}\frac{n_0}{EA_0}(n_0x^3 + 3Nx^2 - 6Vx)\Big]' = 0\,, \tag{1.27b}$$

$$\Big[Bx^2 - 2Hx + R - \frac{1}{8}\frac{n_0}{EA_0}(n_0x^4 + 4Nx^3 - 12Vx^2)\Big]' = 0\,, \tag{1.27c}$$

$$\big[N + n_0x\big]' = 0\,, \tag{1.27d}$$

$$\Big[Nx - V + \frac{1}{2}n_0x^2\Big]' = 0\,. \tag{1.27e}$$

Again, the first three equations (1.27 a-c) express the zeroth-, first- and second-order conservation of *material* force and the last two equations (1.27d and e) express the zeroth- and first-order conservation of *physical* force. To emphasize, for bars, the five conservation laws can always be established for arbitrarily specified functions $EA = E(x)A(x)$ and $n = n(x)$. This is not the case in beam theory (cf. Kienzler and Herrmann, 2000).

All the above expressions for bars remain valid for shafts, if one introduces a change of notation according to the tension-torsion analogy

$$n \rightarrow t\,, \tag{1.28a}$$

$$N \rightarrow T\,, \tag{1.28b}$$

$$EA \rightarrow GI_p\,, \tag{1.28c}$$

$$\varepsilon \rightarrow \gamma\,, \tag{1.28d}$$

$$u \rightarrow \vartheta\,. \tag{1.28e}$$

1.3 Balance and Conservation Laws for Beams

Beams in bending shall be treated employing the same methodology as bars in tension-compression and shafts in torsion. We begin again by considering the derivative of $x^k(W+V)$ to obtain

$$\left[x^k(W+V)\right]' = x^k(W+V)' + kx^{k-1}(W+V)\,, \tag{1.29}$$

which is clearly identical to the corresponding equation for bars and shafts (1.17). It is only when substituting the specific expressions for W and V, equations (1.15) and (1.16), respectively, that differences begin to appear. Indeed,

$$W' = \frac{1}{2}(EI)'\,\psi'^2 + M\psi''\,, \tag{1.30}$$

$$V' = -q'w - qw'\,. \tag{1.31}$$

Since $Q = M'$ from the moment equilibrium condition Q' may be written as

$$Q' = -\left(EIw''\right)'' = -q \tag{1.32}$$

and

$$M'' = -q\,. \tag{1.33}$$

We can write in a first step

$$\left[x^k(W+V)\right]' = x^k\left(\frac{1}{2}(EI)'\,\psi'^2 + M\psi'' - qw' - q'w\right) + kx^{k-1}\left(W+V\right). \tag{1.34a}$$

Repeated partial integration of the term $x^k M\psi''$, together with suitable substitutions, leads to

$$x^k M\psi'' = (x^k M\psi')' + (x^k Qw')' + x^k qw' - kx^{k-1}(M\psi' + Qw') . \tag{1.34b}$$

Further partial integrations and substitutions of the equilibrium equations and the constitutive law (1.14) result finally, omitting intermediate steps,

$$\left[x^k B - kx^{k-1}H + \frac{1}{2}k(k-1)\, x^{k-2}R\right]' =$$

$$= -x^k b - kx^{k-1}h - \frac{1}{2}k(k-1)\, x^{k-2}r - k(k-1)(k-2)\, x^{k-3}EI\left(\psi^2 + \frac{3}{2}\psi' w\right). \tag{1.35}$$

Similarly as in bar theory, the following abbreviations have been introduced for beams

material force	$B = W + V - M\psi' - Qw'$,	(1.36a)
material virial	$H = -\frac{1}{2}M\psi - \frac{3}{2}Qw$,	(1.36b)
second-order virial	$R = 2EI\psi^2 - 3EI\psi' w$,	(1.36c)
material loading	$b = -\frac{1}{2}(EI)'\psi'^2 + q'w$,	(1.36d)
material virial loading	$h = \frac{5}{2}qw$,	(1.36e)
second-order virial loading	$r = 2(EI)'\psi^2$.	(1.36f)

It does not appear possible to generalize the above balance law further, because the combination $\psi^2 + 3/2\psi' w$ cannot be represented as the derivative of an expression, as may be verified.

It is remarkable that the structure of the general balance equation above for beams is the same as that for bars and shafts, except that the six abbreviations (1.36) differ from (1.19) and the last source term in equation (1.35) differs from that of (1.18).

As for bars, we can now, in turn, consider $k = 0, k = 1$ and $k = 2$, realizing again that for $k \geq 3$ no conservation laws would be derivable.

We obtain

$$k=0: \qquad B' = -b, \tag{1.37}$$

$$k=1: \qquad (xB-H)' = -xb-h, \tag{1.38}$$

$$k=2: \qquad (x^2B-2xH+R)' = -x^2b-2xh-r. \tag{1.39}$$

The conservation law

$$B'=0; \qquad B = const. \tag{1.40}$$

is obtained for constant stiffness EI and constant loading q.

The conservation law

$$(xB-H)'=0; \qquad xB-H = const. \tag{1.41}$$

is obtained if, in addition to constant stiffness EI, the loading vanishes, $q=0$.

Finally, the conservation law

$$(x^2B-2xH+R)'=0; \qquad x^2B-2xH+R = const. \tag{1.42}$$

is obtained under the same conditions as the conservation law for $k=1$.

The interpretation of the above balance and conservation laws follows the same lines as for bars and shafts.

1.4 Energy-Release Rates and Stress-Intensity Factors

In this Sub-section, we investigate bars, shafts and beams, with discontinuities in their stiffness. Especially, energy-release rates are calculated due to the translation of discontinuous cross-section. It turns out that remarkably simple formulae evolve to calculate stress-intensity factors for these one-dimensional structural components with cracks. We start within the context of bar theory and extend the results, subsequently, to shafts and beams by analogy considerations.

Consider a bar containing a jump in axial stiffness EA at an arbitrarily fixed position ξ given by

$$E(x)A(x) = \begin{cases} EA^{(1)} = const. \text{ for } x < \xi, \\ EA^{(2)} = const. \text{ for } x > \xi \end{cases} \tag{1.43}$$

(see Figure 1.4, state Ⓐ). The jump in stiffness EA may be due to a jump in E (e. g., phase transition) or to a geometric change in the cross-sectional area A.

At the transition point $x = \xi$, we can distinguish between continuous and discontinuous variables. If it is assumed that the axial load n is smooth, it is obvious that n, N and u are continuous, while EA and u' are discontinuous. The expression for the material force B, given by (1.19a), might be rearranged using (1.2) - (1.5) as

$$B = -\frac{1}{2}\frac{N^2}{EA} - nu \, . \tag{1.44}$$

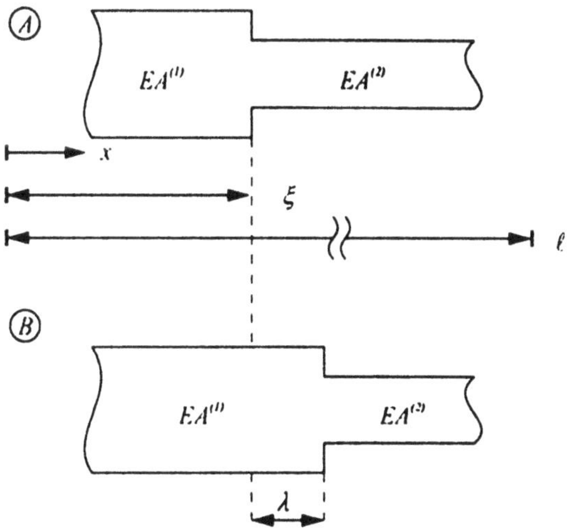

Fig. 1.4 Bar with jump in axial stiffness

Therefore, it follows that the material force B is discontinuous. The jump term $[B]$ is easily calculated to be

$$[B] = B^+ - B^- = -\frac{1}{2}N^2(\xi)[C] \, , \tag{1.45}$$

where [C] is the jump in the compliance $C = \dfrac{1}{EA}$

$$[C] = \frac{1}{EA^{(2)}} - \frac{1}{EA^{(1)}} \, . \tag{1.46}$$

To provide a physical interpretation of [B], a bar (length ℓ) is considered to be composed of two sections, $EA^{(1)}$ (length ξ) and $EA^{(2)}$ (length $\ell - \xi$), subjected (for sake of simplicity) to end forces N_0. The potential Π^i of the internal forces, i. e., the strain energy of the system, is given by

$$\Pi^i = \int_0^\ell W dx = \frac{1}{2}\int_0^\ell \frac{N_0^2}{EA} dx \, , \tag{1.47}$$

and due to Clapeyron's theorem (e. g., Fung, 1965), the potential of the external forces follows to be

$$\Pi^a = -2\Pi^i \, . \tag{1.48}$$

Thus the complete potential energy Π is given by

$$\Pi = \Pi^i + \Pi^a = -\Pi^i \, . \tag{1.49}$$

We wish now to calculate the change of energy when the cross-section $x = \xi$ is translated by a small amount λ (Figure 1.4 state Ⓑ. Using (1.47) and (1.49) the result is

$$\lambda = 0: \qquad \Pi_{Ⓐ} = -\frac{1}{2}\left(\frac{N_0^2 \xi}{EA^{(1)}} + \frac{N_0^2(\ell - \xi)}{EA^{(2)}} \right) , \tag{1.50a}$$

$$\lambda \neq 0: \qquad \Pi_{Ⓑ} = -\frac{1}{2}\left(\frac{N_0^2(\xi + \lambda)}{EA^{(1)}} + \frac{N_0^2(\ell - \xi - \lambda)}{EA^{(2)}} \right) . \tag{1.50b}$$

The change of energy due to this material translation turns out to be

$$\Delta\Pi = \Pi_{Ⓑ} - \Pi_{Ⓐ} = \frac{1}{2}\lambda N_0^2 [C] = -\lambda [B] \, . \tag{1.51}$$

The quantity $\lambda\,[B]$ may be , therefore, interpreted as the work of the material (concentrated) force $[B]$ in the material translation λ. The force $[B]$ acts at the cross-section $x = \xi$ and points horizontally in the direction of the stiffer material (Kienzler, 1993) and not, as mentioned erroneously in Kienzler and Herrmann (1986 a, b), towards the softer material. This force points hence in the direction of larger E or in the direction of the larger cross-sectional area A, thus implying possible removal, not acretion, of material in this latter case.

With

$$\overline{\mathcal{G}} = - \lim_{\lambda \to 0} \frac{\Pi(\xi + \lambda) - \Pi(\xi)}{\lambda} = \frac{\partial \Pi}{\partial \lambda} = +[B] , \tag{1.52}$$

[B] may be identified as the energy.release rate $\overline{\mathcal{G}}$ due to a translation of the cross-section at $x=\xi$ in x - direction. Equation (1.51) and (1.52) are also valid for arbitrary loading and boundary conditions. Here, N_0 has to be merely replaced by $N(\xi)$. For the calculation of [B] it is necessary to know only $N(\xi)$ and [C].

Next, we consider a bar (length ℓ) containing at $x=\xi$ a segment of length $2c$ with reduced stiffness (Figure 1.5 state Ⓐ).

$$E(x)A(x)=\begin{cases} EA^{(1)} = const.\ \text{for}\ 0 \le x < \xi - c\ \text{and}\ \xi + c < x \le \ell, \\ EA^{(2)} < EA^{(1)}\ \text{for}\ \xi - c < x < \xi + c. \end{cases} \tag{1.53}$$

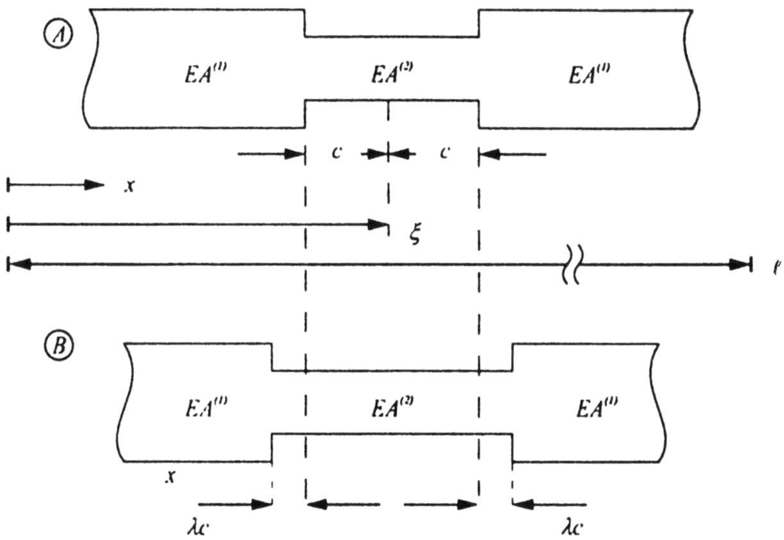

Fig. 1.5 Bar with partly reduced stiffness

Again, the bar is subjected to pure tension $N_0 = const$. The change of total energy, now, due to a self-similar expansion $c \to c(1+\lambda)$ (Figure 1.5 state Ⓑ) is calculated in the same way as above yielding

$$\Delta\Pi = -\lambda c N^2[C] = 2\lambda c[B]. \tag{1.54}$$

We recall that a physical (concentrated) moment is defined as a limit when the lever arm a of two equal and opposite forces F approaches zero while the product aF remains finite. In analogy to the physical moment, we introduce the material (concentrated) moment [H] as

$$[H] = \lim_{c \to 0} c[B]. \tag{1.55}$$

However, c and $[B]$ are both parallel to the x-axis. Thus the quantity $[H]$ is in its nature a scalar moment or "Fliehmoment" or "virial".

The energy-release rate due to the transformation $c \to c(1+\lambda)$ is

$$\overline{\mathcal{G}} = -\frac{\partial \Pi}{\partial(\lambda c)} = -2[B]. \tag{1.56}$$

For shafts in torsion and beams in bending, the jump term $[B]$ is given by

$$[B] = -\frac{1}{2}T^2[C] \qquad \text{with} \qquad [C] = \frac{1}{GI_p^{(2)}} - \frac{1}{GI_p^{(1)}}, \tag{1.57}$$

$$[B] = -\frac{1}{2}M^2[C] \qquad \text{with} \qquad [C] = \frac{1}{EI^{(2)}} - \frac{1}{EI^{(1)}}. \tag{1.58}$$

The energy-release rates due to translation and self-similar expansion can be calculated directly from (1.52) and (1.56), respectively. In the following, we will assume that the length c in Figure 1.5 is small in comparison to the length ℓ, such that the bar, shaft or beam with partly reduced stiffness might be regarded as a structural element with two symmetric edge cracks.

As mentioned in the Introduction, energy-release rates play an essential role in fracture mechanics. The energy-release rate $\mathcal{G}$ due to crack-extension $a \to a + \Delta a$ is closely related in linear elasticity to stress-intensity factors K, valuable parameters to assess a structure with cracks. Via Rice's J integral $J = J_1 = \mathcal{G}$ may be interpreted as the crack-driving force, i. e., material force acting at the crack tip and pointing in the direction of crack extension.

In Kienzler and Herrmann (1986 b) it was assumed that the energy-release rate $\overline{\mathcal{G}}$ for crack extension is equal to that for crack widening, i. e., $c \to c(1+\lambda)$. Although $\overline{\mathcal{G}}$ corresponds to a virial acting perpendicular to the crack-driving force, it was postulated that

$$\mathcal{G} = \overline{\mathcal{G}} \,/\, d. \tag{1.59}$$

For dimensional reasons ($\mathcal{G}$ is defined in plane elasticity, where $\overline{\mathcal{G}}$ is defined in a bar or beam theory) a measure of thickness, d needs to be introduced.

For rectangular cross-sections ($d \times h$) or piecewise rectangular cross-sections (I-, and T-beams) d is the width of the beam. For other cross-sections an effective width d^* was used. Bazant (1990) noted that the hypothesis (1.59) is not exact but merely a good approximation. He provided the following explanation (see also Gao and Herrmann, 1992, and Kienzler, 1993).

Consider a crack with initial length a which may subsequently extend by a small amount Δa to length $a+\Delta a$ as shown in Figure (1.6a) or widen into a fracture band of width $\Delta b \doteq c\lambda$ as shown n Figure (1.6b). The formation of such a crack in a stressed body may be imagined to completely release the strain energy from triangular areas 021 and 023, which are limited by "stress diffusion lines" (e. g., Knott, 1973). The slope β of the triangular area, which is generally of the order of unity, is dependent on the geometry and the crack size. If the crack is widened into a band of width Δb in the direction of constant stress, the size of the triangular zones remains the same so that the stress relief zone is changed from area 1231 to area 45784 (Figure 1.6 b). In comparison, when the crack is extended by Δa, strain energy is released from strips 2683 and 2641 (Figure 1.6 a). It is seen that the stress relief zone 123876541 for crack extension differs from 12387541 for crack widening only by a triangular area 56725. This triangular area is proportional to Δa^2 and may be neglected in comparison with the strips (proportional to Δa), since Δa is small in comparison to a. Since $\Delta b/2 = \beta\Delta a$, Bazant (1990) pointed out that in the limits $\Delta a, \Delta b \to 0$

$$\frac{\partial \Pi}{\partial a} = 2\beta \frac{\partial \Pi}{\partial b}. \tag{1.60}$$

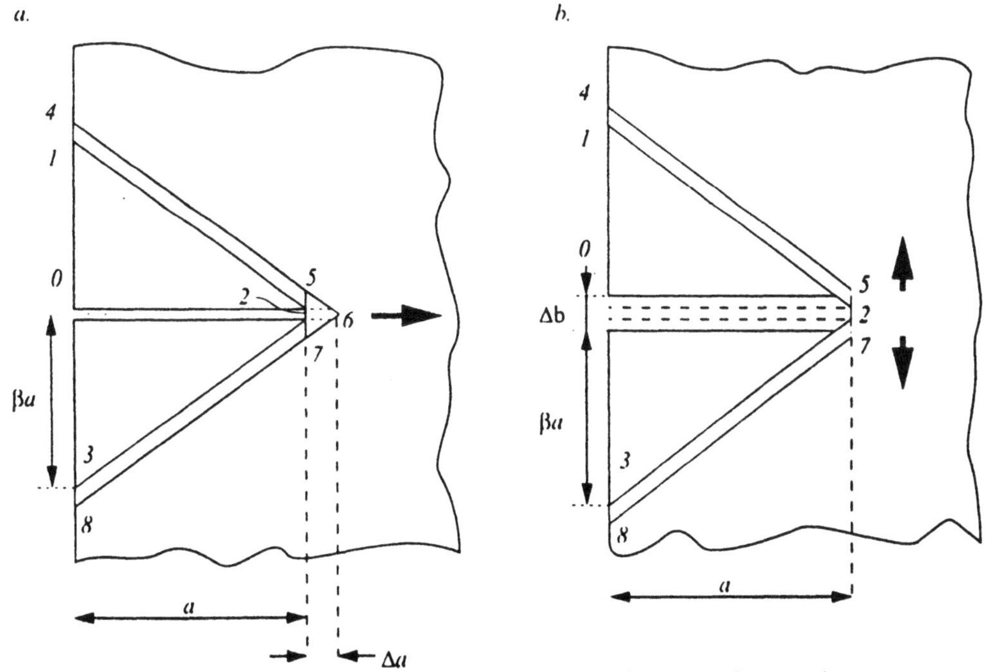

Fig. 1.6 Energy-release zones at a crack tip for (a) crack extension (b) crack widening into a fracture band.

Thus, the energy-release rate $\mathcal{G}$, as a work conjugate to the lengthening of the crack size, is given by

$$\mathcal{G} = -\frac{1}{d}\frac{\partial\Pi}{\partial a} \tag{1.61}$$

It follows from (1.60) and (1.61) that $\mathcal{G}$ can be computed from $\partial\Pi/\partial b$ once β is known. The stress intensity factor $K = K_{\mathrm{I}}$ may subsequently be calculated via the Irwin relation (Irwin, 1957)

$$\mathcal{G} = \frac{K^2}{E^*}, \tag{1.62}$$

where $E^* = E$ for plane stress and $E^* = E/(1-\upsilon^2)$ for plane strain conditions.

Restricting ourselves to plane stress, the expression for K follows with (1.60)-(1.62) as

$$K = \sqrt{-\frac{2\beta}{d}E[B]} \tag{1.63}$$

and, with (1.45), (1.57) and (1.58), the specific expressions are

bars $$K = N\sqrt{\frac{\beta_N}{dA^{(1)}}\left(\frac{A^{(1)}}{A^{(2)}}-1\right)}, \tag{1.64}$$

shafts $$K = T\sqrt{\frac{\beta_T}{d^* I_p^{(1)}}\left(\frac{I_p^{(1)}}{I_p^{(2)}}-1\right)}, \tag{1.65}$$

beams $$K = M\sqrt{\frac{\beta_M}{dI^{(1)}}\left(\frac{I^{(1)}}{I^{(2)}}-1\right)}. \tag{1.66}$$

For beams and bars, K represents the crack-opening mode I, whereas in torsion, K is related to the crack-sliding mode III.

The correction factor depends on the loading conditions, the geometry and the crack size. Bazant (1990) proposed to determine β through optimum fitting of the exact solutions. Gao and Herrmann (1992) showed that this correction factor can be obtained through simple matching with standard asymptotic limiting crack solutions from plane elastostatics.

But, even if one does not adopt these corrections and takes $\beta = 1$, i. e., one is concerned solely with elementary beam and bar theory, the approximation is remarkably accurate. This has been shown by Kienzler and Herrmann (1986 b), Herrmann and Sosa (1986), Gao and Herrmann (1992) and Müller et al. (1993 a and b) with several examples.

1.5 Examples

As a first example, consider a bar/beam of rectangular cross-section $d \times h$ with symmetrical edge cracks of length a under pure tension/bending (Figure 1.7 and 1.8).

The cross-sectional areas are

$$A^{(1)} = dh\,,$$

$$A^{(2)} = d(h-2a),$$

and the moments of inertia are

$$I^{(1)} = \frac{dh^3}{12}\,,$$

$$I^{(2)} = \frac{d}{12}(h-2a)^3\,.$$

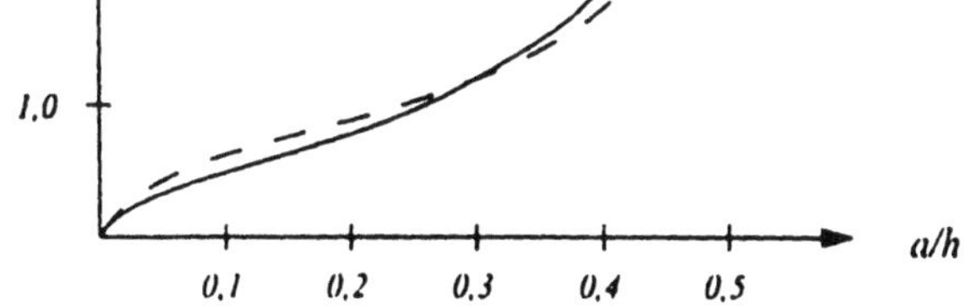

Fig. 1.7 Stress-intensity factor vs. dimensionless crack length for a bar under simple tension with symmetrical edge cracks (— (eq. 1.67),---- Benthem and Koiter (1973))

With (1.64) and (1.66) ($\beta_N = \beta_M = 1$), the stress-intensity factors for the bar under tensile loading are given by

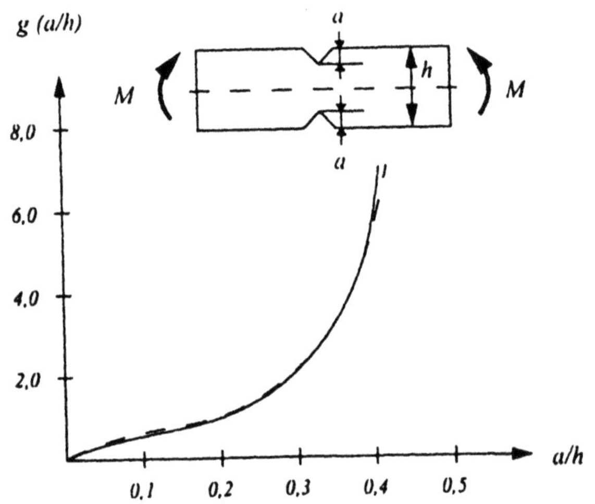

Fig. 1.8 Stress-intensity factor vs. dimensionless crack length for a beam under pure bending with symmetrical edge cracks (— eq. (1.68), ---- Benthem and Koiter (1973))

$$K = \frac{N}{d\sqrt{h}} g_N\left(\frac{a}{h}\right) \tag{1.67a}$$

with

$$g_N\left(\frac{a}{h}\right) = \sqrt{\frac{1}{1-2\frac{a}{h}} - 1} \tag{1.67b}$$

and for the beam under pure bending

$$K = \frac{6M}{d\sqrt{h^3}} g_M\left(\frac{a}{h}\right) \tag{1.68a}$$

with

$$g_M = \sqrt{\frac{1}{3}\left(\frac{1}{(1-2\frac{a}{h})^3} - 1\right)} \tag{1.68b}$$

The result are compared graphically with those of Benthem and Koiter (1973) in Figures 1.7 and 1.8. The agreement is quite satisfactory.

As a second example, consider a bar/beam of rectangular cross-section $d \times h$ with a centered crack of length $2a$ under pure tension/bending (Figure 1.9 and 1.10). As in the example above, the cross-sectional areas are

$$A^{(1)} = dh\,,$$

$$A^{(2)} = d(h - 2a).$$

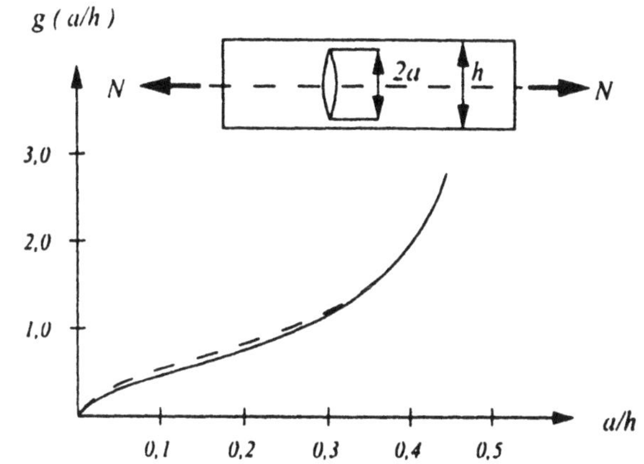

Fig. 1.9 Stress intensity factor vs. dimensionless crack length for a bar under simple tension with center crack (— eq. (1.67), ---- Benthem and Koiter (1973))

It is, therefore, not possible to distinguish between edge cracks and a center crack for tension loading. Equation (1.67) is still applicable. The results are compared graphically with those of Benthem and Koiter (1973) in Figure 1.9.

For bending, the moments of inertia are

$$I^{(1)} = \frac{dh^3}{12}\,,$$

$$I^{(2)} = \frac{d}{12}\left(h^3 - 8a^3\right).$$

The stress-intensity factor for the beam under pure bending is with (1.66) given by

$$K = \frac{6M}{d\sqrt{h^3}}\, g_M\left(\frac{a}{h}\right) \tag{1.69a}$$

with

$$g_M = \sqrt{\frac{1}{3}\left(\frac{1}{1-(2\frac{a}{n})^3}-1\right)}\,. \tag{1.69b}$$

Again, the results are compared graphically with those of Benthem and Koiter (1973) (cf. Figure 1.10) and again, the agreement is satisfactory

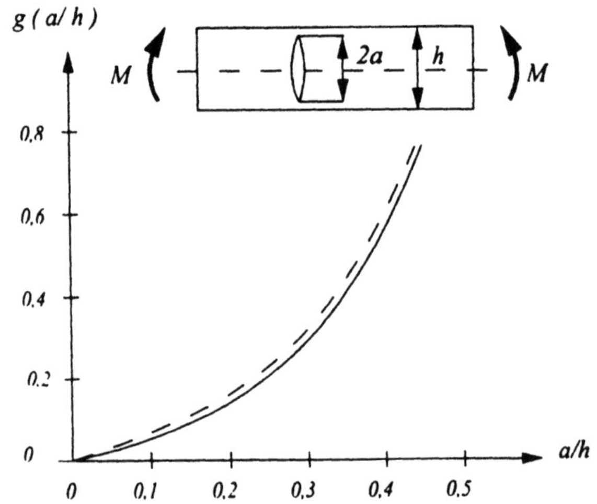

Fig. 1.10 Stress intensity factor vs. dimensionless crack length for a beam under pure bending with center crack (— eq. (1.69), ---- Benthem and Koiter (1973))

As a third example consider a circular shaft with radius R under pure tension with a circumferential edge crack of depth a and a penny-shaped centre crack of radius a (Figures 1.11 and 1.12).

From (1.65) it follows

$$K = \frac{2T}{\sqrt{\pi R^5}}\, g_T\left(\frac{a}{R}\right) \tag{1.70a}$$

with

$$g_T = \sqrt{\frac{1}{(1-2\frac{a}{R})^4}-1} \tag{1.70b}$$

for the edge crack and with

$$g_T = \sqrt{\frac{1}{1-(2\frac{a}{R})^4} - 1} \tag{1.70c}$$

for the penny-shaped crack. The graphical comparison of (1.70) with the results of Benthem and Koiter (1973) is depicted in Figures 1.11 and 1.12.

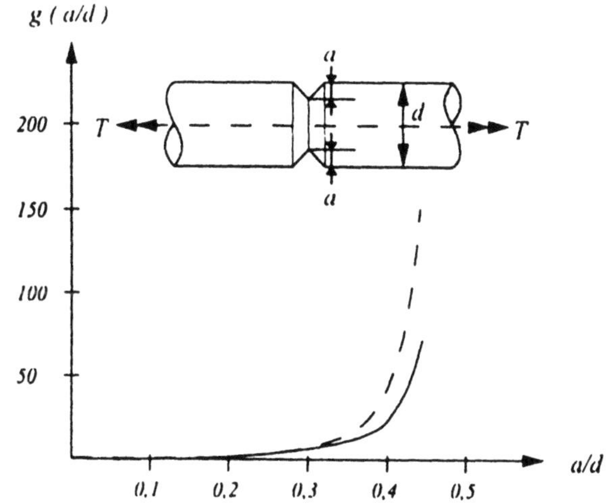

Fig. 1.11 Stress-intensity factor vs. dimensionless crack length for a shaft under pure torsion with circumferential edge crack (— eq. (1.70b), ---- Benthem and Koiter(1973))

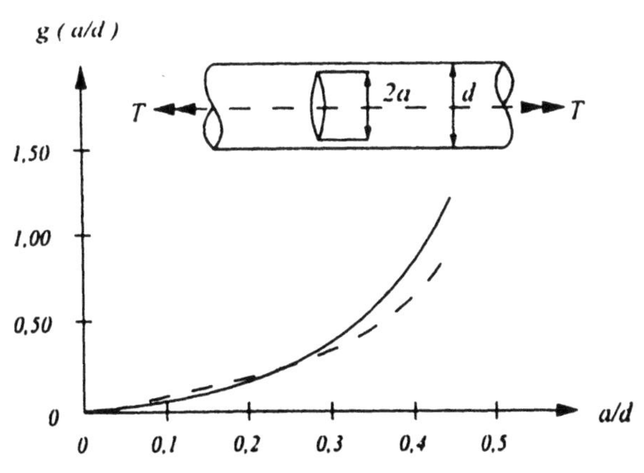

Fig. 1.12 Stress-intensity factor vs. dimensionless crack length for a shaft under pure torsion with penny-shaped centre crack (— eq. (1.70c), ---- Benthem and Koiter(1973))

It is astonishing, that elementary beam theory may be applied with success even for circumferentially cracked cylindrical pipes, cf. Gao and Herrmann (1992). As a fourth example we present here the principal results of that analysis. Consider a pipe segment depicted in Figure 1.13.

The stress intensity factor K is given by

$$K = \frac{M}{\pi R^2 t}\sqrt{\pi R}\, g \tag{1.71 a}$$

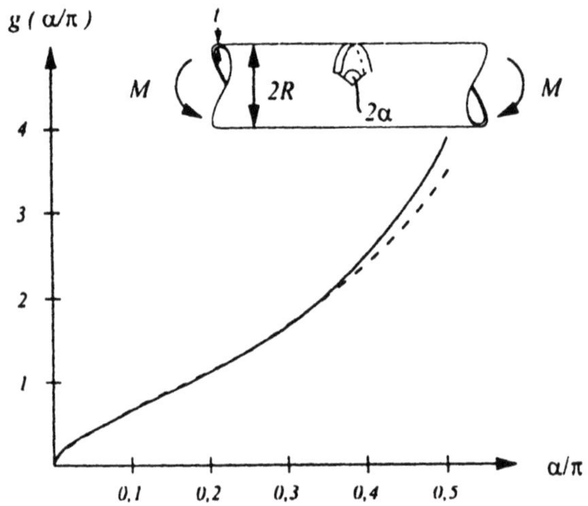

Fig. 1.13 Stress-intensity factor vs. normalized crack angle α for a circumferentially cracked cylindrical pipe (radius R, thickness t) under bending (— eq. (1.71), ---- Sanders (1982 and 1983))

with the function $g(\alpha/\pi)$ defined as

$$g = \sqrt{\frac{\pi}{2}\left(\frac{I^{(1)}}{I^{(2)}} - 1\right)}\,. \tag{1.71b}$$

The bending stiffness of the uncracked pipe is

$$EI^{(1)} = E\pi r^3 t\,,$$

whereas the bending stiffness of the cracked pipe is

$$EI^{(2)} = \frac{EI^{(1)}}{\pi} \int_{\alpha}^{2\pi-\alpha} (\cos\vartheta + \frac{\sin\alpha}{\pi - \alpha})\, d\,\vartheta.$$

Figure 1.13 (taken from Gao and Herrmann, 1992) compares the prediction of beam theory with the result from the complete shell analysis of Sanders (1982 and 1983). Up to a crack angle of $\alpha / \pi = 0.4$, corresponding to a nearly half-cracked cylindrical shell, the results match closely. It is remarkable that the elementary beam theory can actually reproduce the results generated through a complicated shell analysis.

Likewise it is possible to investigate curved cracked beams in bending and tension as sketched in Figure 1.14, cf. Müller at al. (1993a) and (1993b).

In closing this Section, the far-reaching and rather unexpected applicability of one-dimensional theories of strength-of-materials to cracked structural elements of great variety, once more confirmed the power of these theories which rests mainly on their simplicity and accuracy as compared to theories of elastic continua.

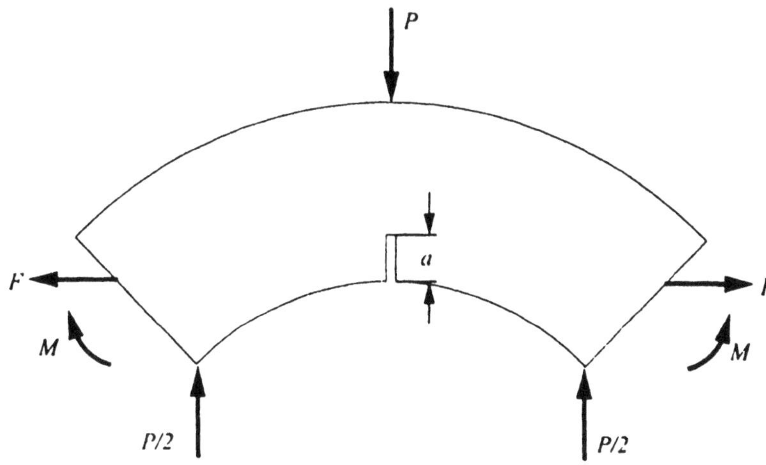

Fig. 1.14 Cracked circular beam subjected to a bending couples M, tensile forces F and three-point bending P

2 Plates and Shells

2.1 General Remarks

Plate and shell theories attempt to describe the three-dimensional state of deformation and stress in terms of the deformed and undeformed configurations of the middle surface. (The middle surface of a plate is plane, that of a shell is curved). In this sense, plate and shell theories are inherently approximative and their quality depends on the validity of the a priori assumptions (cf. Bernoulli's normal hypothesis in beam theory) introduced. The variety of existing plate and shell theories might be viewed as a measure of their degree of approximation. Since no "exact" theory is available, various features that characterize a *consistent* theory have been introduced to estimate possible errors of the results and to indicate the range of applicability, e. g., Koiter (1960), Naghdi (1972), Krätzig (1980).

As for bars and beams, conservation laws for plates and shells might be derived in different ways. Starting from conservation laws in three-dimensional elasticity, the a priori assumptions may be introduced into those expressions and the integration over the plate or shell thickness can be performed. However, it is not guaranteed that the invariant integrals in three-dimensional elasticity are compatible with the plate or shell theory employed. It has been shown by Sosa et al. (1988) that differences occur, especially when dealing with the M integral. As pointed out by Li and Shyy (1997), it would be both technically significant and aesthetically appealing to establish conservation laws within the original theory, because the partial differential equations which govern the motion of a particular class of structures may yield their own invariant integrals. An alternative way would then be, to apply Noether's formalism, the Neutral Action (NA) method, or any other method mentioned so far, directly to the governing Lagrangian or to the governing field equations within the proposed theory.

2.2 Plate Theories

With reference to an (x_1, x_2, x_3)-rectangular coordinate system, a plate is considered as an elastic body bounded by two surfaces $x_3 = \pm h/2$ (usually called the faces of the plate) and by a cylindrical surface perpendicular to the mid-plane $x_3 = 0$ of the plate. The distance h between the two faces is assumed to be small compared with the other linear dimensions of the plate and might be, in general, a function of x_1 and x_2

$$h = h(x_1, x_2) .$$

The most common theory of plates is referred to as the Poisson-Kirchhoff or classical theory. Within this theory it is assumed that rotations are small, i. e., $\partial w/\partial x_1$, $\partial w/\partial x_2 \ll 1$, where w is the transverse displacement of the mid-surface ($x_1, x_2, 0$). It is also assumed that the linear fibers of the plate which are initially perpendicular to the middle surface remain straight and perpendicular to the deformed middle surface (sometimes referred to as Kirchhoff's conditions). Additionally, the shear strains ε_{13} and ε_{23} are assumed to be negligible (analogously to the assumptions of the Bernoulli-Euler beam theory in Section 1). Finally, it is assumed that the normal stress σ_{33} is negligibly small in comparison with the inplane stress components.

Although very useful for many practical problems, this theory leads to some undesirable consequences such as the occurrence of concentrated forces at the corner of simply-supported plates of polygonal shape and inaccurate results in the determination of stress concentrations at the edge of holes when the dimensions of the hole are comparable to the magnitude of the thickness of the plate. These effects were explained by Kirchhoff (1850), who showed that the assumptions underlying this theory were responsible for a contraction on the number of boundary conditions that have to be satisfied at each edge of the plate. For example, along a free edge, three physical boundary conditions must be satisfied. These are the conditions of vanishing bending and twisting moments and of vanishing transverse force. However, in the classical theory the three conditions collapse into two by an appropriate combination of the shear force with the twisting moment, called ersatz-shear force.

Reissner (1944, 1945, 1947) established a theory that accounted for shear deformations, leading to a higher-order theory in which three boundary conditions can and must be prescribed along the edge of the plate. As a consequence, due to the effect of shear deformation, Kirchhoff's conditions do not hold any more.

Mindlin (1951) advanced a dynamic theory, including the effects of transverse shear and rotatory inertia, and showed that the two-dimensional theory of flexural motions of isotropic elastic plates could be deduced from the three-dimensional equations of elasticity.

Both theories are approximate as compared to the theory of elasticity because of imposed constraints. Reissner's and Mindlin's theories are distinguished from each other in that the former takes the effect of linearly weighted averages of the normal stress distribution in thickness direction into account. The difference between both theories is insignificant for thin plates, and in the absence of transverse loading they become identical. Since, in addition, numerical analyses, such as the finite-element method, usually develop their formulation based on a Mindlin-type plate, the following considerations rely on the latter theory. We start with the general elastodynamics of Mindlin plates and introduce later the assumptions of statics and those of the classical theory.

It is postulated that the displacements have the following form

$$u_1 = x_3 \psi_1(x_1, x_2, t), \tag{2.1a}$$

$$u_2 = x_3 \psi_2(x_1, x_2, t), \tag{2.1b}$$

$$u_3 = w(x_1, x_2, t), \tag{2.1c}$$

where u_1, u_2 and u_3 are, respectively, the displacements in the x_1, x_2 and x_3 directions; t is the time. Here, w is the linearly weighted average (taken over the plate thickness) of the transverse displacement; ψ_1 and ψ_2 are, respectively, the linearly weighted averages of the rotation about the x_2 and x_1 axes of fibers normal to the mid-plane before deformation. The variables ψ_1, ψ_2 and w are called *plate displacements* with sign conventions depicted in Figure 2.1.

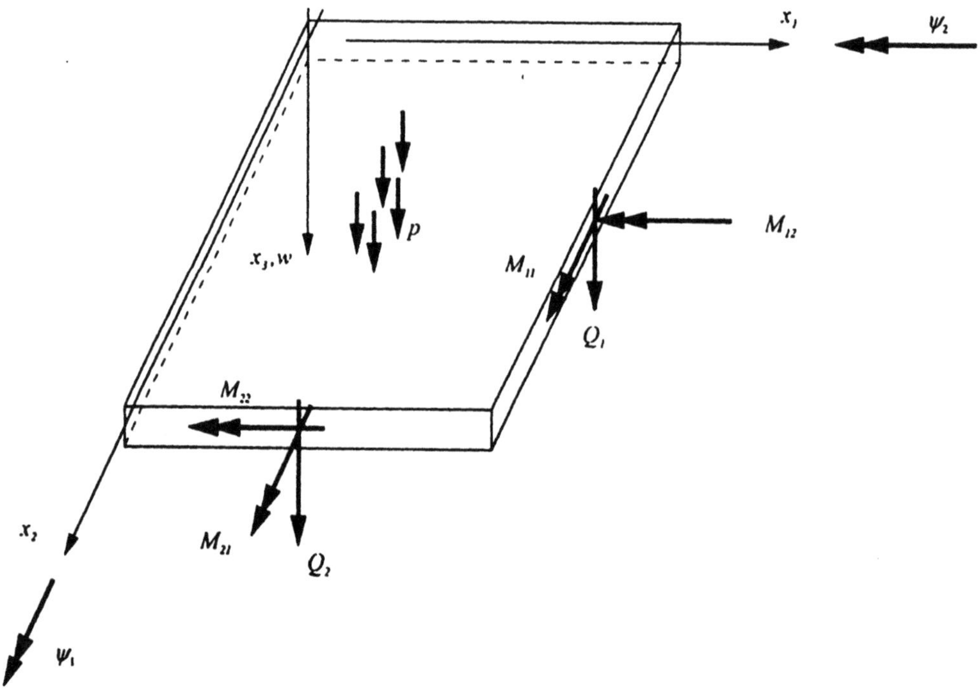

Fig. 2.1 Sign conventions of plate stress and displacement components

The bending and twisting moment resultants, $M_{11}, M_{22}, M_{12}, M_{21}$ and the transverse shear force resultants Q_1 and Q_2 are related to the plate displacements by

$$M_{11} = D(\psi_{1,1} + \upsilon \psi_{2,2}), \tag{2.2a}$$

$$M_{22} = D(\psi_{2,2} + \upsilon\psi_{1,1}), \quad (2.2b)$$

$$M_{12} = M_{21} = \frac{1-\upsilon}{2} D(\psi_{1,2} + \psi_{2,1}), \quad (2.2c)$$

$$Q_1 = C_S(w_{,1} + \psi_1), \quad (2.2d)$$

$$Q_2 = C_S(w_{,2} + \psi_2), \quad (2.2e)$$

where the flexural stiffness D and the shear stiffness C_S of the plate are given by, respectively,

$$D = \frac{Eh^3}{12(1-\upsilon^2)}, \quad (2.3a)$$

$$C_S = k\,Gh \quad (2.3b)$$

with E, G, υ being Young's modulus, shear modulus and Poisson's ratio, respectively, while the shear correction factor k takes the value $5/6$ (based on energy considerations in Reissner's theory) or $\pi^2/12$ (based on dynamic considerations in Mindlin's theory). The quantities $M_{11}, M_{22}, M_{12}, Q_1$ and Q_2 are functions of (x_1, x_2, t) only and are referred to as *plate stresses* with sign conventions depicted also in Figure 2.1.

The equations of motion expressed in terms of plate quantities are obtained by integrating the corresponding 3-D elasticity equations through the thickness to give

$$\Delta_1 = M_{11,1} + M_{21,2} - Q_1 - \rho I\psi_{1,tt} = 0, \quad (2.4a)$$

$$\Delta_2 = M_{12,1} + M_{22,2} - Q_2 - \rho I\psi_{2,tt} = 0, \quad (2.4b)$$

$$\Delta_2 = Q_{1,1} + Q_{2,2} + p - \rho h w_{,tt} = 0, \quad (2.4c)$$

with $p = p(x_1, x_2)$ being the transverse pressure acting in the x_3-direction, ρ being the density per unit of volume and I being the moment of inertia per unit width of the plate. For a inhomogeneous plate, $D, C_S, \upsilon, \rho, h$ and I might be functions of x_1 and x_2.

Finally, the strain energy density W, the kinetic energy T and the potential of external force V per unit of area are given by, respectively,

$$W = \frac{1}{2}\left[M_{11}\psi_{1,1} + M_{22}\psi_{2,2} + M_{12}(\psi_{2,1} + \psi_{1,2}) + Q_1(w_{,1} + \psi_1) + Q_2(w_{,2} + \psi_2)\right], \quad (2.5a)$$

$$T = \frac{1}{2}\left[\rho h w_{,t}^2 + \rho I \psi_{1,t}^2 + \rho I \psi_{2,t}^2\right], \tag{2.5b}$$

$$V = -wp\,. \tag{2.5c}$$

Thus, the Lagrangian density L per unit of area of the plate

$$L = T - (W + V) = L\left(x_1, x_2, t, \psi_1, \psi_{1,1}, \psi_{1,2}, \psi_{1,t}, \psi_2, \psi_{2,1}, \psi_{2,2}, \psi_{2,t}, w, w_{,1}, w_{,2}, w_{,t}\right) \tag{2.6}$$

is of first order, i. e., it does not depend on higher than first derivatives of the three dependent variables w, ψ_1 and ψ_2.

In order to find solutions for a particular plate problem, the governing field equations have to be supplemented by a suitable set of initial and boundary conditions, that are not required in the following considerations. The reader is, therefore, referred to the literature, e. g., Timoshenko and Woinowsky-Krieger (1959) and Eschenauer et al. (1997).

The classical theory is recovered, if the Kirchhoff condition (normal hypothesis) is introduced. With

$$\psi_i = -w_{,i}\,, \tag{2.7}$$

shear strains are neglected. Additionally, an infinite shear stiffness C_S and a vanishing moment of inertia I are assumed. Thus, the shear forces remain finite but constitutive relations for the shear forces do not exist. These are obtainable from the equations of motion.

The equations of the classical theory reduce to

- kinematics

$$u_1 = -zw_{,1}, \tag{2.8a}$$

$$u_2 = -zw_{,2}, \tag{2.8b}$$

$$u_3 = w \tag{2.8c}$$

- constitutive relations

$$M_{11} = -D\left(w_{,11} + \upsilon w_{,22}\right), \tag{2.9a}$$

$$M_{22} = -D\left(w_{,22} + \upsilon w_{,11}\right) \tag{2.9b}$$

$$M_{12} = -(1-\upsilon) D w_{,12} \tag{2.9c}$$

- strain energy density

$$W = -\frac{1}{2}\left(M_{11} w_{,11} + M_{22} w_{,22} + 2 M_{12} w_{,12}\right) \tag{2.10a}$$

- kinetic energy $$T = \frac{1}{2}\rho h w_{,t}^2 \qquad (2.10b)$$

- potential of external forces $$V = -wp\,. \qquad (2.10c)$$

Thus, the Lagrangian density L (per unit of area) of the classical theory

$$L = L(x_1, x_2, t, w, w_{,11}, w_{,12}, w_{,22}, w_{,t}) \qquad (2.11)$$

is of second order and depends on the single dependent variable w. In order to obtain the associated Euler-Lagrange equation we eliminate the shear forces in equations of motion with the result

$$\Delta_1 = M_{11,11} + 2M_{12,12} + M_{22,22} + p - \rho h w_{,tt} = 0\,, \qquad (2.12)$$

or, in term of the deflection w, by introducing the constitutive relations

$$\Delta_1 = D\Delta\Delta w - p + \rho h w_{,tt} = 0\,. \qquad (2.13)$$

Here, the Laplace operator $\Delta() = ()_{,11} + ()_{,22}$ is used. Initital and boundary conditions have to be adjusted analogously (cf., e. g., Timoshenko and Woinowsky-Krieger, 1970).

2.3 Conservation Laws for Elastostatics of Mindlin Plates

As in beam theory, we omit the lengthy derivation of conservation laws for inhomogeneous plates, including elastodynamics. Details may be found in Kienzler and Herrmann (2000). We restrict ourselves to balance laws for statically loaded inhomogeneous plates which will reduce to conservation laws in the absence of inhomogeneities, i. e., dependence of the Lagrangian explicitely on x_i.

In this Sub-section, we are interested mainly in conservation laws which lead in turn to path-independent integrals applicable in fracture mechanics. It is not our intention to strive for completeness. The most efficient way to derive these conservation laws, then, is to apply the vector operations gradient, curl and divergence to the Lagrangian L and its moment xL. In static cases, the Lagrangian is just equal to the negative of the sum of the strain energy density W and the potential of external forces V

$$L = -(W + V) = L(x_k, \psi_i, \psi_{i,j}, w, w_{,i}) \tag{2.14}$$

with

$$W = \frac{1}{2}\left[M_{ij}\psi_{j,i} + Q_i(\psi_i + w_{,i})\right] \qquad (i, j, k = 1,2) \tag{2.15}$$

and V given by (2.5c). For convenience, the equilibrium equations (2.4) are recalled. Neglecting all time-dependent terms in (2.4) we have

$$M_{ji,j} - Q = 0, \tag{2.16a}$$

$$Q_{i,i} + p = 0, \tag{2.16b}$$

where summation convention applies.

In the sequel, the three mathematical operations are applied sequentially.

Gradient. Considering the gradient of the Lagrangian, we find with (2.14), (2.15) and (2.2)

$$\mathbf{grad}\ (W + V) = (W + V)_{,k} = M_{ji}\psi_{i,jk} + Q_i(\psi_{i,k} + w_{,ik}) - pw_{,k} + \frac{\partial(W + V)}{\partial x_k}.$$

The explicit derivative of $W + V$ with respect to the independent variable $x_k (k = 1,2)$ accounts for material inhomogeneities produced, for example, by stiffness distributions $D(x_k)$ and $C_S(x_k)$ smoothly or discontinuously varying with x_k (defects), and for physical inhomogeneities, produced by $p(x_k)$.

Integration by parts yields

$$(W + V)_{,k} = (M_{ji}\psi_{i,k})_{,j} - (M_{ji,j} - Q_i)\psi_{i,k} + (Q_j w_{,k})_{,j} - (Q_{j,j} + p)w_{,k} + \frac{\partial(W + V)}{\partial x_k}. \tag{2.18}$$

Due to (2.16) and ($(W + V)_{,k} = (W + V)_{,j}\delta_{jk}$) it is

$$\left[(W + V)\delta_{ik} - M_{ji}\psi_{i,k} - Q_j w_{,k}\right]_{,j} = \frac{\partial(W + V)}{\partial x_k}. \tag{2.19}$$

The left hand side of (2.19) is equal to the Eshelby tensor of plate theory,

$$B_{jk} = (W + V)\delta_{jk} - M_{ji}\psi_{i,k} - Q_j w_{,k}. \tag{2.20}$$

It corresponds to b_{jk} in three-dimensional elasticity and is the two-dimensional analogue to the material force B (1.36a) in the one-dimensional beam theory.

In the absence of inhomogeneities, the Eshelby tensor (2.20) is divergence-free

$$B_{jk,j}=0. \tag{2.21}$$

If the functions involved are "smooth", integration of (2.19) with respect to the middle-plane surface S (elements of area dA) and application of the divergence theorem yields with (2.20) the $\boldsymbol{J}$ integral of plate theory

$$J_k=\int_\Gamma B_{jk}n_j ds=\int_S \frac{\partial(W+V)}{\partial x_k}dA, \tag{2.22}$$

where Γ is the boundary enclosing S, s is the arc length along Γ and $\boldsymbol{n}$ is the unit outward normal vector.

It can be observed that when the pressure load p is constant and the material is homogeneous within S, equation (2.22) can be written as a single line integral on Γ, and the balance law (2.22) becomes a conservation law.

Curl. This operation is performed on the vector quantity $(W+V)\boldsymbol{x}$. With the permutation tensor ε_{kij} it follows

$$\textbf{curl}\ \left[(W+V)\boldsymbol{x}\right]=\varepsilon_{kij}-\left[(W+V)x_j\right]_{,i}=$$

$$=\varepsilon_{kij}-\left[(W+V)_{,i}x_j+(W+V)\delta_{ji}\right]=\varepsilon_{kij}x_j(W+V)_{,i}. \tag{2.23}$$

Of particular interest from a physical point of view regarding plate problems is the case $k=3$. Use of (2.18) and integration by parts yields

$$\varepsilon_{3ij}\left[(W+V)x_j\right]_{,i}=\varepsilon_{3ij}\left\{\left(x_jM_{mn}\psi_{n,i}\right)_{,m}+\left(x_jQ_mw_{,i}\right)_{,m}+x_j\frac{\partial(W+V)}{\partial x_i}-M_{jn}\psi_{n,i}-Q_jw_{,i}\right\}.$$

We add and substract the term $M_{nj}\psi_{i,n}+Q_j\psi_i$ on the right-hand side and find

$$\varepsilon_{3ij}\left[(W+V)x_j\right]_{,i}=\varepsilon_{3ij}\left\{\left[x_j\left(M_{nm}\psi_{m,i}+Q_nw_{,i}\right)\right]_{,n}-\left[M_{jn}\psi_{n,i}+M_{nj}\psi_{i,n}+\right.\right.$$

$$+Q_j\left(\psi_i+w_{,i}\right)\big]+\left[M_{nj}\psi_{i,n}+Q_j\psi_i\right]+x_j\frac{\partial(W+V)}{\partial x_i}\Big\} \tag{2.24}$$

The term in the second square bracket vanishes on substitution of (2.2). If the material behaviour is not isotropic, i. e., the constitutive equations (2.2) have a different form, this term has to be treated as an extra source term.

Focusing attention on the term in the third square bracket we can write

$$\varepsilon_{3ij}\left(M_{nj}\psi_{i,n}+Q_j\psi_i\right)=\varepsilon_{3ij}\left(-M_{ni}\psi_{j,n}-Q_i\psi_j\right)=$$

$$=\varepsilon_{3ij}\left[\left(-M_{ni}\psi_j\right)_{,n}+\left(M_{ni,n}-Q_i\right)\psi_j\right]=\varepsilon_{3ij}\left[-M_{ni}\psi_j\right]_{,n} \tag{2.25}$$

because of (2.16a). Upon introduction of (2.25) into (2.24), a rearranged leads finally with (2.20) to the desired balance law

$$\varepsilon_{3ij}\left(B_{ni}x_j+M_{ni}\psi_j\right)_{,n}=\varepsilon_{3ji}\frac{\partial(W+V)}{\partial x_i}x_j. \tag{2.26}$$

If the material is isotropic and homogeneous, the source term vanishes giving rise to the divergence-free expression

$$\varepsilon_{3ij}\left(B_{ni}x_j+M_{ni}\psi_j\right)_{,n}=0. \tag{2.27}$$

Integration of (2.26) and application of the divergence theorem leads to the L_3 integral of plate theory

$$L_3=\int_\Gamma\varepsilon_{3ij}\left(B_{ni}x_j+M_{ni}\psi_j\right)n_n ds=\int_S\varepsilon_{3ij}\frac{\partial(W+V)}{\partial x_i}x_j dA. \tag{2.28}$$

Again, if the integrand of the surface integral vanishes, the balance law changes to a conservation law in integral form.

Divergence. This operation is applied to the moment of the sum of strain energy and potential of external forces per unit of mid-plane surface as

$$\textbf{div}\ \left[(W+V)\mathbf{x}\right]=\left[(W+V)x_i\right]_{,i}=2(W+V)+x_i(W+V)_{,i}. \tag{2.29}$$

The factor 2 arises from $x_{i,i} = \delta_{ii}$, which is equal to two for plane problems. With equation (2.15) and (2.18) it is

$$[(W+V)x_i]_{,i} = M_{ij}\psi_{j,i} + Q_{,}(\psi_j + w_{,j}) - 2pw + x_1\left[(M_{jk}\psi_{k,i})_{,j} + (Q_j w_{,i})_{,j} + \frac{\partial(W+V)}{\partial x_i}\right].$$

Integration by parts leads to

$$[(W+V)x_i]_{,i} = (x_i M_{jk}\psi_{k,i})_{,j} + (x_i Q_{,} w_{,i})_{,j} + x_i \frac{\partial(W+V)}{\partial x_i} - 2pw + Q_j\psi_j .$$

The last term may be transformed with (2.2d, e) and, again, integration by parts leads to

$$Q_j\psi_j = \frac{Q_j Q_j}{C_S} - Q_j w_{,j} = \frac{Q_j Q_j}{C_S} - (Q_j w)_{,j} - pw. \qquad (2.30)$$

Introduction of the last equation and rearrangement yields with (2.20) finally

$$(B_{ji}x_i + Q_j w)_{,j} = x_i \frac{\partial(W+V)}{\partial x_i} - 3pw + \frac{Q_j Q_j}{C_S}. \qquad (2.31)$$

The first term on the right-hand side of (2.31) has the same structure as previous balance laws (compare, e. g., with (1.22) in combination with (1.19d) or with (1.38) in combination with (1.36)). The second and third terms are completely different in nature and do not vanish even for a homogeneous material and constant pressure p. It is also noteworthy that the last term is dependent on the plate theory employed, indicated by the presence of the effective shear stiffness C_S.

Integration of (2.31) and application of the divergence theorem leads to the M integral of plate theory

$$M = \int_{\Gamma} (B_{ji}x_i + Q_j w) n_j ds = \int_{S} \left(\frac{\partial(W+V)}{\partial x_i} x_i - 3pw + \frac{Q_j Q_j}{C_S} \right) dA. \qquad (2.32)$$

The surface integral vanishes only under the quite restrictive conditions that the material is homogeneous and the plate is subjected to pure bending. As will be shown below, the third term vanishes in the classical Poisson-Kirchhoff plate theory.

The integral expressions (2.22), (2.28) and (2.32) have been discussed in terms of path- and path-domain independent integrals in Sosa (1986), Sosa et al. (1988) and Sosa and Herrmann (1989). In addition, the physical interpretation of the integrals as energy-release rates due to cavity translation, rotation and self-similar expansion were given. Finally, the relationships of J_k, L_3 and M to fracture parameters, e. g., stress-intensity factors, were established, that make the integrals valuable tools, from a practical point of view, in fracture mechanics.

2.4 Reduction to the Classical Theory

In order to obtain conservation laws, the same mathematical operations introduced in Paragraph 2.3 could be performed within the classical Poisson-Kirchhoff theory governed by equations (2.7)-(2.13). This Section is devoted to show that in the classical plate theory, invariant integrals can be obtained as a particular case of equations (2.22), (2.28) and (2.32).

It should be noted, however, that in general such a reduction does not apply in the presence of a flaw. As it was shown in Sosa (1986), the asymptotic expressions of the elastic fields in the neighbourhood of a crack for the classical plate theory do not follow from the fields corresponding to the transverse shear theory. As a consequence, crack problems need to be solved individually within each plate theory.

For the following, it is useful to refer the plate field quantities in the integrand of the contour integral to components n and s normal and tangential, respectively, to the boundary. The relations between the plate stresses in both coordinate systems then become

$$M_{nn} = M_{11}n_1^2 + M_{22}n_2^2 + 2M_{12}n_1n_2, \tag{2.32a}$$

$$M_{ns} = (M_{22} - M_{11})n_1n_2 + M_{12}(n_1^2 - n_2^2), \tag{2.32b}$$

$$Q_n = Q_1n_1 + Q_2n_2, \tag{2.32c}$$

where

$$n_1 = \cos(n, x_1) = \cos\alpha, \tag{2.33a}$$

$$n_2 = \cos(n, x_2) = \sin\alpha \tag{2.33b}$$

and α is the angle between the coordinate systems (x_1, x_2) and (n, s) as depicted in Figure 2.2.

Likewise, the following relations are useful

$$\psi_n = \psi_1n_1 + \psi_2n_2, \tag{2.34a}$$

$$\psi_s = \psi_2 n_1 - \psi_1 n_2, \tag{2.34b}$$

$$\frac{\partial w}{\partial n} = w_{,1} n_1 + w_{,2} n_2, \tag{2.34c}$$

$$\frac{\partial w}{\partial s} = w_{,2} n_1 - w_{,1} n_2. \tag{2.34d}$$

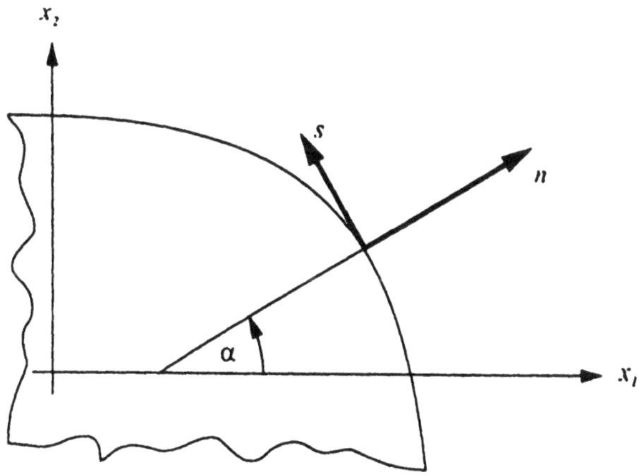

Fig. 2.2 Coordinates along the contour Γ of integration

Under the condition of neglecting the transverse shear strain components, the three plate balance laws are considered in sequence.

J_k **integral**. The analogue of the J integral of the three-dimensional theory of elasticity follows from (2.22) and is rewritten as

$$J_k = \int_\Gamma B_{jk} n_j ds$$

$$= \int_\Gamma \left[(W+V) n_k - M_{nn} \psi_{n,k} - M_{ns} \psi_{s,k} - Q_n w_{,k} \right] ds$$

$$= \int_S \frac{\partial (W+V)}{\partial x_k} dA \tag{2.36}$$

(no summation over n and s).

(no summation over n and s).

The strain-energy density W depends merely on $w_{,ik}$, as indicated in equation (2.10a). With the normal hypothesis (2.7) and with (2.35) one can write

$$-M_{nn}\psi_{n,k} - M_{ns}\psi_{s,k} - Q_n w_{,k} = M_{nn}\frac{\partial w_{,k}}{\partial n} + M_{ns}\frac{\partial w_{,k}}{\partial s} - Q_n w_{,k}$$

$$= M_{nn}\frac{\partial w_{,k}}{\partial n} - \left(Q_n + \frac{\partial M_{ns}}{\partial s}\right)w_{,k} - \frac{\partial}{\partial s}\left(M_{ns}w_{,k}\right).$$

The last term vanishes upon integration along a closed curve, since it is a total differential. The second term is Kirchhoff's so-called ersatz-shear force

$$\overline{Q}_n = Q_n + \frac{\partial M_{ns}}{\partial s}. \tag{2.37}$$

In the (x_1, x_2) coordinate system it is commonly defined as

$$\overline{Q}_1 = Q_1 + M_{12,2}, \tag{2.38a}$$

$$\overline{Q}_2 = Q_2 + M_{21,1}. \tag{2.38b}$$

Equation (2.36) can then be written as

$$J_k = \int_\Gamma \left[(W+V)n_k - M_{nn}\psi_{n,k} - \overline{Q}_n w_{,k}\right]ds = \int_S \frac{\partial(W+V)}{\partial x_k}dA. \tag{2.39}$$

It may be noted, that (M_{nn} and $\overline{Q}_n$) and ($\psi_n = -\partial w/\partial n$ and w) are compatible with classical plate theory, which considers only two components of force and displacement at any boundary of the plate.

***L_3* integral.** The analogue of the L_3 integral of the three-dimensional theory of elasticity follows from (2.28) and is rewritten as

$$L_3 = \int_\Gamma \varepsilon_{3ij}\left(B_{ki}x_j + M_{ki}\psi_j\right)n_k ds$$

$$= \int_{\Gamma} \varepsilon_{3ij} \left\{ x_j \left[(W+V) n_i - M_{nn}\psi_{n,i} - M_{ns}\psi_{s,i} - Q_n w_{,i} \right] + \left(M_{nn}\psi_s - M_{ns}\psi_n \right) \right\} ds$$

$$= \int_{S} \varepsilon_{3ij} x_j \frac{\partial (W+V)}{\partial x_i} dA . \tag{2.40}$$

Manipulations along similar lines as above lead to

$$- \varepsilon_{3ij} x_j \left(M_{nn}\psi_{n,i} + M_{ns}\psi_{s,i} + Q_n w_{,i} \right) =$$

$$= \varepsilon_{3ij} x_j \left[M_{nn} \frac{\partial w_{,i}}{\partial n} - \left(Q_n + \frac{\partial M_{ns}}{\partial s} \right) w_{,i} + \frac{\partial}{\partial s}\left(M_{ns} w_{,i} \right) \right] .$$

Focussing attention on the last term in the above expression we can write

$$\varepsilon_{3ij} x_j \frac{\partial (M_{ns} w_{,i})}{\partial s} = \varepsilon_{3ij} \frac{\partial}{\partial s}\left(x_j M_{ns} w_{,i} \right) - \varepsilon_{3ij} \frac{\partial x_j}{\partial s} M_{ns} w_{,i}$$

$$= \varepsilon_{3ij} \frac{\partial}{\partial s}\left(x_j M_{ns} w_{,i} \right) - M_{ns} \frac{\partial w}{\partial n} . \tag{2.41}$$

The first term will vanish when integrated along a closed path. The last two terms in the line integral of (2.40) become

$$M_{nn}\psi_s - M_{ns}\psi_n = -M_{nn} \frac{\partial w}{\partial s} + M_{ns} \frac{\partial w}{\partial n} . \tag{2.42}$$

Therefore, upon introduction of equations (2.41) and (2.42), the terms $M_{ns} w_{,n}$ cancel and (2.40) is written as

$$L_3 = \int_{\Gamma} \left\{ \varepsilon_{3ij} x_j \left[(W+V) n_i - M_{nn} \frac{\partial \psi_i}{\partial n} - \overline{Q}_n w_{,i} \right] - M_{nn} \frac{\partial w}{\partial s} \right\} ds$$

$$= \int_{S} \varepsilon_{3ij} x_j \frac{\partial (W+V)}{\partial x_i} dA . \tag{2.43}$$

M Integral. Since in this case the shear stiffness $C_S \to \infty$ the last term in (2.32) vanishes and the analogue to the M integral of the three-dimensional theory of elasticity is rewritten as

$$M=\int_\Gamma \left(B_{ji}x_i+Q_jw\right)n_j ds$$

$$=\int_\Gamma \left[\left(W+V\right)x_in_i-\left(M_{nn}\psi_{n,i}+M_{ns}\psi_{s,i}+Q_nw_{,i}\right)x_i+Q_nw\right]ds$$

$$=\int_S \left[\frac{\partial(W+V)}{\partial x_i}x_i-3pw\right]dA. \tag{2.44}$$

Transformations as indicated above and integration by parts where appropriate lead to

$$M=\int_\Gamma \left\{x_i\left[\left(W+V\right)n_i-M_{nn}\frac{\partial\psi_i}{\partial n}-\overline{Q}_nw_{,i}\right]+\overline{Q}_nw\right\}ds$$

$$=\int_S \left[x_i\frac{\partial(W+V)}{\partial x_i}-3pw\right]dA. \tag{2.45}$$

When the applied pressure p is constant, the surface integrals for J_k and L_3 vanish, which is not the case for M.

In concluding this Sub-section, it might be mentioned that the conservation laws presented are useful in the numerical analysis of cracked plates, cf., e. g., Sosa and Eischen (1986).

2.5 Conservation Laws for Shells

It would be beyond the scope of this text, if a detailed derivation of shell theory, and the underlying theory of surfaces were to be presented here. Thus the reader is expected to be familiar with those topics, the aim being a concise discussion of conservation laws.

The internal geometry of the curved middle surface of a shell imposes the necessity of using curvlinear coordinates and related to them, the special features of Riemannian geometry. Covariant and contravariant coordinates have be introduced and partial derivatives are to be replaced by covariant derivatives with the help of Christoffel symbols, cf., e. g., Flügge (1972, 1973), Niordson (1985), Eschenauer et al. (1997). Shell theories of various degrees of approximation have been derived and are used in practice. Independent of a specific shell theory under consideration, however, a severe restriction to establish conservation laws in shell theory becomes apparent if we think in terms of material translations and rotations.

These transformations are introduced by the comparison of two neighbouring volume elements which differ from each other by a *rigid body translation* (or *rotation*) in the undeformed state. Because of the curvature of the shell middle surface, such two elements will not be congruent with each other in general, giving rise to a distortion, except for shells which enjoy a high degree of symmetry, e. g., cylindrical shells in axial direction, shells of revolution in circumferential direction and spherical shells as indicated by Lo (1980) and Kienzler and Golebiewska-Herrmann (1985). In theories of shallow shells the distortion is an effect of second order. Thus, for shallow shells, path-independent integrals might be established for a wider class of mid-surface geometries, cf. Nicholson and Simmonds (1980) and Li and Shyy (1997).

The aim of the present Sub-section is to give a short account of a shell theory and the associate Eshelby tensor. We will not give a detailed discussion but wish merely to acquaint the reader with special features of problems one faces in shell theory.

In the following, it is assumed that the range of greek indices is (*1, 2*) whereas the range of latin indices is (*1, 2, 3*). The displacements of a shell are described commonly by five independent components u_α, ψ_α, w as

$$\mathbf{u} = \left(u_\alpha + x_3 \psi_\alpha\right) \mathbf{a}^\alpha + w \mathbf{a}^3 . \tag{2.46}$$

The state of deformation consists, therefore, in a combination of the displacements of a disc (plane stress) and a plate. The vectors $\mathbf{a}^\alpha$ and $\mathbf{a}^3$ are contravariant base vectors of the shell middle surface and the unit vector perpendicular to the surface, respectively.

The elongation and shear strain of the middle surface are denotes as $\alpha_{\alpha\beta}$, the change of curvature and the twist by $\beta_{\alpha\beta}$ and the transverse shear strain by $\gamma_{3\alpha}$. The change of thickness and warping of the cross-sections are neglected in (2.46) but may be considered in higher-order shell theories (Kienzler, 1982 and Kienzler and Golebiewska-Herrmann, 1985).

The kinematic relations are given by

$$\alpha_{\alpha\beta} = \frac{1}{2}\left(u_\alpha\big|_\beta + u_\beta\big|_\alpha - 2b_{\alpha\beta} w\right), \tag{2.47a}$$

$$\beta_{\alpha\beta} = \frac{1}{2}\left(\psi_\alpha\big|_\beta + \psi_\beta\big|_\alpha - b_\alpha^\rho\, u_\rho\big|_\beta - b_\beta^\rho\, u_\rho\big|_\alpha + 2b_\alpha^\rho\, b_{\rho\beta} w\right), \tag{2.47b}$$

$$\gamma_{3\alpha} = \frac{1}{2}\left(\psi_\alpha + w_{,\alpha} + b_\alpha^\rho\, u_\rho\right). \tag{2.47c}$$

Vertical bars denote covariant differentiation and $b_{\alpha\beta}$ (or b^{α}_{β}) is the curvature tensor of the middle surface.

The tensor of resultant membrane forces, the tensor of moments and the tensor of resultant shear forces are denoted by $N^{\alpha\beta}$, $M^{\alpha\beta} = M^{\beta\alpha}$ and $Q^{\alpha 3} = Q^{3\alpha} = Q^{\alpha}$, respectively. In contradistinction to the tensor of moments, the tensor of membrane forces is not symmetric. In view of (2.47a), i. e., symmetrical measures of strain, it is convenient to introduce the symmetrical "pseudo" tensor of resultant membrane forces (cf. Eshenauer et al., 1997) as

$$\tilde{N}^{\alpha\beta} = N^{\alpha\beta} + b^{\beta}_{\rho} M^{\alpha\rho} . \tag{2.48}$$

With the strain energy density W per unit of area of the shell middle surface, the constitutive equations are given by

Membran forces $$\tilde{N}^{\alpha\beta} = \frac{\partial W}{\partial \alpha_{\beta\alpha}} , \tag{2.49a}$$

Moments $$M^{\alpha\beta} = \frac{\partial W}{\partial \beta_{\beta\alpha}} , \tag{2.49b}$$

Transverse shear forces $$Q^{\alpha} = \frac{\partial W}{\partial \gamma_{3\alpha}} . \tag{2.49c}$$

The external forces p^i follow from the potential V as

$$p^i = -\frac{\partial V}{\partial u_i} , \tag{2.50}$$

and the Lagrangian for statics of shells is given as usual by

$$L = -(W + V) = L\left(x^{\beta}; u_{\alpha}, u_{\alpha}\big|_{\gamma}, w, w_{,\gamma}, \psi_{\alpha}, \psi_{\alpha}\big|_{\gamma}\right). \tag{2.51}$$

The equations of equilibrium follow either from the Euler-Lagrange equations $\mathbf{E}_k(L) = 0$, from considerations of an infinitesimal shell element or from suitable integrations of the three-dimensional equations with respect to the shell thickness.

They are given by

$$N^{\alpha\beta}\big|_{\alpha} - Q^{\alpha}\, b^{\beta}_{\alpha} + p^{\beta} = 0 , \tag{2.52a}$$

$$Q^{\alpha}\big|_{\alpha} + N^{\alpha\beta}\, b_{\alpha\beta} + p^{3} = 0\,, \tag{2.52b}$$

$$M^{\alpha\beta}\big|_{\alpha} - Q^{\beta} \qquad = 0\,. \tag{2.52c}$$

If the shell middle surface is plane, i. e., $b_{\alpha\beta} = b^{\alpha}_{\beta} = 0$, the above equations of shell theory uncouple into those of sheet theory (akin to plane elasticity) and plate theory.

All methods discussed before to establish conservation laws are obviously applicable also in shell theory. We choose the vector operation gradient applied to the Lagrangian as the most convenient approach. Thus

$$\begin{aligned}
&\mathbf{grad}\,(W+V) = (W+V)_{,\beta} \\
&= \frac{\partial(W+V)}{\partial\alpha_{\nu\mu}}\left(\frac{\partial\alpha_{\nu\mu}}{\partial u_{\gamma}|_{\alpha}}u_{\gamma}\big|_{\alpha\beta} + \frac{\partial\alpha_{\nu\mu}}{\partial w}w_{,\beta}\right) + \\
&+ \frac{\partial(W+V)}{\partial\beta_{\nu\mu}}\left(\frac{\partial\beta_{\nu\mu}}{\partial\psi_{\gamma}|_{\alpha}}\psi_{\gamma}\big|_{\alpha\beta} + \frac{\partial\beta_{\nu\mu}}{\partial u_{\gamma}|_{\alpha}}u_{\gamma}\big|_{\alpha\beta} + \frac{\partial\beta_{\nu\mu}}{\partial w}w_{,\beta}\right) + \\
&+ \left(\frac{\partial(W+V)}{\partial\gamma_{3\mu}} + \frac{\partial(W+V)}{\partial\gamma_{\mu 3}}\right)\left(\frac{\partial\gamma_{3\mu}}{\partial\psi_{\alpha}}\psi_{\alpha}\big|_{\beta} + \frac{\partial\gamma_{3\mu}}{\partial w_{,\alpha}}w_{,\alpha\beta} + \frac{\partial\gamma_{3\mu}}{\partial u_{\alpha}}u_{\alpha}\big|_{\beta}\right) + \\
&\frac{\partial(W+V)}{\partial u_{\alpha}}u_{\alpha}\big|_{\beta} + \frac{\partial(W+V)}{\partial w}w_{,\beta} + \frac{\partial(W+V)}{\partial x^{\beta}}\,.
\end{aligned} \tag{2.53}$$

Use of (2.47) - (2.49) leads to

$$\begin{aligned}
(W+V)_{,\beta} \quad &= (N^{\alpha\gamma} + b^{\gamma}_{\rho}M^{\alpha\rho})\, u_{\gamma}\big|_{\alpha\beta} - (N^{\alpha\gamma} + b^{\gamma}_{\rho}M^{\alpha\rho})\, b_{\gamma\alpha}w_{,\beta} + \\
&+ M^{\alpha\gamma}\psi_{\gamma}\big|_{\alpha\beta} - M^{\rho\alpha}b^{\gamma}_{\rho}u_{\gamma}\big|_{\alpha\beta} + M^{\alpha\gamma}b^{\rho}_{\gamma}b_{\rho\alpha}w_{,\beta} + \\
&+ Q^{\alpha}\left(\psi_{\alpha}\big|_{\beta} + w_{,\alpha\beta} + b^{\rho}_{\alpha}u_{\rho}\big|_{\beta}\right) - \\
&- p^{\alpha}\, u_{\alpha}\big|_{\beta} - p^{3}w_{,\beta} + \frac{\partial(W+V)}{\partial x^{\beta}}\,.
\end{aligned}$$

Addition and subtraction of the terms $N_{\alpha\gamma}u_\gamma|_{\beta\alpha}$ and $M^{\alpha\gamma}\psi_\gamma|_{\beta\alpha}$, integration by parts and rearrangement yields

$$(W+V)_{,\beta} = N^{\alpha\gamma}\left(u_\gamma|_{\alpha\beta} - u_\gamma|_{\beta\alpha}\right) + M^{\alpha\gamma}\left(\psi_\gamma|_{\alpha\beta} - \psi_\gamma|_{\beta\alpha}\right) +$$

$$+\left(N^{\alpha\gamma}\, u_\gamma|_\beta\right)\Big|_\alpha + \left(M^{\alpha\gamma}\psi_\gamma|_\beta\right)\Big|_\alpha + \left(Q^\alpha w_{,\beta}\right)\Big|_\alpha -$$

$$-\left(N^{\alpha\gamma}|_\alpha - Q^\alpha\, b_\alpha^\gamma + p^\gamma\right)u_\gamma|_\beta -$$

$$-\left(Q^\alpha|_\alpha + N^{\alpha\gamma}\, b_{\gamma\alpha} + p^3\right)w_{,\beta} -$$

$$-\left(M^{\alpha\gamma}|_\alpha - Q^\gamma\right)\psi_\gamma|_\beta +$$

$$+\frac{\partial(W+V)}{\partial x^\beta}.$$

Finally, with the equations of equilibrium (2.52), we obtain

$$\left[\left(W+V\right)\delta_\beta^\alpha - N^{\alpha\gamma}\, u_\gamma|_\beta - M^{\alpha\gamma}\, \psi_\gamma|_\beta - Q^\alpha w_{,\beta}\right]\Big|_\alpha =$$

$$= N^{\alpha\gamma}\left(u_\gamma|_{\alpha\beta} - u_\gamma|_{\beta\alpha}\right) + M^{\alpha\gamma}\left(\psi_\gamma|_{\alpha\beta} - \psi_\gamma|_{\beta\alpha}\right) + \frac{\partial(W+V)}{\partial x^\beta} \tag{2.54}$$

It should be recalled that covariant differentiation on a surface is, in general, not commutative; rather Gauss' relationship should be used, e. g.,

$$\eta_\gamma|_{\alpha\beta} - \eta_\gamma|_{\beta\alpha} = R^\rho{}_{\gamma\alpha\beta}\, \eta_\rho, \tag{2.55}$$

where

$$R^\rho{}_{\gamma\alpha\beta} = b_{\gamma\beta}\, b_\alpha^\rho - b_{\gamma\alpha}\, b_\beta^\rho \tag{2.56}$$

is the Riemann-Christoffel curvature tensor (Naghdi, 1972).

The left-hand side of (2.54) is the Eshelby tensor of shell theory

$$B_\beta^\alpha = \left(W+V\right)\delta_\beta^\alpha - N^{\alpha\gamma}\, u_\gamma|_\beta - M^{\alpha\gamma}\, \psi_\gamma|_\beta - Q^\alpha w_{,\beta}. \tag{2.57}$$

Thus, (2.54) is rewritten as

$$B^{\alpha}_{\beta}\big|_{\alpha} = R^{\rho}{}_{\gamma\alpha\beta}\left(N^{\alpha\gamma}\,u_{\rho} + M^{\alpha\gamma}\,\psi_{\rho}\right) + \frac{\partial(W+V)}{\partial x^{\beta}} \tag{2.58}$$

To find the condition for conservation laws, i. e., $B^{\alpha}_{\beta}\big|_{\alpha} = 0$, we recall that the Lagrangian of a shell made of an isotropic material has in general the following form

$$L = L\left(E, \upsilon, h, b^{\gamma}_{\alpha}, p^{j}; u_{\alpha}, u_{\alpha}\big|_{\gamma}, w, w_{,\gamma}\, \psi_{\alpha}, \psi_{\alpha}\big|_{\gamma}\right).$$

All parameters up to the semicolon can contribute to the last term in (2.58). The condition to be satisfied in order to ensure $\partial(W+V)/\partial x^{\beta} = 0$ are

$$E, \upsilon, h = const., \tag{2.59a}$$

$$p^{j}\big|_{\beta} = 0, \tag{2.59b}$$

$$b^{\gamma}_{\alpha}\big|_{\beta} = 0. \tag{2.59c}$$

This last condition is satisfied only for shells of constant curvature. In addition $R^{\rho}{}_{\gamma\alpha\beta}$ has to vanish, as seen from (2.58). In shell theory, this condition is satisfied only for developable surfaces. Thus, conservation laws

$$B^{\alpha}_{\beta}\big|_{\alpha} = 0 \tag{2.60}$$

will exist for cylindrical shells in the direction of the generators and for shells of revolution in the circumferential direction, provided the loading p^{j} is constant in the directions indicated and, of course, the constraint equations (2.59) are also satisfied.

For spherical shells under certain restrictions, further conservation laws can be constructed (cf. Lo, 1980 and Kienzler and Golebiewska-Herrmann, 1985).

3 Concluding Remarks

The rather surprising results of the analysis represented in this Chapter of the Lecture Notes is that the various theories of Strength-of-Materials, such as those of bars, shafts and beams can be employed not only for the calculation of deformations and stress, but also for the analysis of cracks. Thus, as it turns out, it is not necessary to rely exclusively on continuum mechanics, such as the theory of elasticity, but it is, under certain restrictive circumstances, sufficient to use Strength-of-Materials to obtain rather accurate data concerning fracture parameters, such as stress-intensity factors at crack tips.

Hence, the range of applicability, as is shown here, of Strength-of-Materials, as well as that of the more advanced theories of plates and shells is much broader than was expected to be the case until now. This completely novel vista on these theories opens up entirely new didactic possibilities.

It permits, already in undergraduate courses on Strength-of-Materials, to introduce the fundamental concepts of linear elastic fracture mechanics without that the students have any knowledge of the theory of elasticity. Thus, also practicing engineers involved in design and not pocessing any particular expertise in continuum mechanics of elastic solids, may in the future take good advantage of the approaches presented here.

References

Bažant, Z.P. (1990). Justification and improvement of Kienzler and Herrmann's estimate of stress intensity factors of cracked beam. *Engineering Fracture Mechanics* 36: 523-525.

Benthem, J.P., and Koiter, W. T. (1973). Asymptotc approximations to crack problems. In: Sih, G.C. (ed.), *Mechanics of Fracture I*. Noordhoff: Leyden. 131-178.

Eschenauer, H.,Olhoff, N., and Schnell, W. (1997). *Applied Structural Mechanics*. Springer: Berlin.

Eshelby, J.D. (1975). The calculus of energy release rates. In: Sih, G.C., van Elst, H. C., and Broek, D. (eds.), *Prospects of Fracture Mechanics*. Noordhoff: Leyden. 69-84.

Flügge, W. (1972). *Tensor Analysis and Continuum Mechanics*. Springer: Berlin.

Flügge, W. (1973). *Stresses in Shells*. Springer: New York.

Fung, Y.C. (1965). *Foundations of solid mechanics*. Prentice-Hall: Englewood Cliffs, N.J.

Gao, H., and Herrmann G. (1992). On estimates of stress intensity factors for cracked beams and pipes. *Engineering Fracture Mechanics* 41: 695-706.

Günther, W. (1962). Über einige Randintegrale der Elastomechanik. *Abhandlungen der Braunschweigischen Wissenschaftlichen Gesellschaft* 14: 53-72.

Herrmann, G., and Sosa H. (1986). On bars with cracks. *Engineering Fracture Mechanics* 24: 889-894.

Irwin, G.R. (1957). Analysis of stresses and strains near the end of a crack traversing a plate. *Journal of Applied Mechanics* 24: 361-364.

Kienzler, R. (1982). Eine Erweiterung der klassischen Schalentheorie; der Einfluß von Dickenverzerrungen und Querschnittsverwölbungen. *Ingenieur Archiv* 52: 311-322.

Kienzler, R. (1986). On existence and completeness of conservation laws associated with elementary beam theory. *International Journal of Solids and Structures* 22: 789-796.

Kienzler, R. (1993). *Konzepte der Bruchmechanik.* Braunschweig: Vieweg.

Kienzler, R., and Golebiewska-Herrmann, A. (1985). Material conservation laws in higher-order shell theories. *International Journal of Solids and Structures* 21: 1035-1045.

Kienzler, R., and Herrmann, G. (1986 a). On material forces in elementary beam theory. *Journal of Applied Mechanics* 53: 561-564.

Kienzler, R., and Herrmann, G. (1986 b). An elementary theory of defective beams. *Acta Mechanica* 62: 37-46.

Kienzler, R., and Herrmann, G. (2000). *Mechanics in Material Space.* Berlin: Springer.

Kirchhoff, G.R. (1850). Über das Gleichgewicht und die Bewegung einer elastischen Scheibe. *Crelles Journal für die reine und angewandte Mathematik.* 40: 51-88.

Knott, J. F. (1973). *Fundamentals of Fracture Mechanics.* London: Butterworths.

Koiter, W.T. (1960). A consistent first approximation in the general theory of thin elastic shells. In: Koiter, W.T. (ed.). *Proceedings of the Symposium on the Theory of Thin Elastic Shells.* Amsterdam: North-Holland. 12-33.

Krätzig, W.B. (1980). On the structure of consistent linear shell theories. In: Koiter, W.T., and Mikhailov, G.K. (eds.). *Proceedings of the 3rd IUTAM Symposium on Shell Theory.* Amsterdam: North-Holland. 353-368.

Li, S., and Shyy, W. (1997). On invariant integrals in the Marguerre - von Kármán shallow shell. *International Journal of Solids and Structures* 34: 2927-2944.

Lo, K.K. (1980). Path independent integrals for cylindrical shells and shells of revolution. *International Journal of Solids and Structures* 16: 701-707.

Mindlin, R.D. (1951). Influence of rotatory inertia and shear on flexural motions of isotropic elastic plates. *Journal of Applied Mechanics* 73: 31-38.

Müller, W.H., Herrmann, G., and Gao, H. (1993 a). A note on curved cracked beams. *International Journal of Solids and Structures* 30: 1527-1532.

Müller, W.H., Herrmann, G., and Gao, H. (1993 b). Elementary strength theory of cracked beams. *Theoretical and Applied Fracture Mechanics* 18: 163-177.

Naghdi, P.M. (1972). The Theory of Shells and Plates. In: Truesdell, C. (ed.). *Flügge's Handbuch der Physik.* Berlin: Springer. IVa/2: 425-640.

Nicholson, J.W., and Simmonds, J.G. (1980). Sanders' energy-release rate integral for arbitrarily loaded shallow shells and its asymptotic evaluation for circular cylinders. *Journal of Applied Mechanics* 47: 363-369.

Niordson, F.I. (1985). *Shell Theory.* Amsterdam: North-Holland.

Reissner, E. (1944). On the theory of bending of elastic plates. *Journal of Mathematics and Physics* 23: 184-191.

Reissner, E. (1945). The effect of transverse shear deformation on the bending of elastic plates. *Journal of Applied Mechanics* 12: A69-A77.

Reissner, E. (1947). On bending of elastic plates. *Quarterly of Applied Mathematics* 5: 55-68.

Rösel, R. (1986). Duality under dependency inversion and Noether theory for second-order Lagrangians. *International Journal of Solids and Structures* 22: 819-832.

Sanders, J.L. (1982). Circumferential through-cracks in cylindrical shells under tension. *Journal of Applied Mechanics* 49: 103-107.

Sanders, J.L. (1983). Circumferential through-crack in a cylindrical shell under combined bending and tension. *Journal of Applied Mechanics* 50: 221.

Sosa, H.A.(1986). On the analysis of bars, beams and plates with defects. Ph. D. Thesis, Stanford University.

Sosa, H.A., and Eischen J.W. (1986). Computation of stress intensity factors for plate bending via a path-independent integral. *Engineering Fracture Mechanics* 25: 451-462.

Sosa, H.A., and Herrmann, G. (1989). On invariant integrals in the analysis of cracked plates. *International Journal of Fracture* 40: 111-126.

Sosa, H.A., Rafalski, P., and Herrmann, G. (1988). Conservation laws in plate theories. *Ingenieur Archiv* 58: 305-320.

Timoshenko, S.P., and Woinowsky-Krieger, S. (1970). *Theory of Plates and Shells*. New York: McGraw-Hill 2nd ed.

Morphological Equilibrium and Kinetics of Two-Phase Materials

D. Gross

Institute of Mechanics, TU Darmstadt Hochschulstr. 1, D-64289 Darmstadt

1 Introduction

Many engineering and natural crystalline materials are microscopically not homogeneous but consist of two ore more phases with different material properties. The morphology, i.e. the geometric shape of the phases, can be regarded as a result of stresses and strains induced by their different crystalline structure and by the loading history. On account of diffusive mass transport, the morphology may change with time and eventually tend to a final equilibrium state. It is self-evident that in such multiphase materials the overall properties are found to depend not only on the properties of each phase, but also on the morphology of the microstructure and its evolution. As an example, Fig. 1a shows the lens-shaped microstructure of Mg-stabilized Zirconia (ZrO_2) formed by a tetragonal in a cubic phase. Similarly, a lamellar structure is found in a geomaterial consisting of a pigeonite in an augite phase, see Fig. 1b. In high temperature applications, such as turbine blades, Ni-base alloys are becoming more widely used. Microscopically they consist of Ni_3X precipitates, the so-called γ'-phase (where X stands for Al, Ti, Si etc.), embedded in a Ni matrix, the γ-phase. In Fig. 2 the microstructure of such materials is displayed for different heat treatment and loading histories. It can be seen that, depending on the specific parameters, the precipitates may be sphere or cube shaped, they may be oblate or prolate. Furthermore they align along the crystallographic axis and show the so-called *rafting* behaviour under elevated temperature and external loads. The binary system $Ni - Ni_3Al$ has been under careful investigation for several years. Materials science has described the material properties very thoroughly. It was observed that the microstructure of those materials is of great importance for their overall properties, see e.g. Hornbogen and Roth (1967), Ardell and Nicholson (1966), Ardell and Meshkinpour (1994). It is characteristic not only for Ni-base alloys that the crystallographic planes of the two phases are continuous. Such phase boundaries are called *coherent*. Beside these empirical aspects, progress has been made in understanding the mechanics and thermodynamics of two-phase materials, e.g. Leo and Sekerka (1989), Gurtin and Voorhees (1993), Gurtin (1995), Schmidt and Gross (1997). Herein the *configurational forces* play a central role. These more theoretical works lay the basis for understanding how the microstructures form and how different loading situations influence the morphology on the micro level. However, pure analytical investigations can only treat very simple phenomena or predict qualitative results, see Johnson and Cahn (1984), Kaganova and Roitburd (1988). Numerical investigation, on the other hand, are capable of overcoming these limitations. The different approaches which are presently in use can be divided into three different categories:

- Atomistic simulations: Interatomic energies are considered in a Boltzmann weighted chain of exchanges between matrix and particle atoms, see Lee (1995), Lee (1996a) and related works.

a) b)

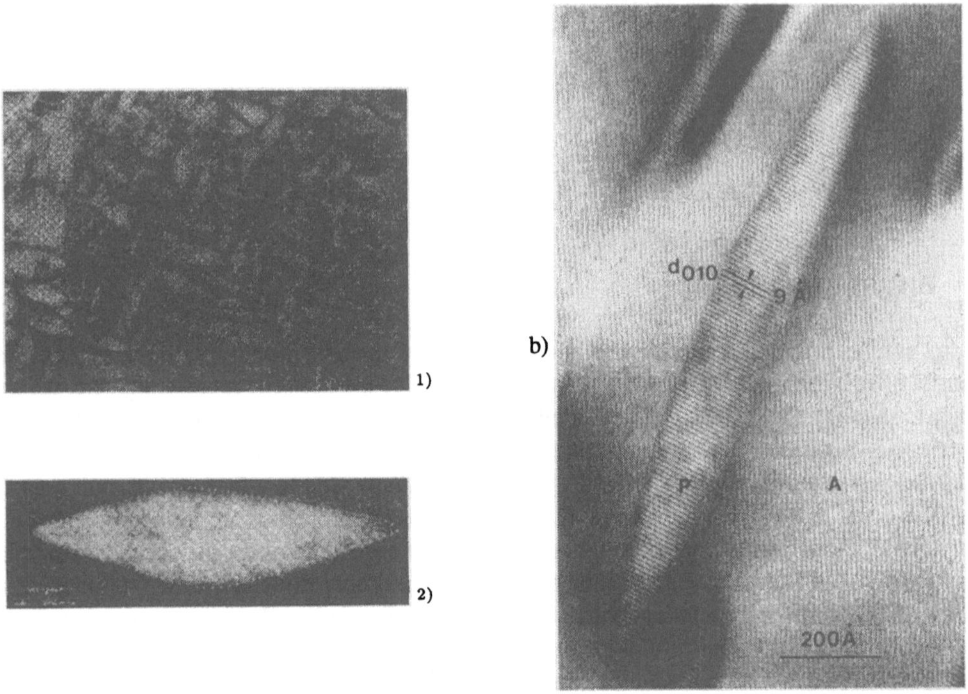

1) 2) 3)

[1] from Schoenlein et al. (1984)
[2] from Rühle and Heuer (1984)
[3] from W.F. Müller, G. Müller (personal communication)

Figure 1. a) Mg-ZrO_2 (tetragonal in cubic phase), b) microstructure in geomaterials (pigeonit in augit)

- Sharp interface models: Matrix and particle phase are separated by a sharp interface, see Voorhees et al. (1992), Thompson et al. (1993), Schmidt and Gross (1995), Schmidt and Gross (1997), Su and Voorhees (1996)
- Continuous models: An order parameter or a concentration is introduced to continuously describe properties of the matrix and the particle, see Fried and Gurtin (1993), Leo et al. (1998) and the literature cited in these works.

However, as the microstructural aspects of these materials and the relevant phenomena are rather complex, accurate models are needed for the computer simulation. In this survey, recent works regarding the first two models are reported. In section 2, a continuum model with a sharp interface is described which can be applied in 3D and in 2D. It is able to simulate the microstructure as accurately as possible, without restrictions to a plane problem, see Mueller and Gross (1998a), Mueller and Gross (1998b). It allows a detailed and accurate description of the local fields being relevant for equilibrium morphologies and their temporal evolution. Since the material properties of the two phases are in general anisotropic and the *interfacial energy* plays an important role, both effects have to be taken into account. A Boundary Element Method (BEM) capable of handling anisotropic materials efficiently is employed. It is then shown how the model can

Two-phase materials

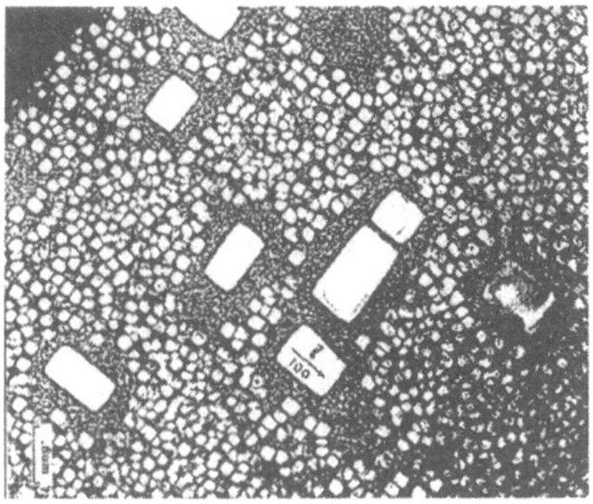

$Ni_3Al - Ni$[1)]

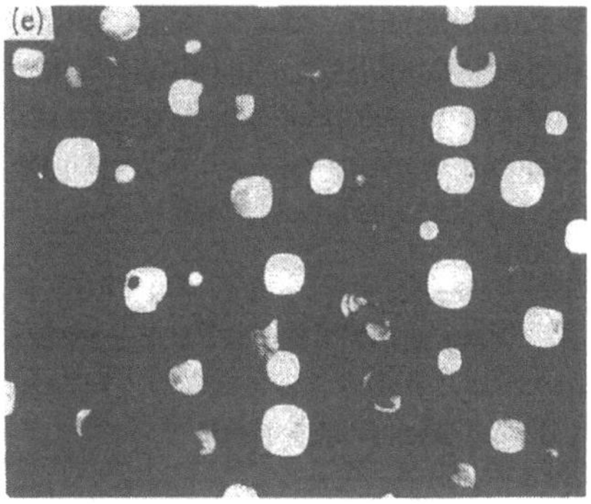

$Ni_3Si - Ni$[2)]

Heat treatment (Ostwald ripening)

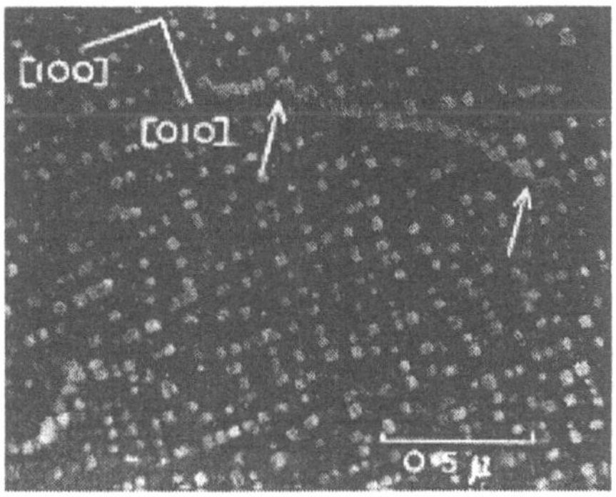

Aged for 8h at 750° C [3)]

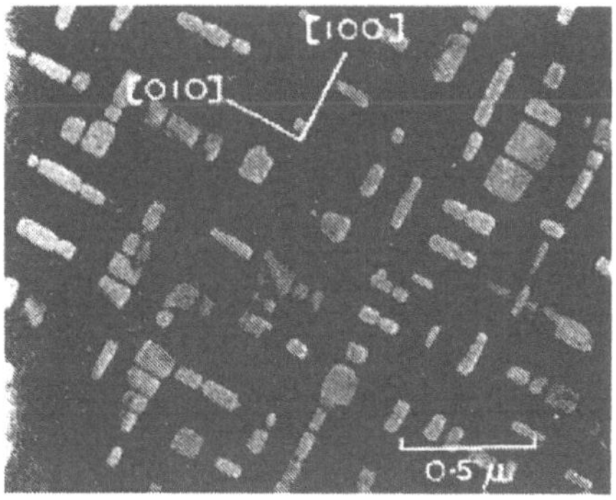

Aged for 96h at 750° C [3)]

Stress induced microstructures [4)]

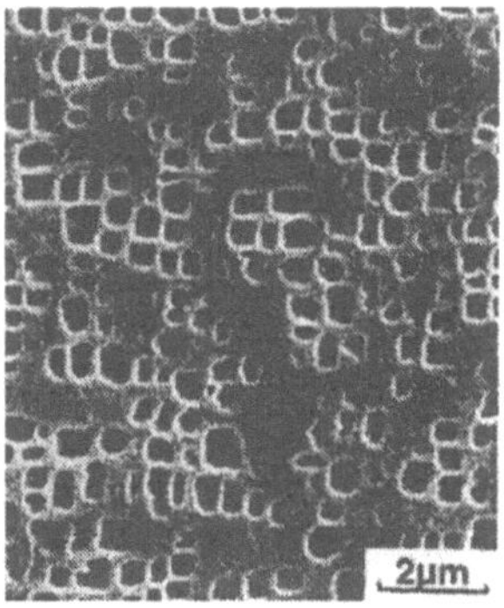

$\varepsilon = 0.0\%$[4)]

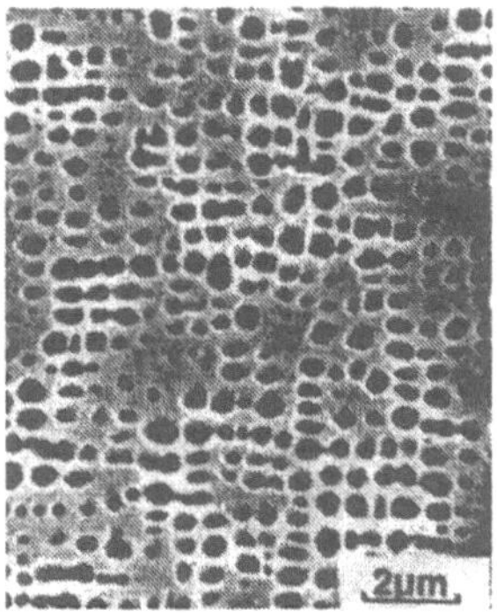

$\varepsilon = 0.5\%$[4)]

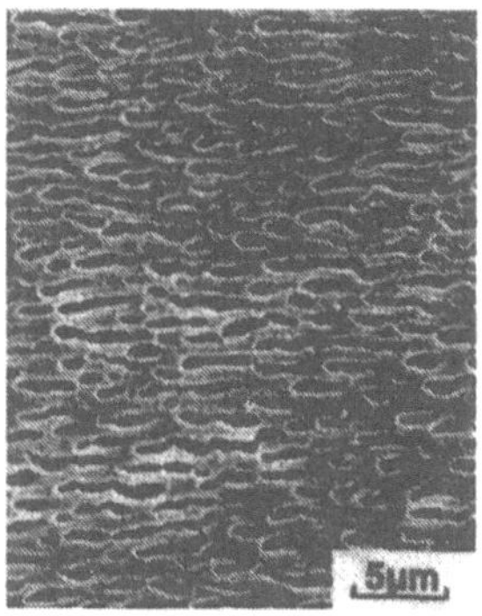

$\varepsilon = 5.0\%$[4)]

[1)] from Johnson et al. (1988)
[2)] from Ardell and Meshkinpour (1994)
[3)] from Ardell and Nicholson (1966)
[4)] from Wahi (1997)

Figure 2. Experimental Observations: Ni–base alloys

be simplified in order to describe the influence of morphology changes on the overall material behaviour within a constitutive law which can be implemented into a FEM program. In section 3, the use of atomistic simulations is discussed. Furthermore, to simulate the morphological diffusion process of typical microstructures with single or several randomly distributed misfitting inclusions, a computational technique in terms of a FE Monte Carlo simulation is presented.

2 The sharp interface model

2.1 Model description

Considered is a single particle $\mathcal{B}^{in}$ fully embedded in an infinite or finite matrix domain $\mathcal{B}^{out}$, see Fig. 3. In case of a finite matrix, the boundary $\partial\mathcal{B}^{out}$ is a closed surface. The two phases are

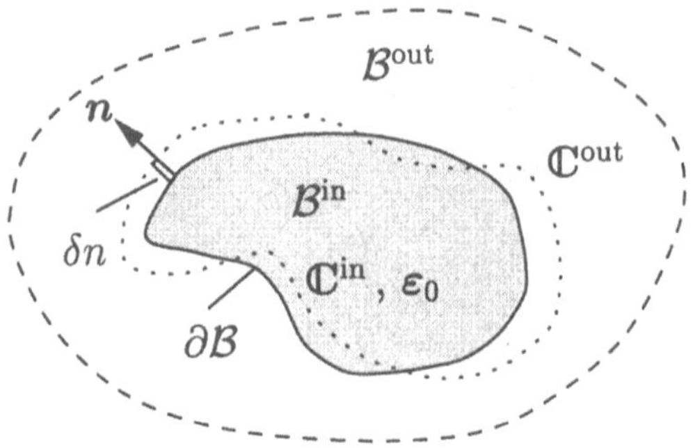

Figure 3. Sharp interface model

separated by a sharp interface $\partial\mathcal{B}$ with the normal vector $\boldsymbol{n}$ pointing into the matrix. The interface is a movable material surface with an interface energy density γ assigned to it. The particle and the matrix are linear elastic materials with the elasticity tensors $\mathbb{C}^{in/out}$ respectively. The particle material and matrix material can be different from each other and completely anisotropic. This is necessary to account for real material behaviour as encountered in e.g. fcc/bcc crystalline materials. The lattice misfit, which arises due to the different lattice parameters between the precipitate and the matrix phase, is taken into account by an inelastic eigenstrain tensor $\boldsymbol{\varepsilon}_0$, which is assumed constant in the particle. The shape of the particle is not restricted to a specific class of morphologies to allow for great generality.

We will recast the theory for small strains only to such an extend to make the present work self-contained. For a detailed discussion the reader is referred to the work of Schmidt (1997), where an extension to large deformations and finite elasticity is given.

2.2 Basic equations

Boundary value problem

We assume that the phase transition occurs at constant temperature and at speeds at which inertia effects can be neglected. Therefore mechanical equilibrium is satisfied during the entire process. In absence of body forces local equilibrium in the bulk material is given by

$$\operatorname{div} \boldsymbol{\sigma} = \mathbf{0}, \tag{1}$$

while on the interface $\partial\mathcal{B}$ the tractions are continuous:

$$[\![\boldsymbol{\sigma}]\!]\, \boldsymbol{n} = \mathbf{0}\,. \tag{2}$$

The stress tensor $\boldsymbol{\sigma}$ is related to the strains $\boldsymbol{\varepsilon}$ by Hooke's law

$$\boldsymbol{\sigma} = \begin{cases} \mathbb{C}^{in}\,(\boldsymbol{\varepsilon} - \boldsymbol{\varepsilon}_0) & \text{in } \mathcal{B}^{in} \\ \mathbb{C}^{out}\boldsymbol{\varepsilon} & \text{in } \mathcal{B}^{out} \end{cases}, \tag{3}$$

where the inelastic eigenstrains $\boldsymbol{\varepsilon}_0$ have been introduced in the precipitate phase. It describes the stress free deformation the precipitate would undergo in the absence of the matrix phase.
The interface $\partial\mathcal{B}$ is a coherent interface, thus the jump of the displacements $\boldsymbol{u}$ has to vanish across the interface $\partial\mathcal{B}$, preventing separation, penetration or slipping of the two domains:

$$[\![\boldsymbol{u}]\!] = \mathbf{0}, \quad \text{where} \quad [\![(\cdot)]\!] = (\cdot)^{out} - (\cdot)^{in}. \tag{4}$$

The complete setting of the problem requires the formulation of boundary conditions on the external boundary $\partial\mathcal{B}^{out}$ which may be at infinity. For prescribed tractions $\boldsymbol{t}$ or prescribed displacements $\boldsymbol{u}_0$ they read

$$\boldsymbol{\sigma n} = \boldsymbol{t} \quad \text{or} \quad \boldsymbol{u} = \boldsymbol{u}_0\,. \tag{5}$$

Energy balance and configurational force

The global energy balance in conjunction with the second law of thermodynamic for a control volume containing a closed interface and loaded at the boundary $\partial\mathcal{B}^{out}$ by a dead load $\boldsymbol{t} = \boldsymbol{\sigma n}$ can be written as

$$\underbrace{\frac{d}{dt}\left[\int_{\mathcal{B}} W\,dV + \int_{\partial\mathcal{B}} \gamma\,dA - \int_{\partial\mathcal{B}^{out}} \boldsymbol{t}\cdot\boldsymbol{u}\,dA\right]}_{\Pi} = -\int_{\partial\mathcal{B}} \tau_n v_n\,dA \leq 0\,. \tag{6}$$

Here W is the strain energy, Π the total potential and v_n the normal interface velocity. The interfacial energy contribution is assumed to be given by an constant density γ on the interface $\partial\mathcal{B}$. This term takes the bonding energy between the different atoms of the two phases into account. Together with the strain energy term it forms the Helmholtz free energy. The quantity

$$\tau_{\mathrm{n}} = \boldsymbol{n}\cdot[\![\boldsymbol{P}]\!]\boldsymbol{n} + \gamma K \tag{7}$$

is the *driving force* or total *configurational force* acting on the interface. Herein the first part takes the change of the elastic energy into account. This part is given by the jump in the normal component of Eshelby's energy-momentum tensor, see Eshelby (1970) or Eshelby's (configurational) "stress tensor"

$$\boldsymbol{P} = W\mathbf{1} - (\nabla\boldsymbol{u})^{\mathrm{T}}\boldsymbol{\sigma}. \tag{8}$$

The second part in (7) is given by twice the mean curvature K, considering creation or annihilation of interface area.

It shall be mentioned that in the considered inclusion problem the generalized driving force τ_n can be expressed by the inner limit of the strains ε at the interface $\partial\mathcal{B}$ (for details the reader is referred to Schmidt and Gross (1995), Schmidt and Gross (1997))

$$\begin{aligned}\tau_{\rm n} = &-\tfrac{1}{2}\left[\left(\mathbb{C}^{in}\varepsilon^0 + [\![\mathbb{C}]\!]\varepsilon\right)\boldsymbol{n}\right]\cdot\boldsymbol{\Omega}^{-1}\left[\left(\mathbb{C}^{in}\varepsilon^0 + [\![\mathbb{C}]\!]\varepsilon\right)\boldsymbol{n}\right]\\ &+\tfrac{1}{2}\varepsilon : \mathbb{C}^{out}\varepsilon - \tfrac{1}{2}\left(\varepsilon - \varepsilon^0\right) : \mathbb{C}^{in}\left(\varepsilon - \varepsilon^0\right) + \gamma K,\end{aligned} \tag{9}$$

where the components of the acoustic-tensor $\boldsymbol{\Omega}$ of the matrix material are given by

$$\Omega_{ik} = C^{\rm out}_{ijkl} n_j n_l . \tag{10}$$

The computation of $\tau_{\rm n}$ requires the solution of the inhomogeneous inclusion problem. In section 2.4 the solution procedure with a Boundary Element Method is explained. It shall be emphasized, that the computation of the driving force on the interface is independent on the specific arrangement of the particles, because only local quantities are involved. The effect of particle interaction e.g. in a two-particle system or in a periodic arrangement enters the calculation of the driving force only through the solution of the elastic field problem, namely through the value of the strains ε.

Kinetic law and morphological equilibrium

From (6) it can be seen that the interface velocity v_n is such that the power of τ_n, i.e. the *dissipation rate*, is non-negative. This condition is fulfilled by assuming an evolution law that relates the normal interface velocity v_n to the driving force τ_n by the linear relation

$$v_n = \alpha(\tau_n + \beta) \tag{11}$$

where α and β are material parameters. The postulated constitutive relation (11) can also be derived by considering the diffusion process in more detail. Using the Gibbs-Thompson equation as given in Voorhees et al. (1992) together with an simplified diffusion model yields a functional dependence as shown in (11).

Furthermore, it can be deduced from (6) that Π takes a minimum when the process has come to an end and the interface does not move. Therefore, the conditions for a stable configurational equilibrium read

$$\delta\Pi = 0\,, \qquad \delta^2\Pi > 0 \tag{12}$$

where the variational symbol denotes a variation of Π on account of an arbitrary admissible displacement of the interface. Alternatively, the necessary condition for equilibrium may be formulated as generalized virtual work principle:

$$\delta\Pi = -\int_{\partial\mathcal{B}} \tau_n \delta n \, dA = 0\,. \tag{13}$$

Here δn describes an infinitesimal morphological displacement of the boundary $\partial\mathcal{B}$ in normal direction.

Mass conservation dictates that the volume of each phase remains constant in the reference configuration, or in other words configurational changes do not affect the total number of atoms of one phase:

$$V_{\mathcal{B}^{\text{in}}} = V_0 \,. \tag{14}$$

Note that elastic volume changes may still happen. In order to incorporate this constraint, an augmented Lagrange function

$$\tilde{\Pi} = \Pi + \lambda(V_{\mathcal{B}^{\text{in}}} - V_0) \tag{15}$$

is defined with the help of the Lagrange multiplier λ. The necessary condition for an equilibrium shape then takes the form

$$\delta\tilde{\Pi} = -\int_{\partial\mathcal{B}} (\tau_{\text{n}} - \lambda)\delta n \, dA + \delta\lambda(V_{\mathcal{B}^{\text{in}}} - V_0) = 0 \,. \tag{16}$$

As the two variations in (16) are arbitrary, the necessary - but not sufficient - conditions for an equilibrium shape are

$$\begin{aligned} \tau_{\text{n}} &= \lambda \qquad \text{on} \quad \partial\mathcal{B}, \\ V_{\mathcal{B}^{\text{in}}} &= V_0 . \end{aligned} \tag{17}$$

Configurational equilibrium therefore requires that the driving force τ_{n} has to be constant along the interface $\partial\mathcal{B}$. Since the Lagrange multiplier λ is only related to the particle volume it is equivalent to prescribe the constant value τ_{n} instead of the particle volume. The method for determining the stability of a solution is described later in the section on shape discretization.

Interaction and periodically arranged particles

In many cases particle interaction cannot be neglected. This interaction may influence the morphology considerably, see Schmidt et al. (1998). It also may give rise to an overall repulsion or attraction of neighboring precipitates resulting possibly in particle coalescence. Such effects can be studied at 2-particle configurations and by periodically arranged particles.

If two interacting particles are considered, the equations of the forgoing sections are still valid. Of course, on account of the two particles, there are now two particle surfaces $\partial\mathcal{B}_1$, $\partial\mathcal{B}_2$ and the constant volume side condition now reads

$$V_{\mathcal{B}_1^{\text{in}}} + V_{\mathcal{B}_2^{\text{in}}} = V_0 \,. \tag{18}$$

The periodic cell arrangement is chosen for two reasons. Firstly due to its physical relevance, as many Ni-base alloys display a rather regular periodic microstructure, secondly for its efficient implementation of 3D interaction effects. In this context, it should be mentioned, that with the computational power of usually available workstations a complex structure of more than two non-periodicly interacting 3D particles presently is not computable. On the other hand, the simulation of a periodic arrangement predicts the effect of the volume fraction of the precipitate phase on the micro morphology. This is generally believed to be an important factor, see Ardell and Rastogi (1971), Ardell and Maheshwari (1992), Ardell and Meshkinpour (1994) together with the particle size and the external loading situation. Again, no important changes appear in the periodic cell model formulation compared with the single particle situation, see Fig.4. Now at the exterior boundary of the cubic cell the displacements $\boldsymbol{u}$ and tractions $\boldsymbol{t}$ only have to satisfy periodicity

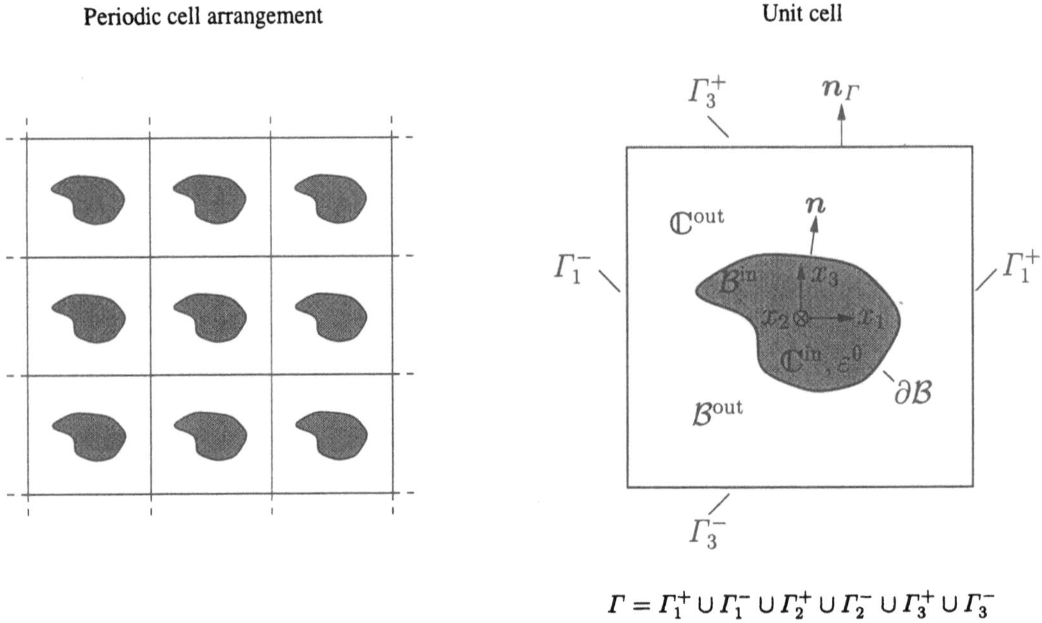

Figure 4. Periodic arrangement of unit cells with precipitate

conditions in order to ensure periodic filling of the entire material space. The cell boundary is denoted by Γ with its associated outward unit vector given by $\boldsymbol{n}_\Gamma$. The displacements $\boldsymbol{u}$ on two opposite cell boundaries $\Gamma_i^\pm$ satisfy

$$\boldsymbol{u}|_{\Gamma_i^+} - \boldsymbol{u}|_{\Gamma_i^-} = \boldsymbol{v}_i \quad \text{for } i = 1, 2, 3, \tag{19}$$

where the three vectors $\boldsymbol{v}_i$ are still undetermined translations. The periodicity conditions for the tractions are given by

$$\boldsymbol{t}|_{\Gamma_i^+} + \boldsymbol{t}|_{\Gamma_i^-} = \boldsymbol{0} \quad \text{for } i = 1, 2, 3. \tag{20}$$

In the absence of external loads the mechanical equilibrium of the periodic cell arrangement leads to three vector equations

$$\int_{\Gamma_i^+} \boldsymbol{t}\, dA = \boldsymbol{0} \quad \text{for } i = 1, 2, 3 \tag{21}$$

to determine the unknown translations $\boldsymbol{v}_i$. Equation (21) needs to be modified in the case of an external load σ^∞. However with the use of an equivalent eigenstrain

$$\varepsilon^0_{\text{eq}} = \varepsilon^0 - [\![\mathbb{C}^{-1}]\!]\sigma^\infty \tag{22}$$

instead of the physical eigenstrains ε^0 equations (3), (4), (19) - (21) can still be used. A detailed discussion of equivalent eigenstrain is given in Mura (1987), Schmidt and Gross (1999) and in section 2.6.

2.3 Shape and time discretization

Shape equilibrium

It is useful to start the numerical treatment of the shape equilibrium problem from equation (16) and recast the variational problem in terms of a shape optimization problem. Describing the interface by a set of M shape parameters $q_1, q_2, \ldots, q_M$ the normal variation is given by:

$$\delta n = \sum_{i=1}^{M} \frac{\partial \boldsymbol{x}}{\partial q_i} \cdot \boldsymbol{n}\, \delta q_i. \tag{23}$$

Inserting (23) in (16) yields $M+1$ nonlinear equations

$$\begin{aligned}
F_1(q_1, \ldots, q_M, \lambda) &= -\int_{\partial\mathcal{B}} (\tau_n - \lambda) \frac{\partial \boldsymbol{x}}{\partial q_1} \cdot \boldsymbol{n}\, dA = 0 \\
&\vdots \\
F_M(q_1, \ldots, q_M, \lambda) &= -\int_{\partial\mathcal{B}} (\tau_n - \lambda) \frac{\partial \boldsymbol{x}}{\partial q_M} \cdot \boldsymbol{n}\, dA = 0 \\
F_{M+1}(q_1, \ldots, q_M, \lambda) &= V_{\mathcal{B}^{\text{in}}} - V_0 = 0
\end{aligned} \tag{24}$$

for the M shape parameters q_i and for the Lagrange multiplier λ, as the variations of the shape parameters δq_i and the Lagrange multiplier $\delta\lambda$ are arbitrary. Here the terms $F_1, \ldots, F_M$ are the work conjugated quantities of the the respective shape parameters $q_1, \ldots, q_M$. To solve this non-linear set of equations, several techniques are available. Unfortunately the efficient Newton-method is not applicable, because the Hessian matrix

$$H_{ij} = \frac{\partial F_i}{\partial q_j} \tag{25}$$

is not easy to compute. Therefore techniques, that iteratively update the Hessian or its inverse, have been implemented. Among the classical methods, Broyden, DFP and BFGS (see Luenberger (1984)), the BFGS-method has proven to be a stable and efficient algorithm for the considered problem and has therefore been used in the calculations. The general setting is equivalent to that of a shape optimization with a geometrical constraint, i.e. constant particle volume. In the terminology of optimization the total potential of the system is the target function.

Stability

The positive or negative definiteness of the Hessian matrix H_{ij} determines the stability of the morphology solution. Since the Hessian matrix is only known approximately, a precise judgment is not possible. In addition, the numerical evaluation of the Hessian is very time-consuming. For this reason stability alternatively can be investigated in a simplified way, by considering solely one second derivative of the total energy Π. It was shown by Johnson and Cahn (1984), Johnson et al. (1988), that in 3D a shape instability manifests in a bifurcation from equi-axed to prolate or oblate shapes. Thus a shape parameter ρ is defined as

$$\rho = \left(\frac{a}{b}\right)^3, \tag{26}$$

where a and b are the principal axis of a shape. The stability of a equi-axed equilibrium shape ($\rho = 1$) with respect to ρ is then judged by the sign of the second derivative:

$$\left.\frac{d^2\Pi}{d\rho^2}\right|_{\rho=1} > 0 \quad \rightarrow \quad \text{stable.} \tag{27}$$

Shape evolution

The solution of the kinetic problem of the temporal evolution of the interface position is found by integrating equation (11) numerically in time. The time step and the time discretization respectively has to be chosen sufficiently small to ensure convergence. Since the parameter β is related to the particle size, β has to be chosen such that the precipitate size remains constant, i.e. that there is no contribution in the velocity v_{n}, that would change the precipitate volume.

As the kinetic system approaches an equilibrium configuration with increasing time, the equilibrium shapes may also be found as a limit case of the temporal evolution. It shall be emphasized here, that there is a physical time (timestep) only if the temporal evolution is considered. In the calculation of equilibrium shapes by using (24) the equilibrium iterations can not be related to any physical time.

2.4 Boundary Integral Equation Method

The solution of the inhomogeneous inclusion problem can be found by applying the Boundary Integral Equation of elasto statics, which is discretized with Boundary Elements. It is a natural choice for the boundary variations considered here. The advantage of this method is, that the unknowns are shifted from the interior of a considered domain to its boundary, which reduces the problem dimension and size. The standard integral equation for linear elastic problems follows from Brebbia and Telles (1985) to be

$$\boldsymbol{K}(\boldsymbol{x})\boldsymbol{u}(\boldsymbol{x}) + \int_A \boldsymbol{T}(\boldsymbol{x}-\boldsymbol{y})\boldsymbol{u}(\boldsymbol{y})\,dA = \int_A \boldsymbol{U}(\boldsymbol{x}-\boldsymbol{y})\boldsymbol{t}(\boldsymbol{y})\,dA, \tag{28}$$

with $\boldsymbol{x}$ representing the source point and $\boldsymbol{y}$ the observation point as well as $\boldsymbol{K} = \frac{1}{2}\mathbf{1}$ for a smooth surface. The fundamental solutions for the tractions and the displacements are expressed through $\boldsymbol{T}$ and $\boldsymbol{U}$ respectively.

Anisotropic fundamental solution

Since the lattice-structure of the different phases in general shows anisotropic behaviour, the analytic fundamental solution for the isotropic case, known as Kelvin-problem, cannot be applied. A method to calculate these quantities numerically is proposed by Schclar (1994) and incorporated to handle anisotropic behaviour. According to Schclar (1994), the fundamental solution for the displacements can be expressed as

$$\boldsymbol{U} = \frac{1}{8\pi^2|\boldsymbol{x}-\boldsymbol{y}|}\,\boldsymbol{G}(\theta_1,\theta_2) \tag{29}$$

with the kernel

$$\boldsymbol{G}(\theta_1,\theta_2) = \int_0^{2\pi} \boldsymbol{M}^{-1}(\boldsymbol{z}(\varphi))\,\varphi \tag{30}$$

and the argument of the integral

$$M_{ik} = \mathbb{C}_{ijkl} z_j z_l \qquad |z| = 1\,. \tag{31}$$

Fig. 5 provides a representation of the geometric quantities in equations (29)-(31). The function $G(\theta_1, \theta_2)$ depends on the orientation of the vector $y - x$ but not on its length. Since G is a smooth

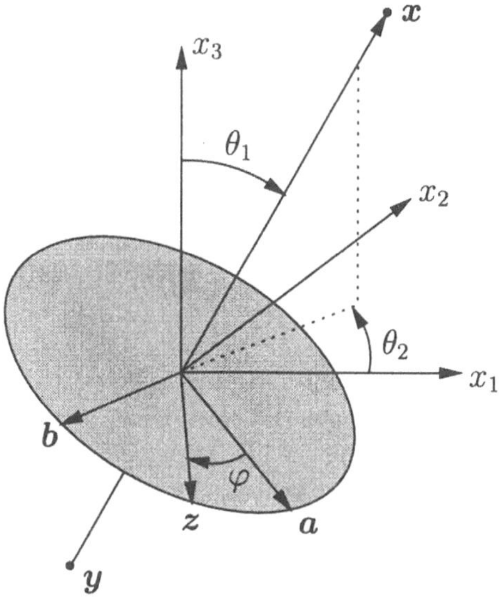

Figure 5. Anisotropic fundamental solution

function, discrete values for the domain $0 \leq \theta_1 \leq \pi$ and $0 \leq \theta_2 \leq 2\pi$ can be calculated. Instead of performing the integration in equation (30) for every pair (x, y) appearing in the problem, quadratic interpolation between the discrete values for the angles θ_1 and θ_2 is utilized. This approach increases the computational speed when integrating the Boundary Integral equations. A similar procedure can be used to find the fundamental solution for the tractions and is described in Schclar (1994).

Boundary Integral equations

Applying the boundary element method to the two-phase problem leads to two distinct sets of equations, one for the matrix and one for the particle-phase. Incorporating the jump equations (2) and (4), the displacement- and the traction-vector are defined as

$$u^{\text{in}} = u^{\text{out}} = u \qquad \text{on} \quad \partial\mathcal{B} \tag{32}$$

and

$$-t^{\text{out}} = t^{\text{in}} = t \qquad \text{on} \quad \partial\mathcal{B}\,. \tag{33}$$

In the following we consider only the case of periodically arranged unit cells since the other cases simply can be constructed by specialization. Here the general procedure to develop the boundary integral equations proposed by Brebbia and Telles (1985) yields three integral equations, two for the matrix boundaries $\partial\mathcal{B}$ and Γ and one for the particle boundary $\partial\mathcal{B}$:

$$\begin{aligned} \boldsymbol{K}^{\rm out}\,\boldsymbol{u} - \int\limits_{\partial\mathcal{B}} \boldsymbol{T}^{\rm out}\,\boldsymbol{u}\,dA + \int\limits_{\Gamma} \boldsymbol{T}^{\rm out}\,\boldsymbol{u}^{\Gamma}\,dA & \\ + \int\limits_{\partial\mathcal{B}} \boldsymbol{U}^{\rm out}\,\boldsymbol{t}\,dA - \int\limits_{\Gamma} \boldsymbol{U}^{\rm out}\,\boldsymbol{t}^{\Gamma}\,dA &= 0 \qquad \text{on} \quad \partial\mathcal{B} \end{aligned} \tag{34}$$

$$\begin{aligned} \boldsymbol{K}^{\Gamma}\,\boldsymbol{u}^{\Gamma} - \int\limits_{\partial\mathcal{B}} \boldsymbol{T}^{\rm out}\,\boldsymbol{u}\,dA + \int\limits_{\Gamma} \boldsymbol{T}^{\rm out}\,\boldsymbol{u}^{\Gamma}\,dA & \\ + \int\limits_{\partial\mathcal{B}} \boldsymbol{U}^{\rm out}\,\boldsymbol{t}\,dA - \int\limits_{\Gamma} \boldsymbol{U}^{\rm out}\,\boldsymbol{t}^{\Gamma}\,dA &= 0 \qquad \text{on} \quad \Gamma \end{aligned} \tag{35}$$

$$\boldsymbol{K}^{\rm in}\,\boldsymbol{u} + \int\limits_{\partial\mathcal{B}} \boldsymbol{T}^{\rm in}\,\boldsymbol{u}\,dA - \int\limits_{\partial\mathcal{B}} \boldsymbol{U}^{\rm in}\,\boldsymbol{t}\,dA = \boldsymbol{K}^{\rm in}\,\boldsymbol{u}^{0} + \int\limits_{\partial\mathcal{B}} \boldsymbol{T}^{\rm in}\,\boldsymbol{u}^{0}\,dA \qquad \text{on} \quad \partial\mathcal{B} \tag{36}$$

To achieve a complete system of equations the periodicity conditions (19), (20) as well as the traction boundary condition for the unit cell (21) have to be considered.

The integral equations are then discretized using 4-node elements with a bilinear interpolation of displacements, tractions and geometry. The discretization leads to a non-symmetric linear system of equations for the displacements and tractions on the interface and on the cell boundary. In the calculation of equilibrium shapes a discretization with 216 elements (218 nodes) for the cell boundary Γ and 864 elements (866 nodes) for the particle boundary $\partial\mathcal{B}$ has been used. This system has to be solved repeatedly with a changed particle geometry in the optimization process. About 25-50 iterations are needed in the quasi-Newton process to obtain convergence. For simulation of the temporal evolution the order of 1000 calculations of the field equations are necessary.

2.5 Some Results

Nondimensionalization

For the numerical implementation it is useful to scale the physical quantities in an appropriate way. According to Mueller and Gross (1998b) the following positive scalars are defined for the 3D case (and similarly in the 2D case):

$$l^* = \sqrt[3]{V_B} \tag{37}$$

$$\varepsilon^* = \max_{i=1,2,3} |\varepsilon_i^0| \tag{38}$$

$$\begin{aligned} \mu^* = \frac{1}{15}[c_{11}^{\rm out} + c_{22}^{\rm out} + c_{33}^{\rm out} - c_{12}^{\rm out} - c_{23}^{\rm out} - c_{13}^{\rm out} & \\ +3(c_{44}^{\rm out} + c_{55}^{\rm out} + c_{66}^{\rm out})] & \end{aligned} \tag{39}$$

The characteristic length of the particle l^* is defined by the particle volume and ε^* describes the maximal absolute value of the eigenvalues of the eigenstrains tensor. The effective shear modulus μ^* characterizes the average stiffness of the anisotropic matrix material, here given by its Voigt constants c_{ij}. Following the procedure in Schmidt and Gross (1997), a dimensionless formulation for all quantities of the system can be derived in conjunction with the dimensionless particle size

$$L = \frac{\mu^* \varepsilon^{*2} l^*}{\gamma} . \tag{40}$$

Since L can be regarded as the ratio of the systems elastic strain energy to the interfacial energy, a simple physical interpretation is possible. If L becomes large the particle shape is determined solely by the minimum of the elastic strain energy (strain energy dominated case). In contrast, for a small value of L the influence of interfacial energy is dominating, which leads to a sphere as an equilibrium morphology.

In order to describe the kinetic problem a dimensionless time T is introduced as

$$T = \alpha \frac{t}{l^*} \left(\mu^* \varepsilon^{*2} + \frac{\gamma}{l^*} \right) . \tag{41}$$

Material parameters

Although many different and practically relevant material parameters can be simulated, only the isotropic case and the case of a cubic anisotropy (orthotropy) are considered here. The last one models the situation in Ni-base alloys. The lattice misfit in most examples is described by positive eigenstrains

$$\varepsilon^0 = \varepsilon^* \mathbf{1} \qquad \text{with} \quad \varepsilon^* = 0.001 . \tag{42}$$

The elastic constants of Nickel in Voigt notation are:

$$\mathbb{C}^{in} = \mathbb{C}^{out} = \begin{pmatrix} 2.465 & 1.473 & 1.473 & 0 & 0 & 0 \\ 1.473 & 2.465 & 1.473 & 0 & 0 & 0 \\ 1.473 & 1.473 & 2.465 & 0 & 0 & 0 \\ 0 & 0 & 0 & 1.247 & 0 & 0 \\ 0 & 0 & 0 & 0 & 1.247 & 0 \\ 0 & 0 & 0 & 0 & 0 & 1.247 \end{pmatrix} \cdot 10^5 \, \frac{\mathrm{N}}{\mathrm{mm}^2} . \tag{43}$$

Single Particle Results in 2D

For a better understanding some basic results for the case of a single inclusion in 2D are discussed. For small L, the interfacial energy prevails and the equilibrium shape is circle-like. With increasing L, the elastic energy becomes more important and the shapes become more square-like, their square sides being oriented along the [001] and [010] direction. At a critical value of L, L_{crit}, these fourfold symmetric shapes become unstable: the stable equilibrium shapes in that regime assume a twofold symmetry and are elongated in the [001] or [010] direction. The situation is illustrated in Fig. 6 which shows the typical diagram of a symmetry breaking bifurcation in terms of the geometric quantity $\bar{\rho} = (\rho - 1)/(\rho + 1)$, where ρ is the aspect ratio of the shape. This shape instability is caused by a redistribution of the energies in matrix, inclusion and interface: the increase of interfacial energy due to the enlarged interface of the elongated shape and the increase of the elastic strain energy in the inclusion is compensated by the saving of elastic strain energy in the matrix. The critical value L_{crit} depends sensitively on the material parameters of both matrix and inclusion. If the symmetry of the system is disturbed by an

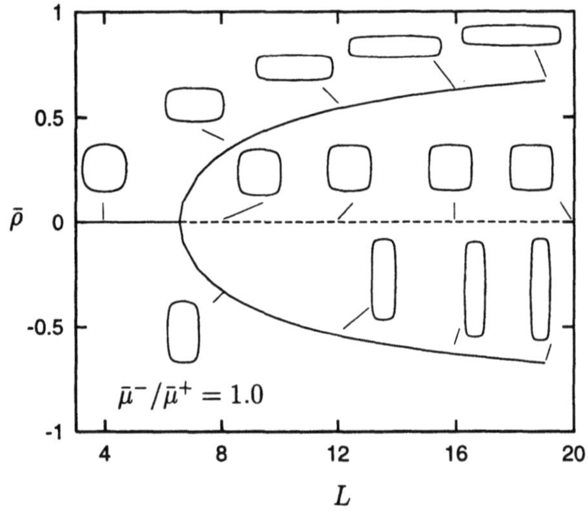

Figure 6. Bifurcation diagram for the single particle case: aspect ratio vs. particle size.

external uniaxial load or a non-isotropic eigenstrain (these two cases are equivalent and can be replaced by one another), the bifurcation diagram assumes the form shown in Fig. 7. One family of solutions exists throughout the whole range of L-values. The respective particle shapes are elongated in the direction of the smaller of the two principal values of the eigenstrain. These shapes represent stable solutions. Two other families of solutions exist above a certain value of L, both being elongated in the perpendicular direction. The upper branch represents unstable-the lower branch stable solutions.

Two interacting Particles in 2D

In the case of a two (or multi-) particle system, each particle experiences a nonuniform "external stressfield" which is induced by the misfit of the other particle. Therefore the results can be expected to be qualitatively similar to the diagram in Fig. 7. Indeed, we find the three different families of shapes described before. The aspect ratio of these shapes is plotted against L in Fig. 8 for the homogeneous case, together with the corresponding shapes. With increasing L, the shapes of the first family become more elongated and move closer together. Above $L \approx 18$ there exist two more solution families. The stability analysis shows that in the homogeneous case all three solution types are unstable in the considered L-interval. This is different for the case of a stiffer inclusion phase where only a single (stable) branch exists, see Fig. 9. While for stiffer inclusions solutions of the second and third type do not exist in the considered L-interval, for soft inclusions

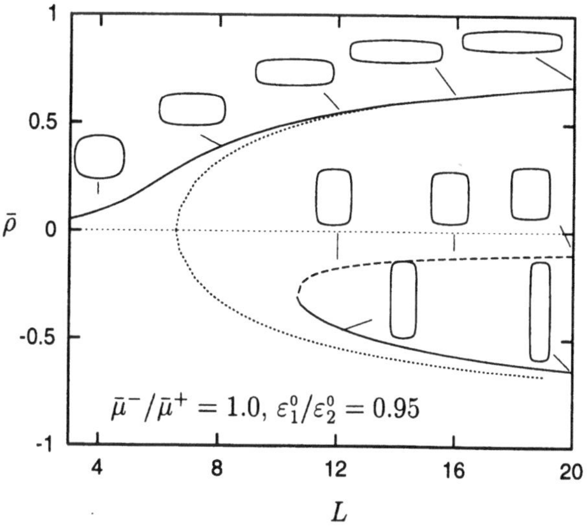

Figure 7. Disturbed bifurcation diagram for tetragonal eigenstrain.

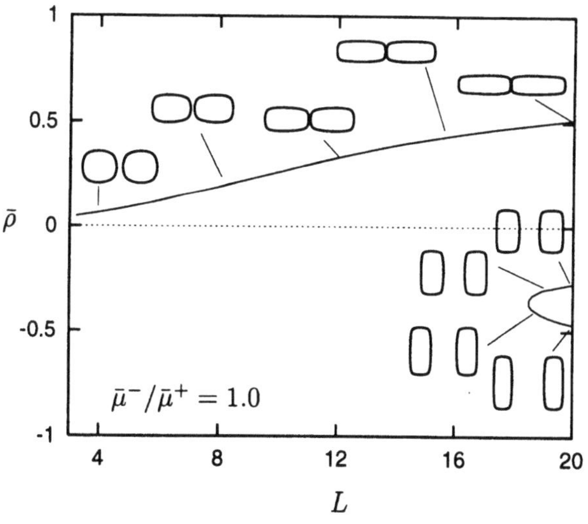

Figure 8. Aspect ratio vs. particle size: Three solution families for the homogeneous case.

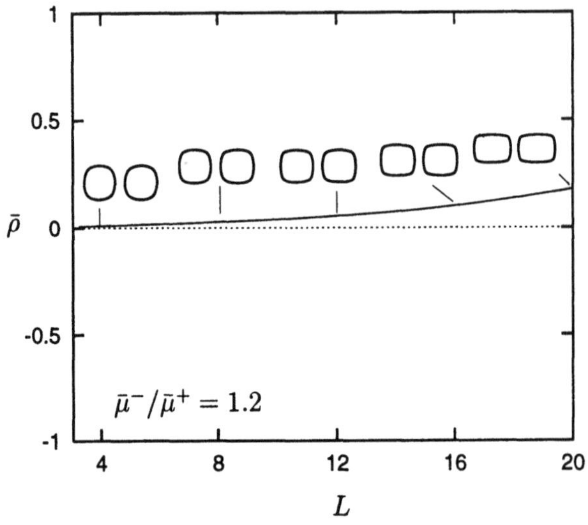

Figure 9. Aspect ratio vs. particle size; Inclusion stiffer than matrix.

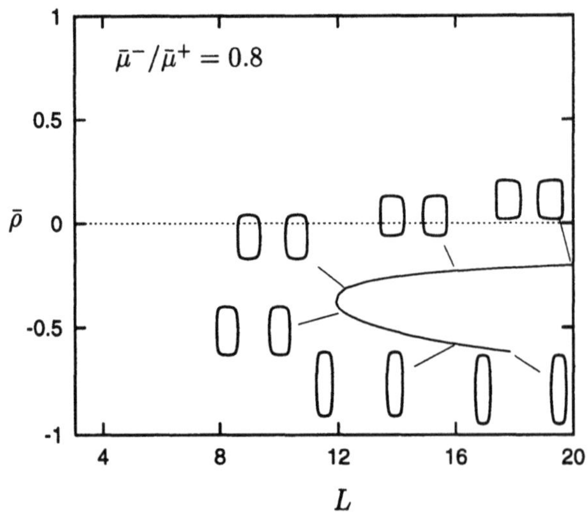

Figure 10. Aspect ratio vs. particle size: Matrix stiffer than inclusion.

these solutions occur at smaller L, while the solutions of the first type do not exist for any value of L (Fig. 10). This is because the particles have the tendency to coalesce. Here, the stability analysis reveals, that all the solutions shown in Fig. 10 are unstable (see also section 3.4.3).

It shall be mentioned that the specific results depend more or less on several more parameters like the "strength" of the anisotropy, its difference between inclusion and matrix or Poisson's ratio. This shall not be discussed in detail.

Single particle in 3D

First, the qualitative different equilibrium shapes that were found with the 3D method shall be discussed. These types of equilibrium shapes are frequently observed in experiments. The simulation revealed three classes of equilibrium shapes:

(1) Cuboids with a convex interface everywhere
(2) Cuboids with a partly concave interface
(3) Prolate or oblate cuboids.

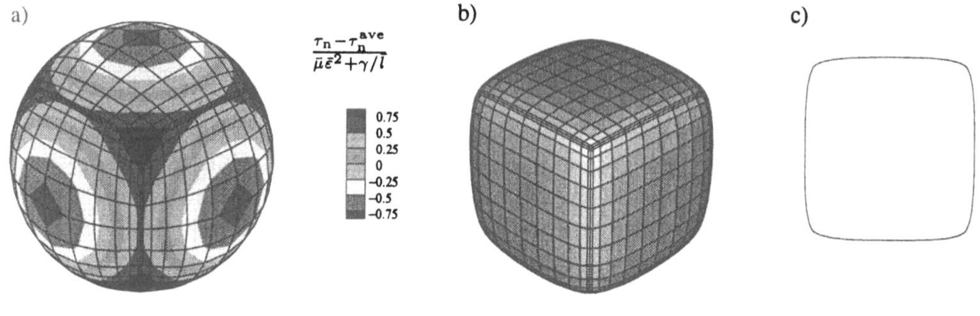

Figure 11. Homogeneous precipitate at size $L = 9$: a) distribution of τ_n for a sphere (in the corner region $\tau_n - \tau_n^{ave}$ is positive), b) equilibrium shape and c) cross-section

It was found that the formation of these microstructures can be explained by basically three important factors, namely the particle size L, the stiffness ratio $\bar{\mu}^{in}/\bar{\mu}^{out}$ and the loading.
Particles of type (2.5) were found by considering a homogeneous set of material parameters of precipitates that are stiffer than the surrounding matrix. Fig. 11 shows the distribution of the generalized driving force τ_n on a homogeneous spherical particle with a dimensionless particle size $L = 9$, together with the equilibrium shape and a cross-section. Fig. 12 shows the equilibrium shapes for different particle sizes. For small precipitates ($L = 1$) the shape is determined by the now dominating interfacial energy. Due to the isotropic interfacial energy density the equilibrium shape becomes almost a sphere. In this morphology the elastic energy contributes only 40% to the total energy. At $L = 5$ the precipitate has grown, meaning that the effect of the elastic energy on the equilibrium shape has increased. In this configuration the elastic strain energy contributes

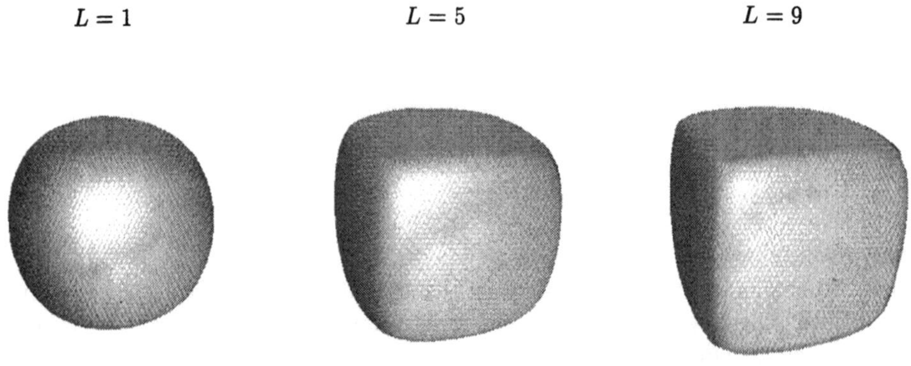

Figure 12. Size effect on equilibrium shape of an homogeneous precipitate in Nickel

76% to the total energy. Due to the orthotropic elastic properties of nickel the equilibrium shape becomes an intermediate configuration between a cube and a sphere. At $L = 9$ the elastic energy dominates (84%) the total energy and the equilibrium shape becomes cube-like.

To obtain equilibrium shapes of type (2.5), the inclusion material was assumed to be softer than the matrix with a shear modulus ratio of $\bar{\mu}^{in}/\bar{\mu}^{out} = 0.7$. The anisotropy was not changed. For two different particle sizes the equilibrium shapes are depicted in Fig. 13 together with a cross-section. In the homogeneous case no concave shapes can be obtained, while making the particle material soft revealed concave shapes at large particle sizes. The occurrence of concave equilibrium shapes is experimentally observered by e.g. Ardell and Meshkinpour (1994). These shapes are usually observed at large particles sizes, proceeding inverse coarsening, i.e. splitting of particles.

Equilibrium shapes of type (2.5) are found when an external load is applied. A brief example of the resulting equilibrium shapes are shown in Fig. 14 where a soft particle is subjected to a compressive and tensile loading.

The stability of the equi-axed equilibrium shapes is treated by a numerical evaluation of the second derivative of the energy with respect to the aspect ratio. For the homogeneous problem the numerical computed values of this derivative are displayed in Fig. 15. At a critical size of $L_{\mathrm{inst}} = 5.35$ the calculated equilibrium shapes become unstable, with respect to this perturbation. The critical size L_{inst} represents the transition from a (local) minimum to a maximum. By directly comparing the energy of perturbed equilibrium shapes, the oblate distortion is energetically favourable. It is therefore reasonable, to expect that stable equilibrium shapes above $L_{\mathrm{inst}} = 5.35$ will have an oblate morphology.

The dependence of the critical size L_{inst} on the stiffness ratio $\bar{\mu}^{in}/\mu^{\bar{o}ut}$ is illustrated in Fig. 16. At a stiffness ratio $\bar{\mu}^{in}/\mu^{\bar{o}ut} = 0.7$ the instability is reached at a dimensionless size $L = 4.5$. This indicates that experimentally observed rafting phenomena should occur at this stiffness ratio for small particles. If the particle becomes stiffer or softer the equi-axed shape remains stable up to higher particle sizes.

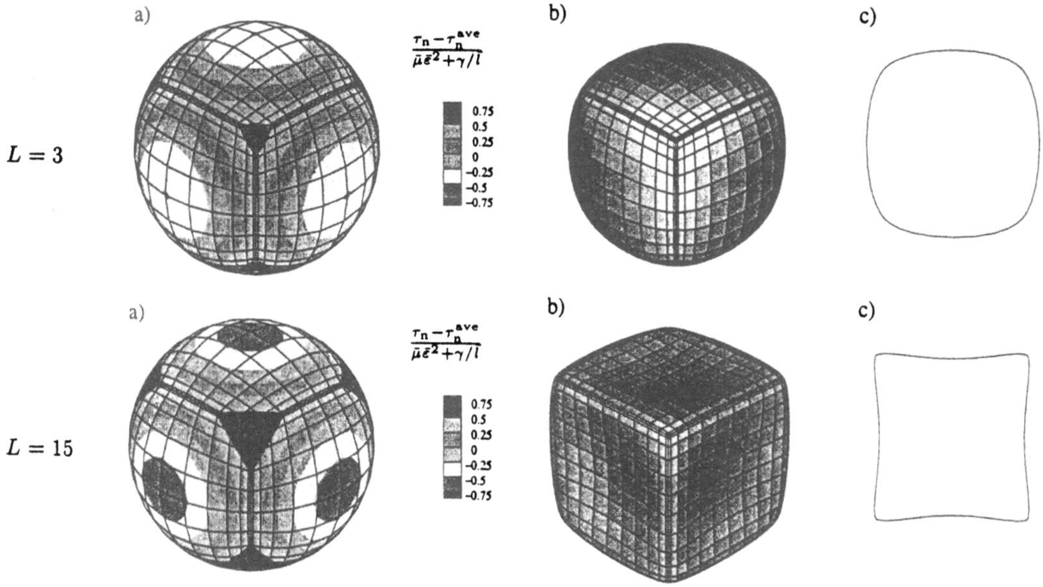

Figure 13. Soft precipitate at size $L = 3$ and $L = 15$: a) distribution of τ_n for a sphere

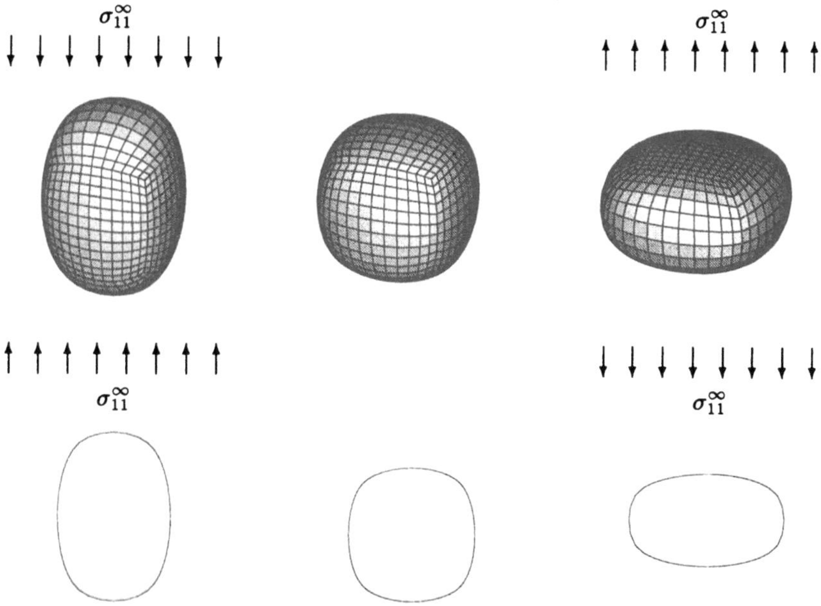

Figure 14. External load: Equilibrium shapes and related cross-sections for $\sigma_{11}^{\infty}/\bar{\mu}\bar{\varepsilon} = \pm 1.06$ and $L = 2$

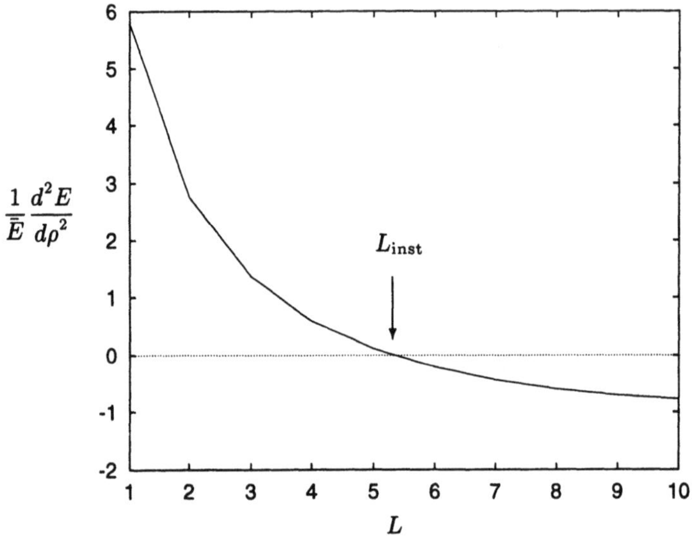

Figure 15. Morphological stability with respect to particle size ($\bar{E} = \mu^* \varepsilon^{*2} l^{*3} + \gamma l^{*2}$)

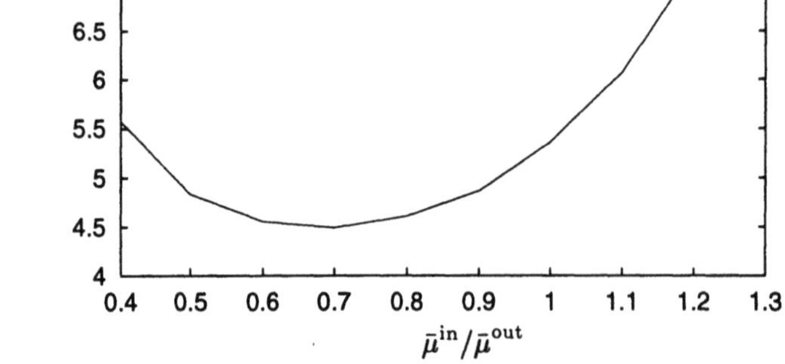

Figure 16. Instability size with respect to stiffness ratio

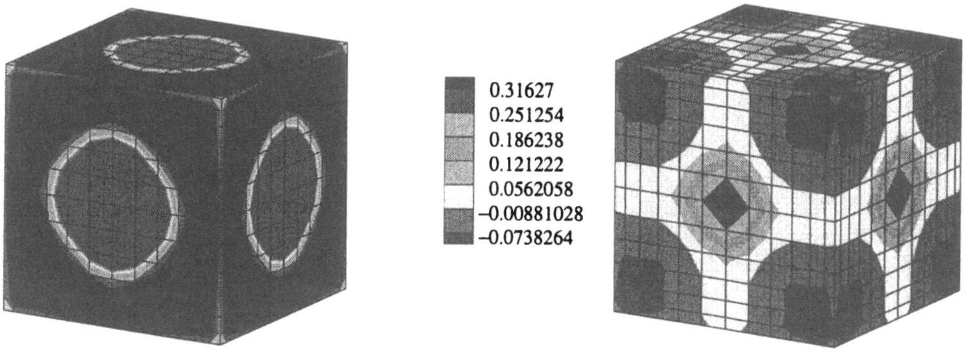

$f = 50\%$ $f = 10\%$

Figure 17. Resolved normal stresses on the matrix boundary for different volume fractions

Periodically arranged precipitates

Before considering particle morphologies, the important influence of the volume fraction f of the precipitate volume on the normal tractions along the cell boundary is shown. In Fig. 17 negative normal tractions for $f = 50\%$ can be found in the center of each matrix face, while in the vicinity of the edges positive values appear. In contrast, for the small volume fraction $f = 10\%$ positive tractions appear at the center, while at the edges negative values can be found. This can be explained by the high stiffness of the orthotropic material in diagonal direction.

The influence of particle size and volume fraction on the equilibrium shape is depicted in Fig. 18, where the shape is characterized by a shape factor $\eta = d/l$. For $\eta = 1$, the shape is a sphere while $\eta = \sqrt{3}$ characterizes a cube. The particle becomes cuboidal if L increases, independent of f. This can be understood by recalling that L denotes the ratio of elastic to interface energy. A large particle minimizes the elastic energy by orienting considerable parts of the interface normal to the soft directions of the material. The interface energy is less dominant at large sizes, thus it rounds off only small regions at the edges and corners to cube-like shapes. When keeping the particle size L constant and increasing the volume fraction f, the particle first tends to a more sphere-like morphology but for even higher values of f, the particle looks more cuboidal. It shall be mentioned that the qualitative dependencies of the equilibrium shapes on the parameters L and f agree well with the 2D results of Schmidt and Gross (1997).

Shape evolution

The treatment of evolution of microstructures is demonstrated by a simple example. Attention is restricted to the shape change only, assuming the volume to remain constant. The particle is softer than the matrix by a factor of 0.7, but has the same anisotropy. The dimensionless particle size is chosen to be $L = 5$. Fig. 19 shows the evolution from a sphere to the cube-like equilibrium shape. In the long time limit the shape approaches the stationary equilibrium morphology. The

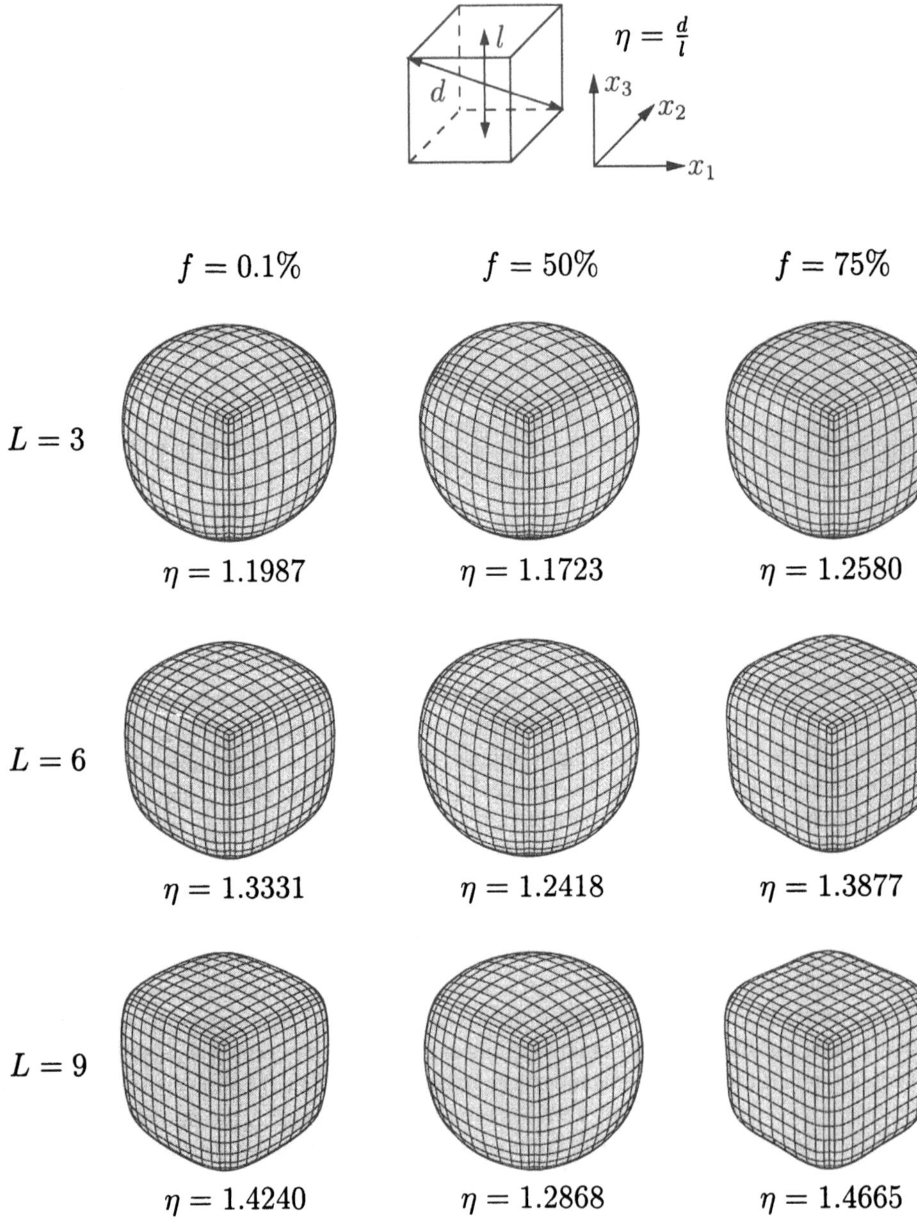

Figure 18. Equilibrium shapes for different values of L and f

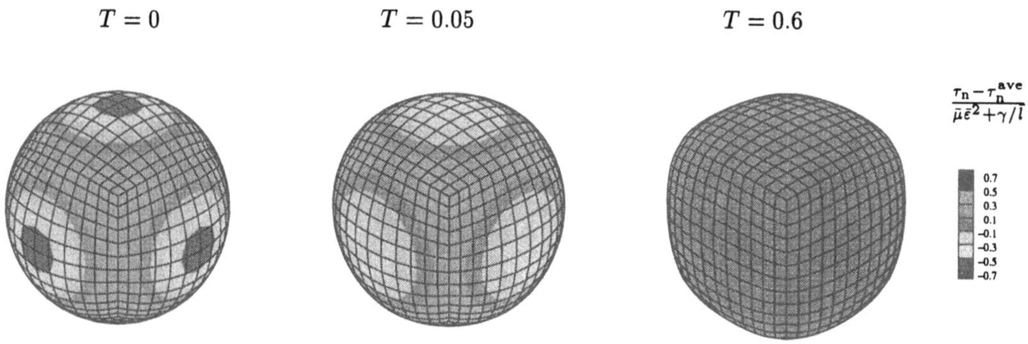

Figure 19. Development from a sphere to a cuboidal equilibrium shape (in the corner region $\tau_n - \tau_n^{ave}$ is positive)

sphere being no energy minimum of the total energy contributes a considerable amount of elastic energy during the initial period, consequently the driving force is not uniform in initially states. As the morphology changes energy is redistributed from elastic to interfacial energy, while the total energy decreases.

2.6 A constitutive law for materials with microstructural evolution

As shown before, materials like Ni-base alloys undergo a microstructural evolution under external loading. The aim of this section is to demonstrate how the effect of microstructural changes on the global constitutive behaviour can be described in a simplified manner. This is done by approximating the real γ'-morphology and its changes by ellipsoidal shapes. Studying first the morphology evolution of a single ellipsoidal precipitate and its dependence on the mechanical loading, macroscopic constitutive equations are subsequently constructed by homogenization. On this macroscopic level, the few morphology parameters of the ellipsoid appear as internal parameters. The overall material behaviour shows a visco-elastic behaviour. But it is revealed that in a loading-unloading processes there exists a hysteresis loop for a certain set of parameters, as observed in visco-plastic materials.

Micromechanical model and basic equations

Considered is (as before) an elastic particle in an elastically different surrounding matrix, see Fig. 20. For simplicity we restrict attention to isotropic material properties. In the particle phase an inelastic eigenstrain $\boldsymbol{\varepsilon}^0$ is introduced to model the lattice misfit between the two phases. The system is loaded by an external stress field $\boldsymbol{\sigma}^\infty$. In the absence of any body forces the elastic field problem is then given by the following set of equations:

$$\operatorname{div}\boldsymbol{\sigma} = \mathbf{0} \qquad \text{where} \quad \boldsymbol{\sigma} \to \boldsymbol{\sigma}^\infty \quad \text{for} \quad |\boldsymbol{x}| \to \infty \tag{44}$$

$$\boldsymbol{\sigma} = \begin{cases} \mathbb{C}^{\mathrm{M}}\boldsymbol{\varepsilon} & \text{in the matrix phase,} \\ \mathbb{C}^{\mathrm{P}}\left(\boldsymbol{\varepsilon} - \boldsymbol{\varepsilon}^0\right) & \text{in the precipitate phase.} \end{cases} \tag{45}$$

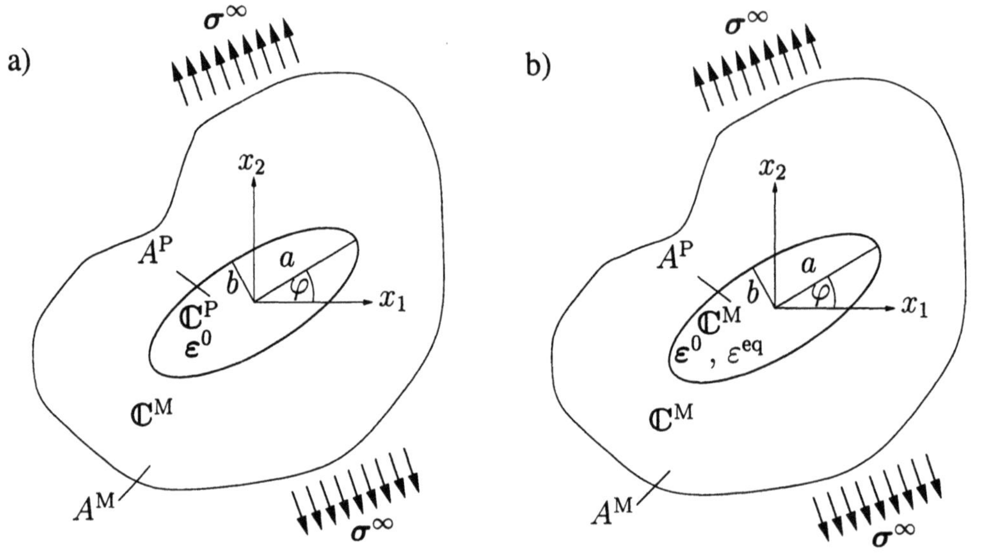

Figure 20. Micromechanical model, with an aspect ratio $m = 1/3$ and orientation $\varphi = 30°$; a) original problem, b) homogeneous comparison problem

The material tensors $\mathbb{C}^P$ and $\mathbb{C}^M$ of the particle (^{P}) and the matrix (^{M}) are isotropic and can be expressed by the Lamé constants λ^P, μ^P and λ^M, μ^M for the precipitate and the matrix respectively.

In real Ni-base alloys the microstructure is usually of a cuboidal shape, but in the following it is assumed that the precipitate A^P can roughly be approximated by an ellipsoidal shape. This assumptions simplifies the analysis as the well known Eshelby solution (Eshelby (1957)) is applicable. Further, for simplicity, in a this approach the analysis is restricted to a two dimensional model, but a three dimensional extension is possible.

The microstructure of the material, i.e. the shape of the ellipse of a given size A^P is determined by the two parameters m and φ. The parameter

$$m = \frac{a - b}{a + b} \tag{46}$$

describes the aspect ratio where a and b are the principal axis. A circle is characterized by $m = 0$. If $m > 0$, then a is the longer principal axis, resulting in an elongated ellipse. The direction of the principal axis a is given by the angle φ. The microstructural evolution is captured by evolution equations for m and φ.

The elasticity problem is solved by using Eshelby's result. For this purpose, a homogeneous comparison problem with the matrix properties $\mathbb{C}^M$ everywhere in the domain is defined. The inhomogeneity is taken into account by an additional eigenstrain ε^{eq} in the precipitate (Fig.20b). A decomposition of the stresses and the strains

$$\boldsymbol{\sigma} = \boldsymbol{\sigma}^\infty + {}'\boldsymbol{\sigma}\,, \quad \boldsymbol{\varepsilon} = \boldsymbol{\varepsilon}^\infty + {}'\boldsymbol{\varepsilon} \tag{47}$$

is related by Hooke's law of the matrix phase

$$\boldsymbol{\sigma}^{\infty} = \mathbb{C}^{\mathrm{M}} \boldsymbol{\varepsilon}^{\infty} . \tag{48}$$

By introducing this new notation the problem to be solved reduces to

$$\operatorname{div} {}'\boldsymbol{\sigma} = 0 \qquad \text{where} \quad {}'\boldsymbol{\sigma} \to 0 \quad \text{for} \quad |\boldsymbol{x}| \to \infty$$

$$ {}'\boldsymbol{\sigma} = \begin{cases} \mathbb{C}^{\mathrm{M}}\, {}'\boldsymbol{\varepsilon} & \text{in the matrix} \\ \mathbb{C}^{\mathrm{M}} \left[{}'\boldsymbol{\varepsilon} - \left(\boldsymbol{\varepsilon}^{\mathrm{eq}} + \boldsymbol{\varepsilon}^{0} \right) \right] & \text{in the precipitate} \end{cases} \tag{49}$$

It is solved by using Eshelby's result, that inside an elliptical inclusion the stresses and strains are constant. In the considered system the strain ${}'\boldsymbol{\varepsilon}$ is determined through the Eshelby tensor $\mathbb{E}^{\mathrm{M}}$ given by the Poisson's ratio ν^{M} of the matrix, the aspect ratio m and the orientation φ (see appendix A for the components of $\mathbb{E}^{\mathrm{M}}$)

$$ {}'\boldsymbol{\varepsilon} = \mathbb{E}^{\mathrm{M}} \left(\boldsymbol{\varepsilon}^{\mathrm{eq}} + \boldsymbol{\varepsilon}^{0} \right) . \tag{50}$$

The equivalent eigenstrain $\boldsymbol{\varepsilon}^{\mathrm{eq}}$ is determined by equating the precipitate stress $\boldsymbol{\sigma}$ of the original problem with the precipitate stress of the comparison problem:

$$\boldsymbol{\sigma} = \mathbb{C}^{\mathrm{P}} \left({}'\boldsymbol{\varepsilon} + \boldsymbol{\varepsilon}^{\infty} - \boldsymbol{\varepsilon}^{0} \right) = \mathbb{C}^{\mathrm{M}} \left[{}'\boldsymbol{\varepsilon} - \left(\boldsymbol{\varepsilon}^{\mathrm{eq}} + \boldsymbol{\varepsilon}^{0} \right) \right] + \boldsymbol{\sigma}^{\infty} . \tag{51}$$

With $\boldsymbol{\varepsilon}^{\infty} = \mathbb{C}^{\mathrm{M}-1} \boldsymbol{\sigma}^{\infty}$ rearranging terms, yields

$$\left[\left(\mathbb{1} - \mathbb{C}^{\mathrm{P}-1} \mathbb{C}^{\mathrm{M}} \right) \mathbb{E}^{\mathrm{M}} + \mathbb{C}^{\mathrm{P}-1} \mathbb{C}^{\mathrm{M}} \right] \left(\boldsymbol{\varepsilon}^{\mathrm{eq}} + \boldsymbol{\varepsilon}^{0} \right) = \boldsymbol{\varepsilon}^{0} - \left(\mathbb{C}^{\mathrm{M}-1} - \mathbb{C}^{\mathrm{P}-1} \right) \boldsymbol{\sigma}^{\infty} \tag{52}$$

to determine $\boldsymbol{\varepsilon}^{\mathrm{eq}}$ as a function of eigenstrains $\boldsymbol{\varepsilon}^{0}$, loading $\boldsymbol{\sigma}^{\infty}$ and the micro geometry m and φ (taken into account by $\mathbb{E}^{\mathrm{M}}$). For later purposes another representation of $\boldsymbol{\varepsilon}^{\mathrm{eq}} + \boldsymbol{\varepsilon}^{0}$ is necessary. Comparing the precipitate stresses ${}'\boldsymbol{\sigma}$ of the original problem

$$\begin{aligned} {}'\boldsymbol{\sigma} &= \mathbb{C}^{\mathrm{P}} \left(\boldsymbol{\varepsilon} - \boldsymbol{\varepsilon}^{0} \right) - \boldsymbol{\sigma}^{\infty} \\ &= \mathbb{C}^{\mathrm{P}} \left(\boldsymbol{\varepsilon} - \boldsymbol{\varepsilon}^{0} \right) - \mathbb{C}^{\mathrm{M}} \left(\boldsymbol{\varepsilon} - {}'\boldsymbol{\varepsilon} \right) \\ &= \mathbb{C}^{\mathrm{M}} \left[{}'\boldsymbol{\varepsilon} + \mathbb{C}^{\mathrm{M}-1} \left(\mathbb{C}^{\mathrm{P}} - \mathbb{C}^{\mathrm{M}} \right) \boldsymbol{\varepsilon} - \mathbb{C}^{\mathrm{M}-1} \mathbb{C}^{\mathrm{P}} \boldsymbol{\varepsilon}^{0} \right] \end{aligned} \tag{53}$$

with those of the comparison problem

$$ {}'\boldsymbol{\sigma} = \mathbb{C}^{\mathrm{M}} \left[{}'\boldsymbol{\varepsilon} - \left(\boldsymbol{\varepsilon}^{\mathrm{eq}} + \boldsymbol{\varepsilon}^{0} \right) \right] , \tag{54}$$

we see that,

$$\boldsymbol{\varepsilon}^{\mathrm{eq}} + \boldsymbol{\varepsilon}^{0} = \mathbb{C}^{\mathrm{M}-1} \left[\mathbb{C}^{\mathrm{P}} \boldsymbol{\varepsilon}^{0} - \left(\mathbb{C}^{\mathrm{P}} - \mathbb{C}^{\mathrm{M}} \right) \boldsymbol{\varepsilon} \right] . \tag{55}$$

Inserting the above result in the Eshelby relation (50) the precipitate strain becomes

$$\boldsymbol{\varepsilon} = \mathbb{G}^{-1} \mathbb{E}^{\mathrm{M}} \mathbb{C}^{\mathrm{M}-1} \boldsymbol{\sigma}^{0} + \mathbb{G}^{-1} \mathbb{C}^{\mathrm{M}-1} \boldsymbol{\sigma}^{\infty} , \tag{56}$$

where

$$\mathbb{G} = \mathbb{1} + \mathbb{E}^{\mathrm{M}} \left(\mathbb{C}^{\mathrm{M}-1} \mathbb{C}^{\mathrm{P}} - \mathbb{1} \right) \quad \text{and} \quad \boldsymbol{\sigma}^{0} = \mathbb{C}^{\mathrm{P}} \boldsymbol{\varepsilon}^{0} . \tag{57}$$

The precipitate stress $\boldsymbol{\sigma}$ is easily obtained from Hooke's law (45) of the original problem to be

$$\boldsymbol{\sigma} = \underbrace{\left(\mathbb{C}^{\mathrm{P}}\mathbb{G}^{-1}\mathbb{E}^{\mathrm{M}}\mathbb{C}^{\mathrm{M}-1} - \mathbb{1}\right)}_{\mathbb{B}^0}\boldsymbol{\sigma}^0 + \underbrace{\mathbb{C}^{\mathrm{P}}\mathbb{G}^{-1}\mathbb{C}^{\mathrm{M}-1}}_{\mathbb{B}^\infty}\boldsymbol{\sigma}^\infty . \tag{58}$$

The forth order tensors $\mathbb{B}^0$ and $\mathbb{B}^\infty$ are the influence tensors, as they describe the precipitate stress caused by eigenstresses $\boldsymbol{\sigma}^0$ or external loads $\boldsymbol{\sigma}^\infty$ respectively.

Total potential

In order to formulate kinetic evolution equations for the morphological variables m and φ the total potential of an inhomogeneity with eigenstrains $\boldsymbol{\varepsilon}^0$ under an external load $\boldsymbol{\sigma}^\infty$ is needed. It is given by the three terms

$$\Pi = \Pi^{\mathrm{int}} + \Pi^{\mathrm{el}} + \Pi^{\mathrm{ext}}, \tag{59}$$

where Π^{int} represents the interface energy, Π^{el} the strain energy, and Π^{ext} the potential of the external forces. The interface energy Π^{int} takes the form

$$\Pi^{\mathrm{int}} = \gamma L, \tag{60}$$

where L is the arc length of the interface. The strain energy Π^{el} of the system is given by

$$\Pi^{\mathrm{el}} = \frac{1}{2}\int_A \sigma : \varepsilon^{\mathrm{el}}\, dA. \tag{61}$$

which after some manipulation can be shown to be

$$\Pi^{\mathrm{el}} = \frac{1}{2}\int_A \boldsymbol{\sigma}^\infty : \boldsymbol{\varepsilon}^\infty\, dA + \frac{A^{\mathrm{P}}}{2}\left(\boldsymbol{\sigma}^\infty : \boldsymbol{\varepsilon}^{\mathrm{eq}} - {}'\boldsymbol{\sigma} : \boldsymbol{\varepsilon}^0\right). \tag{62}$$

The first term is infinite as the integral covers the infinite domain, however this term is not related to the specific microstructure, i.e. it is independent on m and φ. The second term is related to precipitate morphology thus depending on m and φ. Finally, the potential of the external load applied at the remote boundary is given by the tractions $\boldsymbol{t}^\infty$ and the displacements $\boldsymbol{u}$ at infinity:

$$\Pi^{\mathrm{ext}} = -\int_{\partial A} \boldsymbol{t}^\infty \cdot \boldsymbol{u}\, dA\, . \tag{63}$$

In conjunction with a decomposition of the displacement field $\boldsymbol{u}$ and some manipulations the total potential is finially derived as (for details see appendix B)

$$\Pi = -A^{\mathrm{P}}\left(\frac{1}{2}\boldsymbol{\sigma}^\infty : \boldsymbol{\varepsilon}^{\mathrm{eq}} + \frac{1}{2}{}'\boldsymbol{\sigma} : \boldsymbol{\varepsilon}^0 + \boldsymbol{\sigma}^\infty : \boldsymbol{\varepsilon}^0\right) + \gamma L + \Pi^0\, , \tag{64}$$

where

$$\Pi^0 = \frac{1}{2}\int_A \boldsymbol{\sigma}^\infty : \boldsymbol{\varepsilon}^\infty\, dA - \int_{\partial A} \boldsymbol{t}^\infty \cdot \boldsymbol{u}^\infty\, ds. \tag{65}$$

Note, that Π^0 is not dependent on the variables m and φ, so that it will be sufficient to consider only $\tilde{\Pi} = \Pi - \Pi^0$ in the following. Relating $\tilde{\Pi}$ to the cell area A, the density

$$\tilde{\psi} = \frac{\tilde{\Pi}}{A} \tag{66}$$

is obtained, which on the macroscopic level represents Gibbs' free enthalpy for isothermal processes.

Kinetic law for morphology evolution

The second law of thermodynamics for an isothermal process, which is assumed in this derivation, leads to a dissipation inequality. Allowing solely morphological changes (m,φ) in an elastic media the dissipation inequality takes the form

$$\tau_m \, \dot{m} + \tau_\varphi \, \dot{\varphi} \geq 0\,, \tag{67}$$

as in an internal variable approach, see Lubliner (1990). The two terms τ_m and τ_φ represent the driving forces of the two morphology variables m and φ respectively. They are defined by the partial derivatives of the potential with respect to the morphology variables:

$$\tau_m = -\frac{\partial \tilde{\psi}}{\partial m} \quad \text{and} \quad \tau_\varphi = -\frac{\partial \tilde{\psi}}{\partial \varphi}\,. \tag{68}$$

The morphology changes occur due to diffusion processes, which are described by first order differential equations with respect to time. Thus it seems reasonable to propose as the simplest kinetic relation between the evolution $\dot{m}$ and $\dot{\varphi}$ and the respective driving forces a linear relation:

$$\dot{m} = \alpha \tau_m \quad \text{and} \quad \dot{\varphi} = \beta \tau_\varphi \quad \text{with} \quad \alpha \geq 0 \quad \text{and} \quad \beta \geq 0\,. \tag{69}$$

Homogenization and global material behaviour

Using the notation

$$< \cdot >= \frac{1}{A} \int_A (\cdot)\, dA \tag{70}$$

for the mean values in the entire cell, in the precipitate and in the matrix respectively and introducing the volume fraction

$$f = \frac{A^{\mathrm{P}}}{A} \tag{71}$$

the following relation can be deduced from equation (45)

$$< \varepsilon >= \mathbb{C}^{\mathrm{M}-1} < \sigma > + f \left(\mathbb{C}^{\mathrm{P}-1} - \mathbb{C}^{\mathrm{M}-1} \right) < \sigma >^{\mathrm{P}} + f \varepsilon^0 . \tag{72}$$

Introducing the constant precipitate stress (58)

$$< \sigma >^{\mathrm{P}} = \mathbb{B}^\infty \sigma^\infty + \mathbb{B}^0 \sigma^0 \tag{73}$$

and

$$< \sigma >= \sigma^\infty \quad \text{with} \quad \boldsymbol{\sigma n} = \boldsymbol{\sigma}^\infty \boldsymbol{n} = \boldsymbol{t}^\infty \quad \text{at} \quad \partial A \tag{74}$$

yields a relation between $< \sigma >$ and $< \varepsilon >$ of the form

$$< \varepsilon >= \mathbb{C}^{*-1} < \sigma > + e^0 \qquad \text{or} \qquad < \sigma >= \mathbb{C}^* \left[< \varepsilon > - e^0 \right], \tag{75}$$

where

$$\mathbb{C}^{*-1} = \mathbb{C}^{\mathrm{M}-1} + f \left(\mathbb{C}^{\mathrm{P}-1} - \mathbb{C}^{\mathrm{M}-1} \right) \mathbb{B}^\infty \tag{76}$$

and

$$e^0 = f \left[\left(\mathbb{C}^{\mathrm{P}-1} - \mathbb{C}^{\mathrm{M}-1} \right) \mathbb{B}^0 \sigma^0 + \varepsilon^0 \right] . \tag{77}$$

Note that the influence tensors $\mathbb{B}^\infty$ and $\mathbb{B}^0$ depend through the Eshelby tensor $\mathbb{E}^M$ on the micro morphology. In the literature this choice of influence tensors is known as the "dilute distribution" method (see Nemat-Nasser and Hori (1993)) or the "Eshelby method" (see Mura (1987)). Its main draw-back is the negligence of particle interaction. Thus the presented method in principle is only reliable in a regime in which the precipitate volume fraction f is relatively small. On the other hand, the dilute distribution method in 3D delivers satisfactory results in a much wider range since interaction effects in this case are known to be smaller compared with 2D.

Some results

For the calculations, the following material parameters were taken which shall mimic Ni-base alloys. For Nickel matrix the two Lamé constants

$$\lambda^M = 1.0975 \cdot 10^5 \,\mathrm{N/mm^2}, \qquad \mu^M = 0.8715 \cdot 10^5 \,\mathrm{N/mm^2} \tag{78}$$

were taken which were reduced by the factor 0.7 for the softer precipitates. The values for the interface energy density γ and the dilatational eigenstrains ε^0 are taken from Voorhees et al. (1992):

$$\gamma = 10^{-5} \,\mathrm{N/mm^2}, \quad \varepsilon^0 \doteq 0.003 \cdot \begin{bmatrix} 1 & 0 \\ 0 & 1 \end{bmatrix} \tag{79}$$

The area of the representative cell element A is assumed to be

$$A = 10^{-7} \,\mathrm{mm^2}, \tag{80}$$

while the size effect of the precipitate is studied by varying the precipitate area in the range

$$A^P = 1 \cdot 10^{-8} \ldots 8 \cdot 10^{-8} \,\mathrm{mm^2}. \tag{81}$$

The evolution parameters α and β were chosen to be

$$\alpha = 0.1 \,\mathrm{mm^2/Nh}, \quad \beta = 0.1 \,\mathrm{mm^2/Nh}, \tag{82}$$

which seem to be realistic under elevated temperatures.

As first example, the evolution of m (starting from $m = 0$) under uniaxial tension is displayed in Fig.21. In the first time period there is a rapid change but for times larger than $t \approx 150\,\mathrm{h}$ the parameter m reaches a stationary value -0.65. This corresponds to an "equilibrium shape", which is defined as a shape that extremises the total potential with respect to the internal variables. The overall strain response $< \varepsilon >$ is shown in Fig. 22 as a function of time. The change in micro morphology causes an increase of the strain in the loading direction while in the lateral direction the strain decreases. In both direction there is an expansion, that is caused by the dilatational eigenstrains. In the next example the test is modified by an additional applied shear loading. The evolution of the geometry parameter m and φ is shown in Fig. 23. The evolution of the aspect ratio m is similar to the case of uniaxial tension. However, due to the shear loading the orientation of the microgeometry also changes. The final orientation coincides with the principal axis of the applied stress state. The overall strain response $< \varepsilon >$ is depicted in Fig. 24. In the initial period the formation of an elongated ellipse causes $< \varepsilon_{11} >$ to increase rapidly, as the softer precipitate carries less load. In a second phase the strain $< \varepsilon_{11} >$ decreases slightly, as the

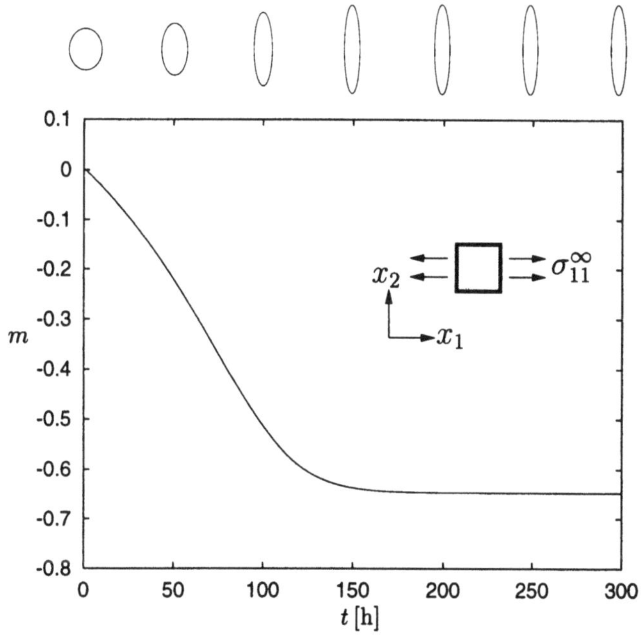

Figure 21. Evolution of m in uniaxial tension ($\sigma_{11} = 100\,\mathrm{N/mm}^2$, $A^{\mathrm{P}} = 5 \cdot 10^{-8}\,\mathrm{mm}^2$)

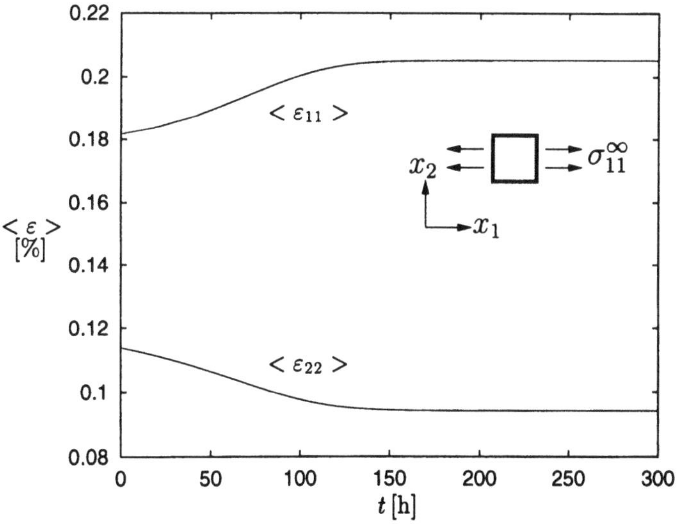

Figure 22. Evolution of $< \varepsilon >$ in uniaxial tension ($\sigma_{11} = 100\,\mathrm{N/mm}^2$, $A^{\mathrm{P}} = 5 \cdot 10^{-8}\,\mathrm{mm}^2$) .

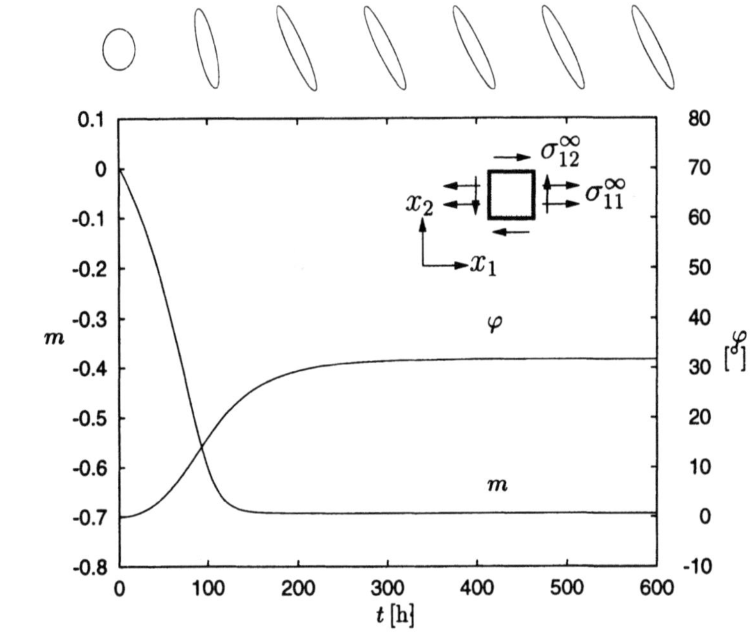

Figure 23. Evolution of m and φ in tension and shear ($\varphi = 0°$ at $t = 0$, $\sigma_{11} = 100\,\mathrm{N/mm^2}$, $\sigma_{12} = 100\,\mathrm{N/mm^2}$, $A^{\mathrm{P}} = 5 \cdot 10^{-8}\,\mathrm{mm^2}$)

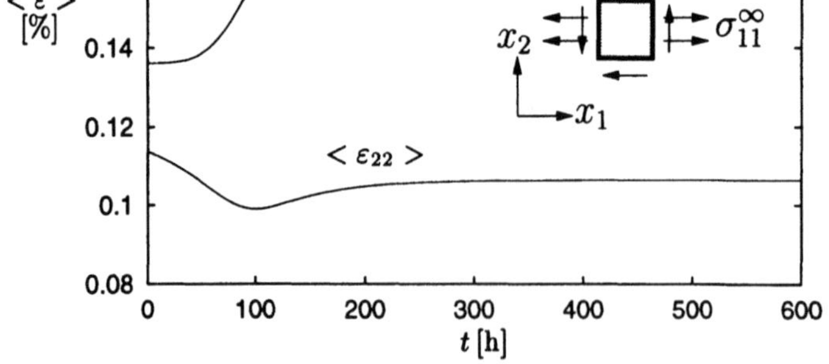

Figure 24. Evolution of $< \varepsilon >$ in tension and shear ($\varphi = 0°$ at $t = 0$, $\sigma_{11} = 100\,\mathrm{N/mm^2}$, $\sigma_{12} = 100\,\mathrm{N/mm^2}$ $A^{\mathrm{P}} = 5 \cdot 10^{-8}\,\mathrm{mm^2}$)

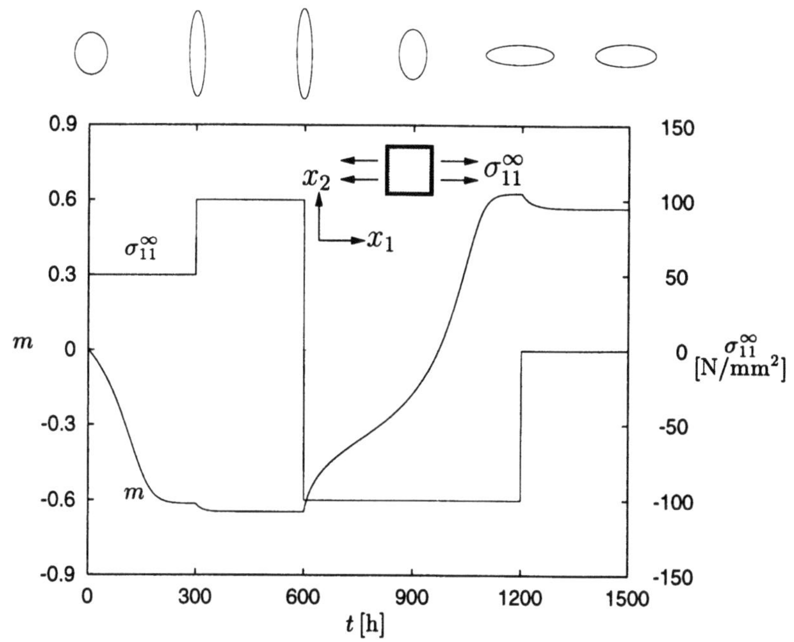

Figure 25. Evolution of m under tensile and compressive loading history ($\sigma_{11}^{\infty}(t)$, $A^{\mathrm{P}} = 5 \cdot 10^{-8}\ \mathrm{mm}^2$)

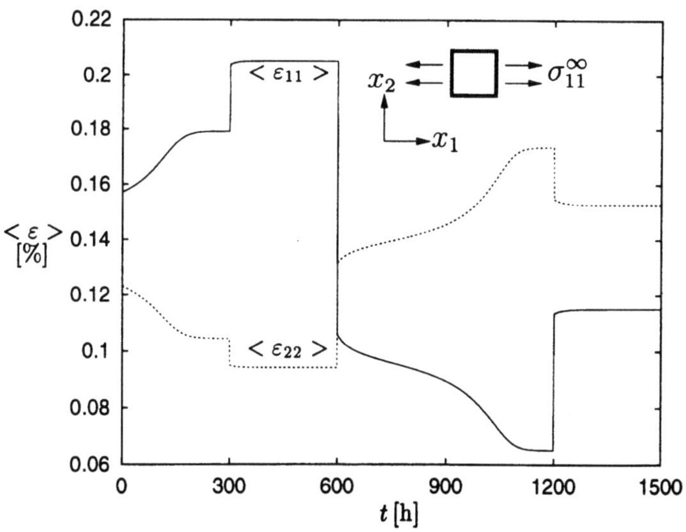

Figure 26. Evolution of $< \varepsilon >$ under tensile and compressive loading history ($\sigma_{11}^{\infty}(t)$,$A^{\mathrm{P}} = 5 \cdot 10^{-8}\ \mathrm{mm}^2$)

area of the precipitate projected onto the x_2-axis reduces due to the rotation. To demonstrate that rather complex loading histories can be simulated, Fig. 25 shows a uniaxial tension-compression experiment. The accompanied overall strain response $< \varepsilon >$ is plotted in Fig. 26. It can be seen how external load changes influence the morphology evolution and by this macroscopic time dependent reaction of the material. It is worth to emphasize, when removing the load σ_{11}^{∞} the microstructure does not return to its initially circular shape. A elongated ellipse remains in the long time limit leading to residual strains and stresses in the material. This in more detail can be understood if the stability of the solution is considered. It is well known that for a soft

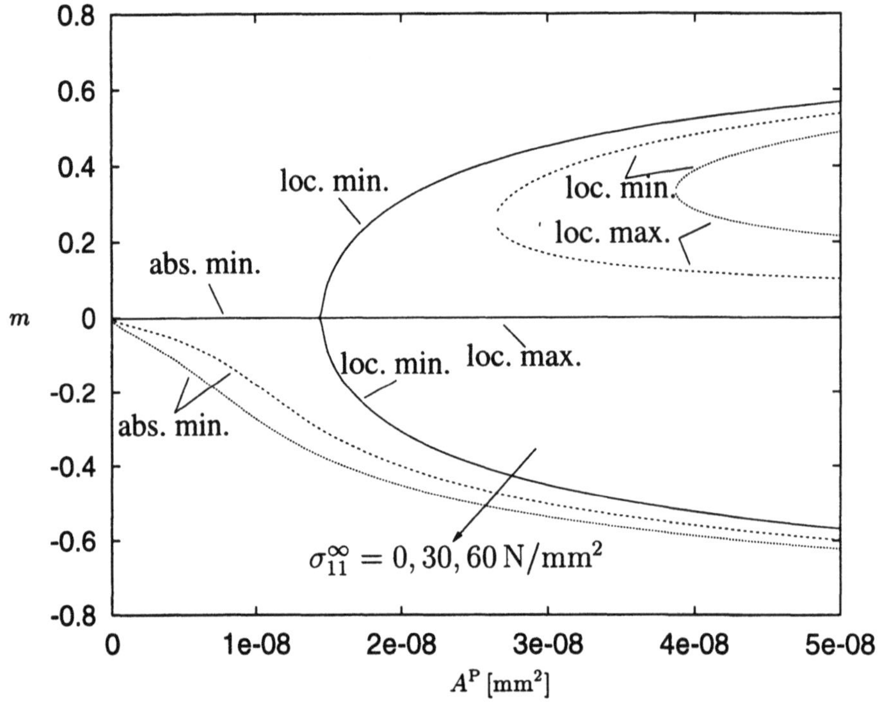

Figure 27. Size instability with respect to m for different load levels, loc. min.: local minimum, abs. min.: absolute minimum, loc. max.: local maximum

precipitate shape bifurcations occur as the particle size increases. This is shown in Fig. 27. For precipitates larger than $A_{\mathrm{crit}}^{\mathrm{P}} = 1.45 \cdot 10^{-8}\,\mathrm{mm}^2$ the circle is no longer an stable shape. A slight distorsion will cause the system to branch in an elongated shape with $m \neq 0$. In the following the term equilibrium shape will be used as a synonym for the shape in the long term limit. Both shapes with $m \neq 0$ are energy minimizers while the circle $m = 0$ is a local energy maximum. In the present setting a local energy maximum can be detected by formally setting $\alpha < 0$. This

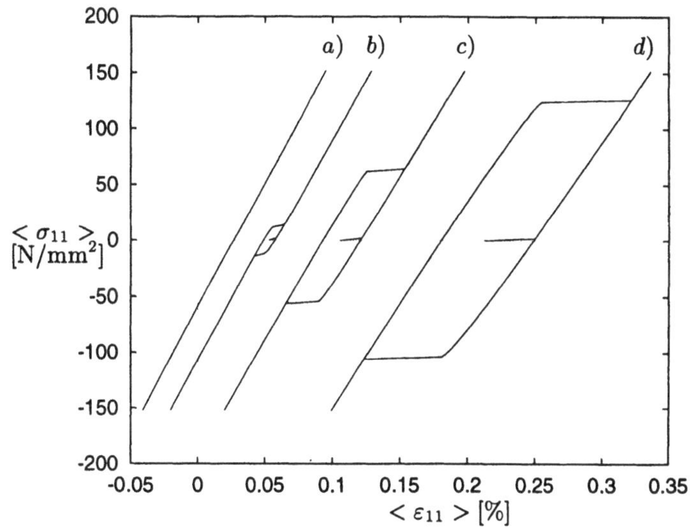

Figure 28. Stress strain curves for different precipitate sizes: a) $A^{\mathrm{P}} = 1 \cdot 10^{-8}\,\mathrm{mm}^2$, b) $A^{\mathrm{P}} = 2$ $10^{-8}\,\mathrm{mm}^2$, c) $A^{\mathrm{P}} = 4 \cdot 10^{-8}\,\mathrm{mm}^2$, d) $A^{\mathrm{P}} = 8 \cdot 10^{-8}\,\mathrm{mm}^2$

obviously violates the dissipation inequality (69). An external load acts as an imperfection to the system. In Fig. 27 also the effect of an uniaxial external load is plotted. In the following the overall stress strain response is considered in Fig. 28. Starting with a perfect circle as the precipitate shape, the macro stress $< \sigma_{11} >$ is increased incrementally. In each load step the microstructural variables m and φ are relaxed to their stationary limit. The macro strain $< \varepsilon_{11} >$ is then observed. At a load level of $< \sigma_{11} >= 150\,\mathrm{N/mm}^2$ the loading direction is reversed until $< \sigma_{11} >= -150\,\mathrm{N/mm}^2$ is reached. Then the loading is again increased until $< \sigma_{11} >= 150\,\mathrm{N/mm}^2$. From the plot d) in Fig. 28 it is obvious that for a precipitate of size $A^{\mathrm{P}} = 8 \cdot 10^{-8}\,\mathrm{mm}^2$ exists a hysteresis loop. For example when unloading from $< \sigma_{11} >= 150\mathrm{N/mm}^2$ the microstructure with $m < 0$ remains almost unchanged until a certain compressive stress level is reached and the microstructure "snaps through" to values of $m > 0$ which are favoured in the compressive stress region. The same happens if this microstructure is loaded again. The size of the hysteresis loop depends on the size of the precipitate. For particles sizes below the before mentioned critical particle size $A^{\mathrm{P}}_{\mathrm{crit}} = 1.45 \cdot 10^{-8}\,\mathrm{mm}^2$ there exits no shape bifurcation, so that no "snap through" in the microstructure can cause a hysteresis loop. These configurations experience also a microstructural change during the loading history, but the influence on the overall behaviour is rather weak. For such material parameters the stress strain relation is only weakly non linear.

3 Atomistic simulations

3.1 Introduction

An alternative to the sharp interface model to simulate equilibrium morphologies are Monte Carlo type atomistic simulations, see Binder (1986), based on a combination of statistical mechanics and linear elasticity, see Lee (1995). Within the Monte Carlo process, the two-phase structure is considered as a lattice structure with matrix and inclusion atoms on the lattice places. Exchanging atoms randomly generates a Boltzmann-weighted chain of configurations. Considering interatomic energies, allow for elastic strain and interfacial parts, these changes are tested with a certain probability and are either accepted or rejected. In this section, a generalization of this discrete model will be discussed and applied to different problems in 2D.

3.2 Model and basic equations

Again, considered is a misfitting isolated precipitate $\mathcal{B}^{in}$ with the elasticity tensor $\mathbb{C}^{in}$ in a finite matrix $\mathcal{B}^{out}$ with an elasticity tensor $\mathbb{C}^{out}$ loaded by an eigenstrain ε^0 caused by the lattice misfit. Both domains are separated by an interface $\partial\mathcal{B}^{in}$.

Breaking the two-phase structure into a large number n of elementary particles, according to classical statistical mechanics, normalization of the probability for a particle state yields the partial function

$$Z(\Theta,\varepsilon,c) = \sum_{r=1}^{n} \mathrm{e}^{-\frac{\Phi_r(\varepsilon,c)}{k_B\Theta}}, \tag{83}$$

where Φ_r is the Hamiltonian for a particle state r, k_B Boltzmann's constant, Θ is the absolute temperature and c is the concentration. Comparison of the total differential of $\ln(Z)$ with macroscopic thermodynamics (see Appendix C1) leads to the configurational free energy

$$\Psi(\Theta,\varepsilon,c) = \Phi - \Theta S = -k_B\Theta\ln(Z(\Theta,\varepsilon,c)) \tag{84}$$

where $\Phi = \Pi$ is the Hamiltonian. According to the laws of thermodynamics, this energy takes a minimum in the equilibrium state:

$$\Psi \to \text{Minimum}\,. \tag{85}$$

It shall be emphasized that minimizing Ψ not necessarily means that Φ is minimized. This is approximately only the case below a certain temperature Θ.

In the continuum formulation, the Hamiltonian $\Phi = \Pi$ is given by the sum of elastic strain energy, the potential of the external load and the interfacial energy. Also all further continuum equations regarding the equilibrium of the system remain unchanged and can be taken from section 2.2.

3.3 Numerical realization

Discrete atom method (DAM)

Within the 2D DAM-model the material is represented by an elastic triangular homogeneous Bravais-lattice $\mathcal{B} = \{n_1\mathbf{a}_1 + n_2\mathbf{a}_2 \mid n_1, n_2 \in \mathbf{Z}\}$ with the primitive unit vectors $\mathbf{a}_1 = (1\ 0)$,

$\mathbf{a}_2 = \left(\frac{1}{2}\ \frac{1}{2}\sqrt{3}\right)$ and the molecular spring constant k between the lattice places and their six nearest neighbours. In the isotropic case, k is proportional to the Lamé-constants: $\lambda = \mu = \frac{1}{4}\sqrt{3}\,k$, see Hoover et al. (1974). Directional spring constants make the triangular crystal elastically anisotropic, restricted to cubic symmetry, see Lee (1996b). The Hamiltonian again is given as the sum of interfacial and elastic strain energy which can be written as

$$\Phi = \sum_{l=1}^{n} \gamma S_l(m) + \frac{1}{2}\bigcup_{e=1}^{n_e} \boldsymbol{v}_e \cdot (\mathbf{k}_e \boldsymbol{v}_e - 2\mathbf{p}_e)\,. \tag{86}$$

Here, the second term describes the elastic energy and a possible potential of an external load at the boundary. In the first term, the interfacial part, n is the total number of inclusion atoms, m the number of unlike nearest neighbours and $S_l(m)$ the equivalent continuum surface of the l-th inclusion atom, see Lee (1995). The elastic strain energy part can be calculated via molecular statics using n_e finite spring elements, with spring constant k_{ij} between atom i and atom $j \neq i$, lattice parameter a_{ij} and the eigenstrain ε^* caused by the misfit-atoms. The equilibrium conditions for the spring model under eigenstrain and an external load F, see Kolling and Gross (2000), lead to a linear system of equations for the atom displacement vector $\boldsymbol{v}_e$

$$\boldsymbol{k}_e \boldsymbol{v}_e = \boldsymbol{p}_e \tag{87}$$

with the local stiffness matrix

$$\boldsymbol{k}_e = k_{ij}\begin{pmatrix} 1 & -1 \\ -1 & 1 \end{pmatrix} \tag{88}$$

and the load vector

$$\boldsymbol{p}_e = \frac{1}{2} a_{ij}(F - k_{ij}\varepsilon^*)\begin{pmatrix} 1 \\ 1 \end{pmatrix}. \tag{89}$$

In appendix C2 it is explained that minimizing the energy difference $\Delta\Phi$ of the system appearing by exchanging arbitrary atoms, leads to a minimum of the free energy Ψ. Thus, as a strategy for determining equilibrium shapes, a Monte Carlo test in the energy difference $\Delta\Phi$

$$f(\Delta\Phi) = e^{-\frac{\Delta\Phi}{k_B\Theta}} > z_n, \quad z_n \in (0,1) \tag{90}$$

after exchange of two atoms with unlike neighbours in comparison with random numbers z_n is used, see Binder (1987). Of course, the side condition of a prescribed constant number of atoms n has to be fulfilled, which is equivalent to a constant inclusion area. It shall be mentioned that the Monte Carlo method is a non-gradient method.

Generalization: Gauss Point Method (GPM)

One disadvantage of the discrete atom method, is the restriction to a single type of isotropic ($\lambda = \mu$) or to purely orthotropic materials. A generalized continuum formulation is capable to overcome these limitations. For this purpose, the field equations will be solved by n_{el} regular arranged isoparametric finite elements with discontinuities inside, assigning elastic material properties to the Gauss points in dependence of their location (Fig. 29).

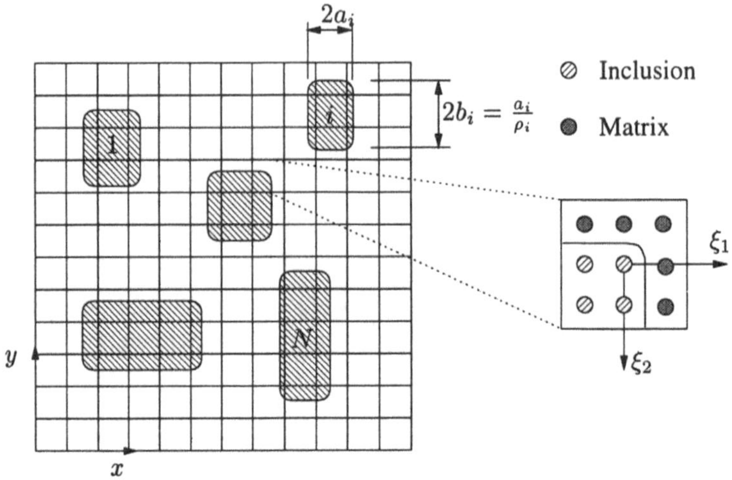

Figure 29. Gauss point method (GPM)

The equilibrium shape is obtained analogous to inequality (90). Using statistical definition of entropy, an exchange of the local position of two Gauss points will be either accepted or rejected, depending on the difference of the Hamiltonian associated with this exchange. In the following this is called the Gauss point method (GPM).

In classical finite element formulation, starting from the weak form of mechanical equilibrium

$$\delta\Phi = \int_{\mathcal{B}} \boldsymbol{\sigma} : \delta\boldsymbol{\varepsilon}\, dV - \int_{\partial\mathcal{B}_\sigma} \mathbf{t} \cdot \delta\mathbf{u}\, dA = 0, \tag{91}$$

the local fields in the quadrilateral element Ω_e are approximated isoparametrically

$$\mathbf{u}(\xi,\eta) = \sum_{I=1}^{4} N_I(\xi,\eta)\hat{\boldsymbol{u}}_I = \boldsymbol{N}\hat{\boldsymbol{u}} \tag{92}$$

$$\boldsymbol{\varepsilon}(\xi,\eta) = \sum_{I=1}^{4} \boldsymbol{B}_I(\xi,\eta)\hat{\boldsymbol{u}}_I = \boldsymbol{B}\hat{\boldsymbol{u}} \tag{93}$$

by bilinear shape functions $N_I \in C^0$

$$N_I(\xi,\eta) = \frac{1}{4}(1+\xi\xi_I)(1+\eta\eta_I) \tag{94}$$

and their derivatives assembled under plane strain conditions in the matrix

$$\boldsymbol{B}_I = \begin{pmatrix} \frac{\partial N_I}{\partial x} & 0 \\ 0 & \frac{\partial N_I}{\partial y} \\ \frac{\partial N_I}{\partial y} & \frac{\partial N_I}{\partial x} \end{pmatrix}. \tag{95}$$

With the approximations $\delta \mathbf{u} = \boldsymbol{N}\delta\hat{u}$ and $\delta\varepsilon = \boldsymbol{B}\delta\hat{u}$ equation (91) leads to the global residual

$$\bigcup_{e=1}^{n_{el}} \left\{ \int_{\Omega_e} \boldsymbol{B}^T \boldsymbol{\sigma}\, d\Omega - \int_{\partial\Omega_e^\sigma} \boldsymbol{N}\cdot\boldsymbol{t}\, d\Gamma \right\} = \mathbf{0}. \tag{96}$$

where $\boldsymbol{t} = \boldsymbol{\sigma}^\infty \boldsymbol{n}$ denotes an external load vector acting on the element boundary $\partial\Omega_e^\sigma$. The occurring integrals are approximated by the Gaussian quadrature formula

$$\int_{\Omega_e} \boldsymbol{B}^T \boldsymbol{\sigma}\, d\Omega_e \approx \sum_{p=1}^{n_p} \boldsymbol{B}^T \boldsymbol{\sigma} det \boldsymbol{J}_e(\xi_p, \eta_p) w_p, \tag{97}$$

where the index p designates a Gauss point, n_p the number of integration points per element, $\boldsymbol{J}_e$ the Jacobian and w_p the weight coefficients.

Contrary to an usual aligned meshing approach, which requires the finite element boundaries to coincide with material interfaces, an unaligned approach of meshing the microstructure is used as depicted in Fig. 29. Within the element Ω_e, material properties are assigned to the integrating points

$$\boldsymbol{\sigma} = \begin{cases} \mathbb{C}^{in}\left(\boldsymbol{B}\,\hat{\mathbf{u}} - \varepsilon^0\right) & \forall\ (x_p, y_p) \in \mathcal{B}_i^{in} \\ \mathbb{C}^{out}\boldsymbol{B}\,\hat{\mathbf{u}} & \forall\ (x_p, y_p) \in \mathcal{B}^{out}, \end{cases} \tag{98}$$

with

$$\mathcal{B}_i^{in} = \{(x, y)\,|\,(x - x_i^0)^p + \rho_i^p (y - y_i^0)^p \le a_i^p\}. \tag{99}$$

This enables a simple and rapid generation of the mesh considering $i = 1, \ldots, N$ inclusions with aspect ratio $\rho_i = a_i/b_i$ simultaneously. Furthermore, high element distortion, which leads to a possible element instability is avoided consequently. As suitable start configurations, elliptical ($p = 2$) inclusions in isotropic and cubic inclusions ($p \to \infty$) in anisotropic materials are recommended. However, the represented method usually requires more finite elements or, alternatively, more integrating points than the aligned approach to achieve the same accuracy. Validity tests at different examples have shown that the GPM especially for global quantities leads to results with a satisfactory high accuracy, see Kolling and Gross (accepted for publication in 2001).

3.4 Applications

To demonstrate the applicability, some examples solved with the GPM are discussed in the following. Considered are isotropic and anisotropic materials where the already introduced dimensionless quantities are used. In addition, the anisotropy factor

$$A = \frac{2C_{44}}{C_{11} - C_{12}} \tag{100}$$

is used to characterize the anisotropy influence in 2D, compare Schmidt and Gross (1997).

a)

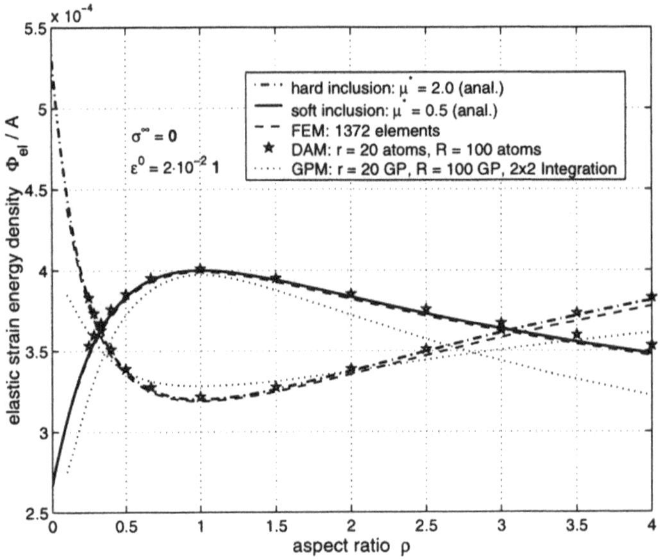

b)

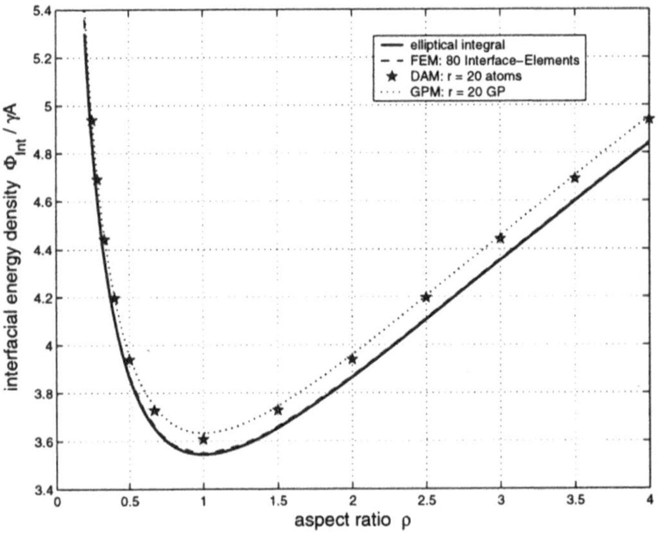

Figure 30. Validity test of the elastic strain and the interfacial energy

Single isotropic particles

A comparison of the results from different methods is shown in Figs. 30 and 31 for the equilibrium and stability of an isotropic inclusion. The elastic strain energy density (Fig. 30a) of a soft ($\mu^* = 0.5$) and a hard ($\mu^* = 2.0$) inclusion is maximum or minimum for the aspect ratio $\rho = 1$. In combination with the interfacial energy density (Fig. 30b), which is always minimum for $\rho = 1$, the equilibrium shape of a soft inclusion depends on the particle size L, for which exists a critical value L_{crit}. As it is shown in the stability diagram (Fig. 31), if L is small enough, the minimum of the total free energy is always given by a circular shape $\rho^* = \frac{a-b}{a+b} = 0 \ \forall \ L \leq L_{crit}$. For overcritical values, the equilibrium shape is getting elliptical $\rho^* \neq 0 \ \forall \ L > L_{crit}$, i.e. prolate for $\rho^* < 0$ and oblate for $\rho^* > 0$.

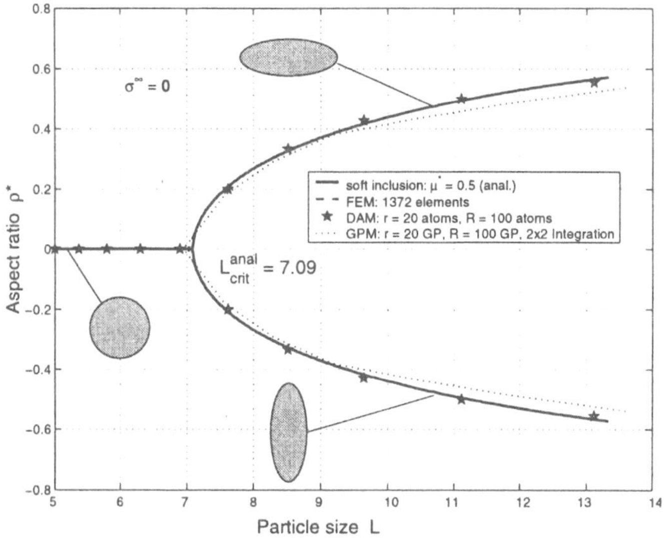

Figure 31. Bifurcation diagram

Fig. 32 shows the morphological evolution of a soft particle from this bifurcation diagram, calculated with the GPM-technique. Starting from an initial circular shape with an overcritical particle size $L = 10 > L_{crit} = 7.09$ (unstable equilibrium), the morphological transition starts with a formation of a wavy interface induced by coherency strain. Then, the waves coarsen more and more into a few larger lobs, which disappear during the simulation. After 30000 Monte Carlo steps (MCS), a stable equilibrium shape (minimum of the total energy) has been found. For comparable results, see Lee (1995).

Single orthotropic particles

The influence of stiffness ratio at same particle size is shown in Fig. 9. Starting from a circular shape as an initial shape, the Eshelby stress part of the configurational force τ_n acting on the interface is depicted (without the interfacial energy). It has the same nature for both hard and soft

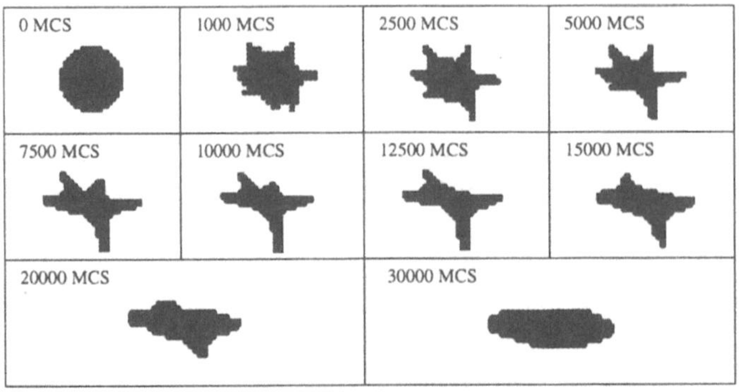

Figure 32. Shape evolution of a soft inclusion

particles. While the homogeneous particle ($\mu^* = 1.0$) becomes cubic, which is also observed for hard particles, the very soft inclusion is not convex anymore and gets a typical pillow-shaped contour, see Ardell and Meshkinpour (1994) for experiments.

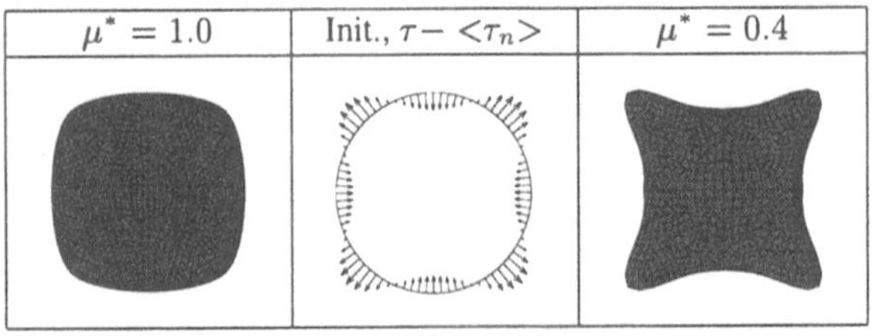

Figure 33. Equilibrium shapes at load free state

The mechanism of morphological development of isotropic and anisotropic coherent precipitates is quite comparable with respect to the rafting process. The first column of Fig. 34 shows the initial equilibrium shape at load free state and the direction of the external load σ^∞ acting on it. In the second column the respective driving force and in the last column the final equilibrium shapes are depicted. As an example for orthotropic materials, a CMSX-6 Ni-base superalloy is considered with -0.38% dilatational eigenstrain, an average Poisson's ratio of $\bar{\nu}^{in/out} = 0.28$ same for matrix and inclusion and an average Young's modulus $\bar{E}^{in} = 246$GPa and $\bar{E}^{out} = 230$GPa, see Göken and Kempf (1999) for nanoindentation results. Under pressure the average driving force acts as tension in [010] and as compression in [100]-direction, which leads to an equilibrium shape elongated in load direction. For a reverse load (in the second row), the elongation of the particle occurs perpendicular to it, which also is observed experimentally. The last row illustrates the influence of shear stresses on the morphology, which effects both an elongation and an orientation in $[1\,-1\,0]$-direction coinciding with the principal axis of the stress. In the GPM-technique, the temporal evolution from an initial to an equilibrium shape is described in terms of Monte Carlo steps (MCS), which are proportional to time t in dependence of the total

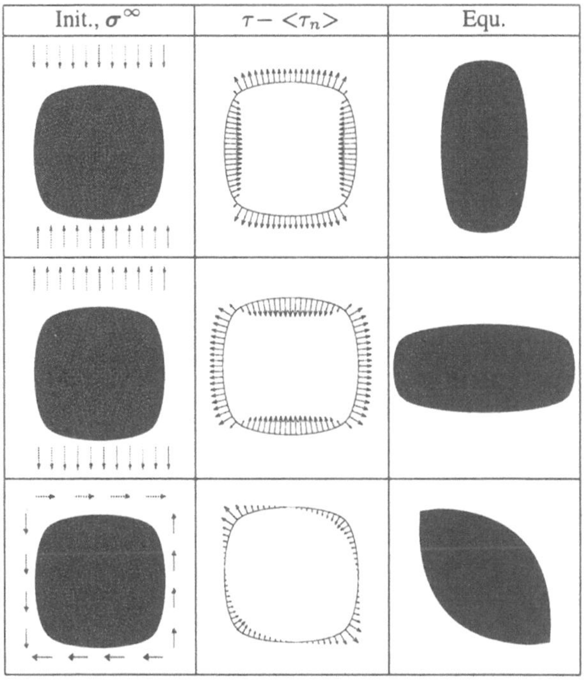

Figure 34. Equilibrium shapes due to external loads

number of Gauss Points:

$$t = \beta \, \mathrm{MCS} \tag{101}$$

where $\beta > 0$ is a material parameter. In Fig. 35 results of the Monte Carlo simulation and a linear constitutive law between the driving force τ_n and the outward normal velocity v_n are compared. They show a good agreement.

As a further Monte Carlo study, Fig. 36 illustrates the influence of the temperature on the equilibrium shape. At $T = 375K$, a smooth transition of the energy term from the initial towards the equilibrium shape takes place. In contrast, at $T = 900K$, a distinct equilibrium shape is not detectable anymore. Although the global tendency is similar as at the low temperature, the particle's interface is always in thermo-activated movement.

Multi particles

For simulating rafting processes due to external loads in real microstructures of superalloys with high second phase volume fraction it is necessary to take particle interaction into account. The interaction energy of n particles Φ_{inter} is defined as the difference of the total energy of the n-particle system and the total energy of n single particles:

$$\Phi_{inter} = \sum_{i=1}^{n} \Phi_{\mathcal{B}_i^{in}} + \Phi_{\mathcal{B}^{out}} - n\Phi_{\mathcal{B}^{in} \cup \mathcal{B}^{out}}. \tag{102}$$

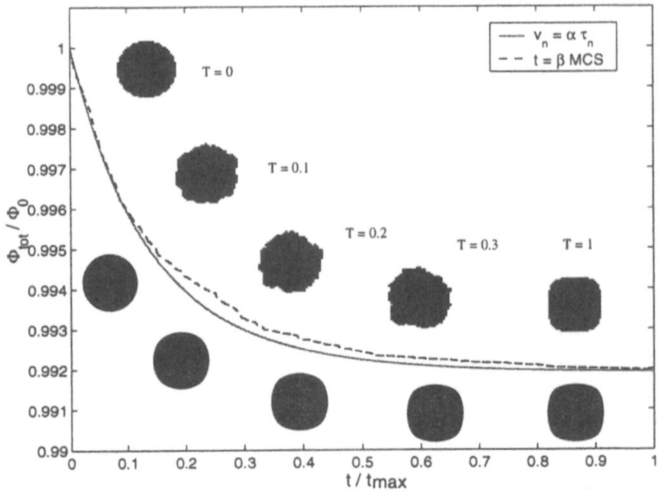

Figure 35. Time dependence of the evolution

As an example, Fig. 37 shows the interaction energy of two 2%-eigenstrained elliptical isotropic inclusions as a function of their distance d/a. For hard inclusions, the interaction energy takes a minimum at $d/a = 2.1$ indicating a stable equilibrium position. In contrast, for soft inclusions, it exists a maximum of Φ_{inter} at the same distance. For any disturbance from this unstable position, the particles either diverge ($d/a \to \infty$) or converge ($d/a \to 0$) which implies a tendency to particle-coalescence.

Using the GPM, the temporal evolution of the rafting process under tension or pressure is illustrated in Fig. 38. The morphology change of regular periodical aligned inclusions of a CMSX-6 superalloy with a volume fraction of $c = 43\%$ at a temperature of $T = 375K$ is shown. The tensile load leads to a directional mass transport orthogonal to the load direction while under pressure the opposite happens. This is qualitatively in accordance with the results for single particles.

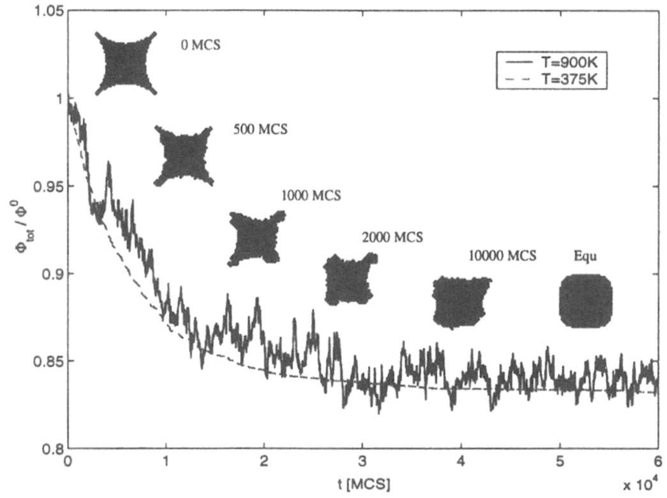

Figure 36. Temperature influence on equilibrium shape

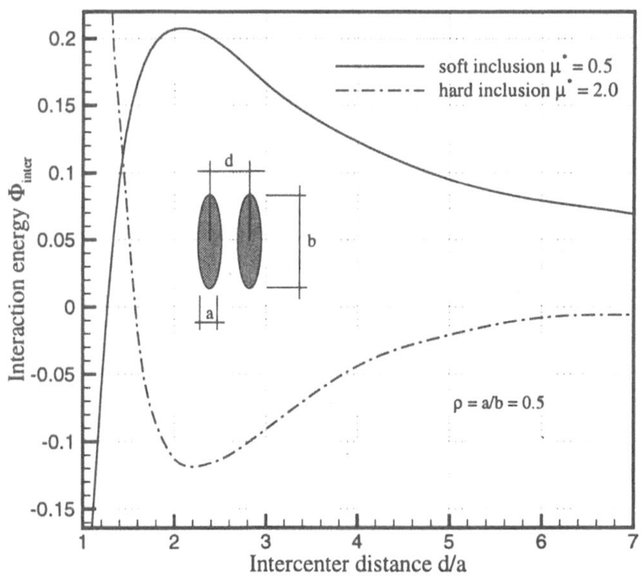

Figure 37. Particle interaction

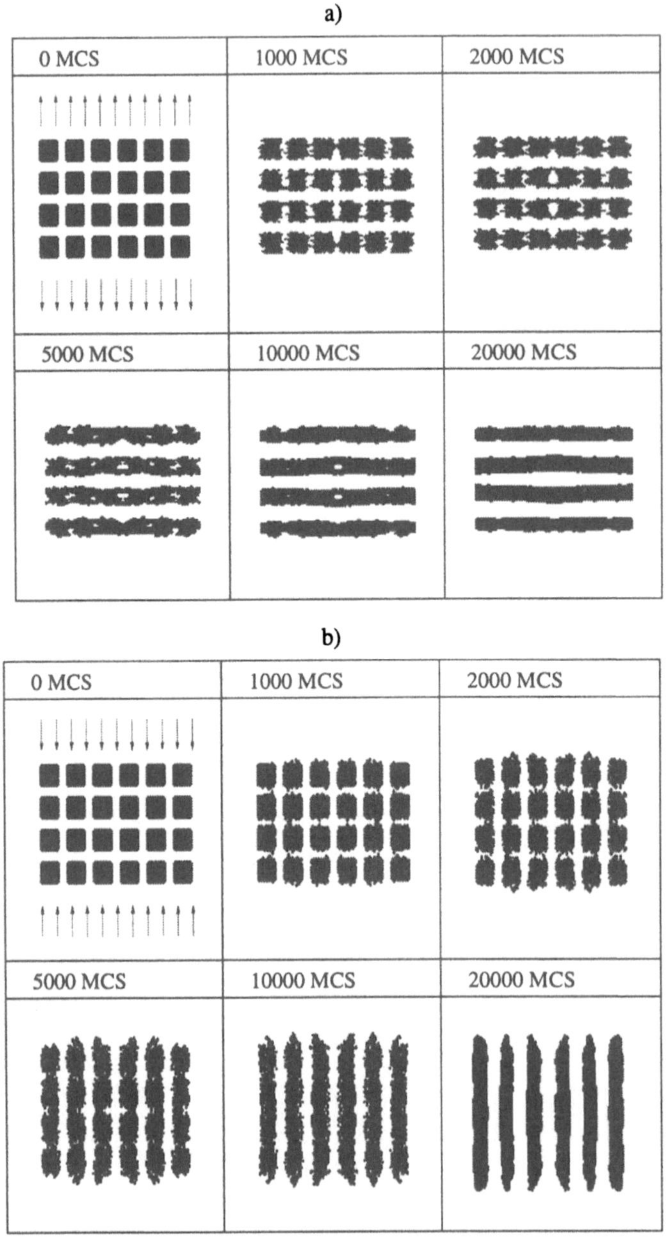

Figure 38. Rafting process of aligned particles

A Eshelby's tensor

The principal axis of an ellipse with a given area A^{P} and a given aspect ratio m are calculated by:

$$a=\sqrt{\frac{A^{\mathrm{P}}}{\pi}\frac{1+m}{1-m}}, \quad b=\sqrt{\frac{A^{\mathrm{P}}}{\pi}\frac{1-m}{1+m}}$$

Using Voigt's notation the Eshelby tensor $\mathbb{E}$ is given by

$$\mathbb{E}=\begin{bmatrix} E_{11} & E_{12} & 0 \\ E_{21} & E_{22} & 0 \\ 0 & 0 & P_{33} \end{bmatrix}$$

in the principal directions where

$$\begin{aligned}
E_{11} &= \frac{1}{2\,(1-\nu)}\left[\frac{b^2+2ab}{(a+b)^2}+(1-2\nu)\,\frac{b}{a+b}\right] \\
E_{12} &= \frac{1}{2\,(1-\nu)}\left[\frac{b^2}{(a+b)^2}-(1-2\nu)\,\frac{b}{a+b}\right] \\
E_{21} &= \frac{1}{2\,(1-\nu)}\left[\frac{a^2}{(a+b)^2}-(1-2\nu)\,\frac{a}{a+b}\right] \\
E_{22} &= \frac{1}{2\,(1-\nu)}\left[\frac{a^2+2ab}{(a+b)^2}+(1-2\nu)\,\frac{a}{a+b}\right] \\
E_{33} &= \frac{1}{2\,(1-\nu)}\left[\frac{a^2+b^2}{2\,(a+b)^2}+\frac{1-2\nu}{2}\right]
\end{aligned}$$

with $\nu=\frac{\lambda}{2(\lambda+\mu)}$ the Poison's ratio of the material. $\mathbb{E}$ is rotated by the angle φ using tensor transformation rules.

B Mechanical energy contributions

The total elastic energy Π^{el} is given by

$$\begin{aligned}
\Pi^{\mathrm{el}} &= \frac{1}{2}\int_A \boldsymbol{\sigma}:\boldsymbol{\varepsilon}^{\mathrm{el}}\,dA=\frac{1}{2}\int_A \left(\boldsymbol{\sigma}^{\infty}+{}'\boldsymbol{\sigma}\right):\left(\boldsymbol{\varepsilon}^{\infty}+{}'\boldsymbol{\varepsilon}-\boldsymbol{\varepsilon}^{0}\right)dA \\
&= \frac{1}{2}\int_A \boldsymbol{\sigma}^{\infty}:\boldsymbol{\varepsilon}^{\infty}\,dA-\frac{1}{2}\int_{A^{\mathrm{P}}} {}'\boldsymbol{\sigma}:\boldsymbol{\varepsilon}^{0}\,dA+\underbrace{\frac{1}{2}\int_A {}'\boldsymbol{\sigma}:\left(\boldsymbol{\varepsilon}^{\infty}+{}'\boldsymbol{\varepsilon}\right)dA}_{I_1} \\
&\quad \underbrace{+\frac{1}{2}\int_A \boldsymbol{\sigma}^{\infty}:\left({}'\boldsymbol{\varepsilon}-\boldsymbol{\varepsilon}^{0}\right)dA}_{I_2},
\end{aligned}$$

where I_1 can be integrated by parts to give

$$\begin{aligned} I_1 &= \int_A {'\boldsymbol{\sigma}} : (\boldsymbol{\varepsilon}^\infty + {'\boldsymbol{\varepsilon}})\, dA \\ &= -\int_A \underbrace{\operatorname{div}{'\boldsymbol{\sigma}}}_{=0} \cdot (\boldsymbol{u}^\infty + {'\boldsymbol{u}})\, dA + \int_{\partial A} \underbrace{{'\boldsymbol{\sigma}}\boldsymbol{n}}_{=0} \cdot (\boldsymbol{u}^\infty + {'\boldsymbol{u}})\, dA\,, \end{aligned}$$

thus I_1 vanishes identically due to equilibrium and boundary conditions of the comparison problem (49).

The second integral I_2 can be rewritten

$$\begin{aligned} I_2 &= \int_A \boldsymbol{\sigma}^\infty : ({'\boldsymbol{\varepsilon}} - \boldsymbol{\varepsilon}^0)\, dA = \int_A \mathbb{C}^{\mathrm{M}}\boldsymbol{\varepsilon}^\infty : ({'\boldsymbol{\varepsilon}} - \boldsymbol{\varepsilon}^0)\, dA \\ &= \int_{A^{\mathrm{M}}} {'\boldsymbol{\sigma}} : \boldsymbol{\varepsilon}^\infty\, dA + \int_{A^{\mathrm{P}}} \mathbb{C}^{\mathrm{M}}\boldsymbol{\varepsilon}^\infty : ({'\boldsymbol{\varepsilon}} - \boldsymbol{\varepsilon}^0 - \boldsymbol{\varepsilon}^{\mathrm{eq}} + \boldsymbol{\varepsilon}^{\mathrm{eq}})\, dA \\ &= \int_{A^{\mathrm{M}}} {'\boldsymbol{\sigma}} : \boldsymbol{\varepsilon}^\infty\, dA + \int_{A^{\mathrm{P}}} {'\boldsymbol{\sigma}} : \boldsymbol{\varepsilon}^\infty\, dA + \int_{A^{\mathrm{P}}} \boldsymbol{\sigma}^\infty : \boldsymbol{\varepsilon}^{\mathrm{eq}}\, dA\,, \end{aligned}$$

where the constitutive relation of the comparison problem (49) has been used. Collecting terms and integrating by parts yields

$$I_2 = \underbrace{\int_A {'\boldsymbol{\sigma}} : \boldsymbol{\varepsilon}^\infty\, dA}_{=0 \quad \text{as in } I_1} + \int_{A^{\mathrm{P}}} \boldsymbol{\sigma}^\infty : \boldsymbol{\varepsilon}^{\mathrm{eq}}\, dA\,.$$

The above relations are generally valid, but for the special case of an ellipse we have

$$\Pi^{\mathrm{el}} = \frac{1}{2}\int_A \boldsymbol{\sigma}^\infty : \boldsymbol{\varepsilon}^\infty\, dA + A^{\mathrm{P}}\left(\frac{1}{2}\boldsymbol{\sigma}^\infty : \boldsymbol{\varepsilon}^{\mathrm{eq}} - \frac{1}{2}{'\boldsymbol{\sigma}} : \boldsymbol{\varepsilon}^0\right).$$

In the potential of the external load

$$\Pi^{\mathrm{ext}} = -\int_{\partial A} \boldsymbol{t}^\infty \cdot \boldsymbol{u}\, ds = -\int_{\partial A} \boldsymbol{t}^\infty \cdot \boldsymbol{u}^\infty\, ds - \underbrace{\int_{\partial A} \boldsymbol{t}^\infty \cdot {'\boldsymbol{u}}\, ds}_{I_3}$$

the last term I_3 is reformulated

$$\begin{aligned} I_3 &= \int_{\partial A} \boldsymbol{\sigma}^\infty\boldsymbol{n} \cdot {'\boldsymbol{u}}\, dA = \int_A \boldsymbol{\sigma}^\infty : {'\boldsymbol{\varepsilon}}\, dA \\ &= \int_A \boldsymbol{\sigma}^\infty : ({'\boldsymbol{\varepsilon}} - \boldsymbol{\varepsilon}^0 - \boldsymbol{\varepsilon}^{\mathrm{eq}})\, dA + \int_A \boldsymbol{\sigma}^\infty : (\boldsymbol{\varepsilon}^0 + \boldsymbol{\varepsilon}^{\mathrm{eq}})\, dA \\ &= \int_{A^{\mathrm{P}}} \mathbb{C}^{\mathrm{M}}\boldsymbol{\varepsilon}^\infty : ({'\boldsymbol{\varepsilon}} - \boldsymbol{\varepsilon}^0 - \boldsymbol{\varepsilon}^{\mathrm{eq}})\, dA + \int_{A^{\mathrm{M}}} \mathbb{C}^{\mathrm{M}}\boldsymbol{\varepsilon}^\infty : {'\boldsymbol{\varepsilon}}\, dA \\ &\quad + \int_{A^{\mathrm{P}}} \boldsymbol{\sigma}^\infty : (\boldsymbol{\varepsilon}^0 + \boldsymbol{\varepsilon}^{\mathrm{eq}})\, dA \\ &= \underbrace{\int_A {'\boldsymbol{\sigma}} : \boldsymbol{\varepsilon}^\infty\, dA}_{=0 \quad \text{as in } I_2} + \int_{A^{\mathrm{P}}} \boldsymbol{\sigma}^\infty : (\boldsymbol{\varepsilon}^0 + \boldsymbol{\varepsilon}^{\mathrm{eq}})\, dA\,. \end{aligned}$$

Thus for an ellipse this yields

$$\Pi^{\text{ext}} = -\int_{\partial A} \boldsymbol{t}^{\infty} \cdot \boldsymbol{u}^{\infty}\, ds - A^{\text{P}} \boldsymbol{\sigma}^{\infty} : \left(\boldsymbol{\varepsilon}^{0} + \boldsymbol{\varepsilon}^{\text{eq}}\right) .$$

Therefore the mechanical contribution $\Pi^{\text{el}} + \Pi^{\text{ext}}$ to the total potential Π becomes

$$\Pi^{\text{el}} + \Pi^{\text{ext}} = -A^{\text{P}} \left(\frac{1}{2} \boldsymbol{\sigma}^{\infty} : \boldsymbol{\varepsilon}^{\text{eq}} + \frac{1}{2}{}' \boldsymbol{\sigma} : \boldsymbol{\varepsilon}^{0} + \boldsymbol{\sigma}^{\infty} : \boldsymbol{\varepsilon}^{0} \right) + \Pi^{0}$$

where Ψ^0 is given in (65).

C HELMHOLTZ'S free energy

For a two-phase structure, broken into a large number of elementary particles, the basic equations defining Helmholtz's free energy are described in the following.

C.1 Total free energy

The total free energy density is defined as

$$\psi(\Theta, \varepsilon, c) = \phi - \Theta s, \tag{103}$$

where ϕ, Θ and s are the internal energy density, temperature and entropy respectively. Furthermore, $\boldsymbol{\varepsilon} = \nabla^{s} \boldsymbol{u}$ is the strain tensor and c is the concentration of foreign particles. For later purposes, first the total differentials of entropy s and free energy density ψ in terms of macroscopic thermodynamics are given:

$$d\psi(\Theta, \varepsilon, c) = d\phi - \Theta\, ds - s\, d\Theta = \underbrace{\frac{\partial \psi}{\partial \Theta}}_{=-s} d\Theta + \underbrace{\frac{\partial \psi}{\partial \varepsilon}}_{=\sigma} d\varepsilon + \underbrace{\frac{\partial \psi}{\partial c}}_{=\mu} dc. \tag{104}$$

The chemical potential μ is a generalized force which corresponds to the concentration c of particles. With this relation, we obtain:

$$\Theta\, ds = d\phi - \boldsymbol{\sigma}\, d\boldsymbol{\varepsilon} - \mu\, dc. \tag{105}$$

C.2 Statistical definition of the free energy

Now we consider a N-particle system with a particle concentration $c = N/\rho$ (ρ is the density of the matrix material). According to classical statistical mechanics, the probability P_r for a micro-state r in a canonical ensemble is represented by the Gauss-type distribution

$$P_r(\Theta, \varepsilon, c) = \frac{1}{Z(\beta, \varepsilon, c)}\; e^{-\beta \phi_r(\varepsilon, c)}, \quad \text{with} \quad \beta = \beta(\Theta) = \frac{1}{k_B \Theta} \tag{106}$$

where Φ_r is the energy corresponding to the state r, Z the yet undetermined partial function and $k_B = 1.380658 \cdot 10^{-23}\ J/K$ is Boltzmann's constant. Normalizing P_r by

$$\sum_r P_r = \sum_r \frac{1}{Z(\beta,\varepsilon,c)}\ e^{-\beta\phi_r(\varepsilon,c)} = 1 \tag{107}$$

the partial function results from (106) as

$$Z(\beta,\varepsilon,c) = \sum_r e^{-\beta\phi_r(\varepsilon,c)}. \tag{108}$$

The total differential of its logarithm yields

$$d\,lnZ(\beta,\varepsilon,c) = \frac{\partial lnZ}{\partial\beta}\,d\beta + \frac{\partial lnZ}{\partial\varepsilon}\,d\varepsilon + \frac{\partial lnZ}{\partial c}\,dc. \tag{109}$$

where the different terms can be formulated and rearranged as

$$\begin{aligned}
\frac{\partial lnZ(\beta,\varepsilon,c)}{\partial\beta} &= \frac{1}{Z(\beta,\varepsilon,c)}\sum_r -\phi_r(\varepsilon,c)e^{-\beta\phi_r(\varepsilon,c)} \\
&= -\sum_r \phi_r(\varepsilon,c)\underbrace{\frac{1}{Z(\beta,\varepsilon,c)}e^{-\beta\phi_r(\varepsilon,c)}}_{P_r} \\
&= -\sum_r \phi_r(\varepsilon,c)P_r = -\langle\phi_r(\varepsilon,c)\rangle = -\phi(\varepsilon,c),
\end{aligned} \tag{110}$$

$$\begin{aligned}
\frac{\partial lnZ(\beta,\varepsilon,c)}{\partial\varepsilon} &= \frac{1}{Z(\beta,\varepsilon,c)}\sum_r -\beta\frac{\partial\phi_r(\varepsilon,c)}{\partial\varepsilon}e^{-\beta\phi_r(\varepsilon,c)} \\
&= -\beta\sum_r \frac{\partial\phi_r(\varepsilon,c)}{\partial\varepsilon}P_r = -\beta\frac{\partial}{\partial\varepsilon}\underbrace{\sum_r \phi_r(\varepsilon,c)P_r}_{\phi(\varepsilon,c)} \\
&= -\beta\underbrace{\frac{\partial\phi(\varepsilon,c)}{\partial\varepsilon}}_{=\sigma} = -\beta\sigma,
\end{aligned} \tag{111}$$

$$\begin{aligned}
\frac{\partial lnZ(\beta,\varepsilon,c)}{\partial c} &= \frac{1}{Z(\beta,\varepsilon,c)}\sum_r -\beta\frac{\partial\phi_r(\varepsilon,c)}{\partial c}e^{-\beta\phi_r(\varepsilon,c)} \\
&= -\beta\sum_r \frac{\partial\phi_r(\varepsilon,c)}{\partial c}P_r = -\beta\frac{\partial}{\partial c}\sum_r \phi_r(\varepsilon,c)P_r \\
&= -\beta\frac{\partial\phi(\varepsilon,c)}{\partial c} = -\beta\mu,
\end{aligned} \tag{112}$$

with $\phi = \langle \phi_r \rangle$ being the observable macroscopic internal energy. With this, equation (109) can be written as

$$d\,lnZ(\beta,\varepsilon,c) = -\underbrace{\phi\,d\beta}_{=d(\phi\beta)-d\phi\beta} -\beta\sigma d\varepsilon - \beta\mu dc \tag{113}$$

or after rearrangement

$$\begin{aligned} d[lnZ(\beta,\varepsilon,c) + \phi\beta] &= \beta d\phi - \beta\sigma d\varepsilon - \beta\mu dc \\ &= \frac{1}{k_B\Theta}\,(d\phi - \sigma d\varepsilon - \mu dc\,.) \end{aligned} \tag{114}$$

Now, comparing the right hand side with the corresponding macroscopic quantities in equation (105) leads to

$$ds = k_B d(lnZ + \beta\phi) \quad \Rightarrow \quad s = k_B\,(lnZ + \beta\phi) + C\,. \tag{115}$$

The integration constant C can be determined by the boundary condition of the zero state entropy. For this purpose, we consider a system with energy eigenvalues

$$\phi_0 < \phi_1 \leq \phi_2 \leq \cdots \tag{116}$$

and the partial function

$$Z = e^{-\beta\phi_0} + e^{-\beta\phi_1} + \cdots = e^{-\beta\phi_0}\left(1 + e^{-\beta\Delta\phi} + \cdots\right). \tag{117}$$

At low temperatures dominates the energy difference $\Delta\phi = \phi_1 - \phi_0 >> k_B\Theta$ and we obtain approximately:

$$Z \approx e^{-\beta\phi_0} \quad \Rightarrow \quad lnZ \approx -\beta\phi_0. \tag{118}$$

Likewise the macroscopical internal energy can be approximated by

$$\phi = \phi_0 + \phi_1 + e^{-\beta\Delta\phi} + \cdots \approx \phi_0. \tag{119}$$

Thus the constant C yields:

$$C = s - k_B\,(-\beta\phi_o + \beta\phi_o) = s. \tag{120}$$

As a consequence of the third law of thermodynamics

$$\lim_{\Theta\to 0} s = 0 \tag{121}$$

the constant C has to vanish. Thus, the relation between classical and statistical formulation reads

$$\psi(\Theta,\varepsilon,c) = \phi(\varepsilon,c) - \Theta s = -k_B\Theta lnZ(\Theta,\varepsilon,c)\,. \tag{122}$$

C.3 Minimization of the free energy

As can be seen from equation (122), the minimum of the total free energy ψ is affine to the minimum of $\ln Z$. Since the partial function Z is monotonous, it is sufficient to minimize the single terms of the sum (108), respectively the difference of the energy $\Delta\Phi$ after a change in the state of the system. The new state will then be accepted with a probability of $\exp(-\beta\Delta\Phi)$. This is realized in the numerical implementation by comparing this exponential function with a random number between 0 and 1:

$$f(\Delta\Phi) = e^{-\beta\Delta\Phi} > z_n, \quad z_n \in (0,1)\,. \tag{123}$$

Only if the random number is less, the new state will be accepted otherwise rejected.

One problem with large-scale simulations are the aforementioned random numbers. They can be obtained by multiplying an odd-numbered random initial integer z_0 (e.g. $z_0 = 1$) with 16807 again and again on a 32-bit computer. This procedure leads to about 10^9 different pseudo-random numbers before their sequence is repeated. Then, in order to obtain numbers within the interval in $(0,1)$, the random number have to be be normalized by 2^{-31}.

In contrast to the continuum formulation, minimization of the free energy Ψ of the discrete system does not correspond necessarily with a minimization of the internal energy Φ. This depends strongly on the temperature of the system, which influences the entropy:

$$\lim_{\Theta\to 0} e^{-\beta\Delta\Phi} = 0 \quad \Rightarrow \quad \text{no exchange accepted}$$

$$\lim_{\Theta\to \infty} e^{-\beta\Delta\Phi} = 1 \quad \Rightarrow \quad \text{every exchange accepted}$$

The last case occurs e.g. for temperatures above the melting temperature Θ_m.

References

Ardell, A., and Maheshwari, A. (1992). Anomalous coarsening behaviour of small volume fractions of Ni_3Al precipitates in binary Ni-Al alloys. *Acta metall. mater.* 40(10):2661–2667.

Ardell, A., and Meshkinpour, M. (1994). Role of volume fraction in the coarsening of Ni_3Si precipitates in binary Ni-Si alloys. *Material Science and Engineering* A 185:153–163.

Ardell, A., and Nicholson, R. (1966). On the modulated structure of aged Ni-Al alloys. *Acta metall.* 14:1205–1309.

Ardell, A., and Rastogi, P. (1971). The coarsening behaviour of the γ' precipitate in Nickel-Silicon alloys. *Acta metall.* 19:321–330.

Binder, K., ed. (1986). *Monte Carlo methods in statistical physics.* Berlin, Heidelberg, New York, Barcelona, Hong Kong, London, Milan, Paris, Singapore, Tokyo: Springer.

Binder, K., ed. (1987). *Applications of the Monte Carlo method in statistical physics.* Berlin, Heidelberg, New York, Barcelona, Hong Kong, London, Milan, Paris, Singapore, Tokyo: Springer.

Brebbia, C., and Telles, J. (1985). *Boundary Element Techniques.* Springer Verlag.

Eshelby, J. (1957). The determination of the elastic field of an ellipsoidal inclusion and related problems. *Proc. R. Soc. Lond.* A 241:376–396.

Eshelby, J. (1970). *Energy relations and the energy-momentum tensor in continuum mechanics,* In Kanninen (1970). 77–115.

Fried, E., and Gurtin, M. (1993). Continnum theory of thermally induced phase transitions based on an order parameter. *Physica D* 68:326–343.

Göken, M., and Kempf, M. (1999). Microstructural properties of superalloys investigated by nanoindentations in an atomic force microscope. *Acta Mater.* (47):1043–1052.

Gurtin, M., and Voorhees, P. (1993). The continuum mechanics of coherent two-phase elastic solids with mass transport. *Proc. R. Soc. Lond.* A 440:323–343.

Gurtin, M. (1995). The nature of configurational forces. *Arch. Rational Mech. Anal.* 131:67–100.

Hoover, W. G., Ashurst, W. T., and Olness, R. J. (1974). Two–dimensional computer studies of crystal stability and fluid viscosity. *J. Chem. Phys.* 60(10):4043–4047.

Hornbogen, E., and Roth, M. (1967). Die Verteilung hohärenter Teilchen in Nickellegierungen. *Z. Metallkde* 58:842–855.

Johnson, W., and Cahn, J. (1984). Elastically induced shape bifurcations of inclusions. *Acta metall.* 32(11):1925–1933.

Johnson, W., Berkenpas, M., and Laughlin, D. (1988). Precipitate shape transitions during coarsening under uniaxial stress. *Acta metall.* 36(2):3149–3162.

Kaganova, I., and Roitburd, R. (1988). Equilibrium between elastically-interacting phases. *Sov. Phys. JETP* 67(4):1173–1183.

Kanninen, M., ed. (1970). *Inelastic Behaviour of Solids.* New York: McGraw Hill.

Kolling, S., and Gross, D. (2000). Description of two–phase materials using discrete atom method. *ZAMM* 80:S385–S386.

Kolling, S., and Gross, D. (accepted for publication in 2001). Simulation of microstructural evolution in materials with misfitting precipitates. *Journal of Probabilistic Engineering Mechanics.*

Lee, J.K. (1995). Coherency strain analysis via discrete atom method. *Scr. Met. Mat.* 32(4):559–564.

Lee, J.K. (1996a). Effects of applied stress on coherent precipitates via a disctrete atom method. *Metals and Materials* 2(3):183–193.

Lee, J.K. (1996b). A study on coherency strain precipitate morphology via a discrete atom method. *Met. Mat. Trans.* 27A:1449–1459.

Leo, P., and Sekerka, R. (1989). The effect of surface stress on crystal-melt and crystal-crystal equilibrium. *Acta metall.* 37(12):3119–3138.

Leo, P., Lowengrub, J., and Jou, H. (1998). A diffuse interface model for microstructural evolution in elastically stressed solids. *Acta mater.* 46(6):2113–2130.

Lubliner, J. (1990). *Plasticity Theory.* New York: Macmillan Publishing Company.

Luenberger, D. (1984). *Linear and nonlinear programming.* Addison-Wesley, 2 edition.

Mueller, R., and Gross, D. (1998a). 3D equilibrium shapes in two-phase materials. *ZAMM* 78(2):635–636.

Mueller, R., and Gross, D. (1998b). 3D simulation of equilibrium morphologies of precipitates. *Comp. Mat. Sci.* 11:35–44.

Mura, T. (1987). *Micromechanics of Defects in Solids.* Martinus Nijhoff Publishers.

Nemat-Nasser, S., and Hori, M. (1993). *Micromechanics: Overall properties of heterogeneous materials.* Amsterdam, London, New York, Tokyo: North Holland.

Schclar, N. (1994). *Anisotropic Analysis using Boundary Elements*, volume 20 of *Topics in Engineering.* Southampton UK and Boston USA: Computational Mechanics Publications.

Schmidt, I., and Gross, D. (1995). A strategy for determinig the equilibrium shape of an inclusion. *Arch. Mech.* 47(2):379–390.

Schmidt, I., and Gross, D. (1997). The equilibrium shape of an elastically inhomogeneous particle. *J. Mech. Phys. Solids* 45(9):1521–1549.

Schmidt, I., and Gross, D. (1999). Directional coarsening in Ni-base superalloys: analytical results for an elasticity based model. *Proc. R. Soc. Lond.* 455:3085–3106.

Schmidt, I., Mueller, R., and Gross, D. (1998). The effect of elastic inhomogeneity on equilibrium and stability of a two particle morphology. *Mechanics of Materials* 30:181–196.

Schmidt, I. (1997). *Gleichgewichtsmorphologien elastischer Einschlüsse*. Ph.D. Dissertation, Technische Hochschule Darmstadt, D-64289 Darmstadt.

Su, C., and Voorhees, P. (1996). The dynamics of precipitate evolution in elastically stressed solids - I, inverse coarsening. *Acta mater.* 44(5):1987–1999.

Thompson, M., Su, C., and Voorhees, P. (1993). The equilibrium shape of a misfitting precipitate. *Acta metall. mater.* 42(6):2107–2122.

Voorhees, P., McFadden, G., and Johnson, W. (1992). On the morphological development of second-phase particles in elastically-stressed solids. *Acta metall. mater.* 40(11):2979–2992.

Schoenlein, L.H., Rühle, M., and Heuer, A.H. (1984). In Situ Straining Experoments of Mg-PSZ Single Crystals *Adv. in Ceramics* 12, Science and Technology of Zirkonia II, eds: Claussen, N., Rühle, M. and Heuer, A.H., The American Ceramic Society:275–282.

Rühle, M., and Heuer, A.H. (1984). Phase Transformations in ZrO_2-Containing Ceramivs: II The Martensitic Reaction in t-ZrO_2 *Adv. in Ceramics* 12, Science and Technology of Zirkonia II, eds: Claussen, N., Rühle, M. and Heuer, A.H., The American Ceramic Society:1–32.

Wahi, R.P. (1997). Nickel base superalloys: Deformation characteristics at elevated temperatures *Adv. in Comp. Eng. Sci.*, Tech. Sci. Press, eds. Atluri, S.N, Yagawa, G.:85–90

Variational Formulation of a Material Ageing Model

A. Chudnovsky[1] and S. Preston[2]

[1] CEMM,College of Engineering, University of Illinois at Chicago, Chicago, IL, 60680.
[2] Department of Mathematical Sciences, Portland State University, Portland, OR, 97207-0751.

Abstract. A material ageing parameter, i.e. an additional internal variable, is introduced as the conjugate to the Canonical Energy Momentum Tensor. Ageing is manifested in variation of basic material characteristics such as density, moduli of elasticity, yeald stress, strength and toughness.
Four-dimensional intrinsic (material) Riemannian metric G of the material 4D space-time continuum P is utilized as the characteristic of the processes developing in the material. The ageing process is modeled as the evolution of the metric G (most importantly of its time related component) of the material space-time P embedded into 4D Absolute (Newtonian) space-time with Euclidian metric.
The kinetic equation for material ageing parameter is derived by following the classical variational approach. The formulation of a Lagrangian for an ageing elastic media and the derivation of a system of coupled elastostatics and ageing equations constitute the central part of the work. The external and internal Balance Laws associated with certain symmetries of internal (material) and external (Absolute) space-time geometries are briefly reviewed from a new viewpoint presented in the paper. A few examples that provide a basis for experimental examinations of the proposed model are also discussed.

1 Introduction

A long list of strength criteria has been proposed during the last three centuries. Numerous attempts to verify the strength criteria in the form of critical stress, critical strain or critical energy density values led to the recognition that the knowledge of stress and strain states are not sufficient, in general, for the modeling of failure. The formulations of deformation and fracture mechanisms maps provide an insight to the problem: for the same stress-strain states the fracture occurs by nucleation, growth and coalescence of different types of micro-defects depending on temperature, strain rate, geometrical constraints, etc. (Ashby et al. 1979). In particular, there is a strong manifestation of time dependence in fracture, i.e., fracture is usually the end result of a long process of material ageing. Here we use the term "ageing" in a general sense that implies any variation in the chemical make-up, i.e., chemical degradation, structural and morphological changes such as phase transformation, phase corsering, nucleation growth of microdefects such as dislocations and voids, shear bands, crazes, micro-cracks etc. Thus it becomes apparent that in addition to the stress and strain tensors, a parameter of state (that can be named an "ageing" parameter) is needed to represent on a continuum level the sub-micro and micro-structural changes of material. A kinetic equation for the evolution of the ageing parameter shall represent the process of material ageing.

* Authors express their gratitude to Dan Streeter for his help with the preparation of the manuscript

An ageing parameter, considered in a narrow sense, that accounts only for micro-defects such as micro-cracks is usually called a "damage" parameter. Various damage parameters (scalars, second rank tensors, a distribution on the unit radius sphere etc.) and the corresponding kinetic equations for those parameters have been widely discussed within Continuum Damage Mechanics (CDM) originated by L. Kachanov (1958). There was noticeable progress in CDM in the seventies and eighties. However, the main obstacles for further development of CDM are the difficulties in the direct observations and measurements of the damage parameter and its evolution. Thus, the introduction of a physically sound damage (ageing) parameter and deriving, rather than *postulating*, a kinetic equation for it.are still a challenge. In this paper we propose a variational approach in addressing this challenge.

Almost a decade prior to the first formulation of CDM (Kachanov, 1958), J. Eshelby (1951) in his pioneering work introduced the concept of the Energy Momentum Tensor (EMT) of elasticity and emphasized the direct connection between EMT and generalized forces acting on the defects. Since the generalized forces associated with the defects or morphological transformations are related to EMT, it is expected that the conjugate parameter to the EMT can be employed to characterize the state of damage or ageing. It is interesting to notice that the field theory approach employed by Eshelby in his work was perfectly suitable for the formulation of a geometrical parameter conjugate to the EMT as well as the corresponding kinetic equation. However, the particular realization of the field theory and the form of the EMT selected by Eshelby for quasi-static elasticity made it difficult to recognize the conjugate as an age (damage) parameter.

There was a significant development of Classical Field Theory during the last twenty years triggered by the studies in Gauge Field Theories and Hamiltonian Mechanics. It resulted in the development of powerful mathematical methods of analysis and geometry and revived interest in studying geometrical structures of various theories in continuum mechanics, astrophysics and solid state physics.

Developments of the microstructural theory of continuous media pioneered in the works of Eshelby (1951,1970), Kondo and further developed by E.Kroner, C.-C.Wang, W.Noll, F. Bloom, G.Mougin, M.Epstein and others (see exposition and bibliography in (Maugin 1993)) , demonstrated the importance of intrinsic geometrical structures such as "material metric" and "material connections," that reflects the inhomogeneity of the material properties and allows an elegant "dual" space-material presentation of Elasticity Theory. Especially interesting is the relationship between the Eshelby Tensor and the **material connection** proposed by G.Mougin and M.Epstein (Epstein and Maugin 1990) and their work on the material evolution laws (Epstein 1999; Epstein and Maugin preprint 2000).

In a series of works by Pak and Herrmann (1986), Herrmann (1981), Golebiewska-Herrmann (1981) and their coworkers the relationship between the "laboratory" and "material" conservation laws (balance equations) of Elasticity and Thermoelasticity have been studied in detail.

At the same time in the works by J. Marsden (1983), J. Marsden and T.Hughes (1983), the classical non-linear Elasticity Theory was reformulated as a Lagrangian Theory where the Lagrangian was considered as a functional of two metrics: an inner material metric and a metric induced via deformation from the Euclidean metric of the laboratory space. J.Marsden and J.Simo have developed the covariant approach to the balance equations of Elasticity Theory (Marsden 1983); Simo and Marsden 1984)) and have studied the relations between the "space" and "material" (rotated) stress tensors for the **material manifold endowed with two metrics**, getting the **rotated stress tensor** as the parameter conjugate to one of these two material metrics, (Simo,

Marsden 1984). Along the same lines an interesting work of N. Rosen (1980), on the General Relativity Theory for a space endowed with **two** metrics, one serving as the "reference frame" for the theory should be mentioned.

The objective in this paper is to employ a 4D inner material metric tensor G (MMT) as the ageing (damage) parameter. The equations of evolution for the MMT G are introduced as the Euler-Lagrange equations resulting from a Variational Principle. The driving force of the evolution of metric G (and of the related quantities, characterizing the properties of the material) is the Canonical Energy-Momentum Tensor closely related with the Eshelby Tensor.

A justification of such a choice of ageing parameter is based on the observation of a dilatation or shrinkage associated with material ageing as well as a variation with ageing of the resonance atomic frequencies on the atomic scale and characteristic relaxation times measured in macroscopical studies. In other terms, the internal length and time scales change with ageing when compared with the corresponding absolute (laboratory) scales. More specifically, a variation of an intrinsic material time scale is the most sensitive indicator of ageing. "The measurement of time in the laboratory as well as in material (intrinsic time) can be accomplished by several methods. One is the use of oscillating processes as in clocks with pendulum or crystal-based timepieces. Another way of measuring time is the use of a uni-directional evolution of state. For instance, in King Arthur's eras, time was measured by the burning of a candle which had numbered and colored beeswax strips. The third well-known way of measuring time is associated with relaxation processes. An example is an electronic relaxation generator which utilizes discharge of a capacitor. Fading of a phosphorus luminescence of the coating on the watch hands and numbers is another example. Any relaxation measurement essentially requires an exitation input which enables a response. The exitations are placement of electric charge on a capacitor and illumination of a watch face. the responses are the exponential discharge of a capacitor through a resistor and dimming of luminiscence of the phosphorous. Such processes are well suited for measuring intrinsic time scale changes because they reflect atomic or interatomic events.

Let us consider an external exitation of a material which responds with a specific change in its state. the decay or fading of the response constitutes the relaxation process. Usually, the decay can be described by an exponential function (within certain limits) e^{t/τ_0} where t is time and τ_0 is the time constant characterizing the rate of relaxation. Usually τ_0 becomes smaller with increase in temperature or decrease in pressure. For instance, fading in phosphorous glow runs slower at colder temperatures (Chudnovsky 1996)." The stress relaxation or strain recovery process of organic glasses are strongly effected by temperature and most importantly by ageing. Thus the time constant τ_0 may serve as the material scale parameter characterizing an intrinsic time.

The kinematics of a media with a variable Riemannian metric G in a 4D material space-time P, embedded into 4D Absolute (Newton's) space-time M^4 with the Euclidean metric H is discussed in Section II. A conventional elastic strain tensor E^{el} and a 3D "ground state" metric tensor are introduced in this section as a measure of deformation and a natural analog of the "unstrained state" respectively. A variational formulation of ageing theory and the Euler-Lagrange equations (equations of elasticity coupled with the ageing equations) are considered in Section III. In Section IV we discuss some properties of Ageing Equations and present several examples. In Section V we present the Space and Material Balance Laws, corresponding to the material and laboratory symmetries. The Energy-Momentum Balance Law and the decomposition of the Energy Momentum Tensor into components, including the Eshelby Tensor and terms related to the ageing processes are studied in Section VI. The relationship between the Canonical Energy-

Momentum Tensor of elasticity with ageing and the Eshelby tensor are analyzed in Section VII. We conclude with the summary in Section VIII.

2 Kinematics of media with a variable metric.

In this section we introduce the basic elements of the kinematics of a continuum with a variable metric, including material space-time P, 4D material metric G, 4D deformations φ, slicing of the material space-time by the surfaces of constant laboratory time $B_{\phi,t}$, and total, elastic and irreversible deformation tensors. The mass conservation law is also formulated.

2.1 Laboratory and Material Space-Time

Let us consider 4D Euclidean vector space $M = \mathbb{R} \times \mathbb{R}^3$ (**laboratory space-time**) with the standard Euclidean metric H. There exists the volume form d^4v generated by this metric.

We select global coordinates $x^i, i = 1, 2, 3$, in the laboratory space $\mathbb{R}^3$ and $x^0 = t$ on the time axes $\mathbb{R}$.

Hyperplanes $t = c$ are endowed with the 3D Euclidean metric h induced by H. One can extend tensor h to the degenerate covariant tensor $\hat{h}$ in M, taking $\hat{h}_{0i} = \hat{h}_{i0} = 0$.

A material body is considered here, in a conventional way as a 3D manifold B, i.e. a set of "idealized" material points (with the coordinates $X^I, I = 1, 2, 3$ induced by a (fixed) **reference configuration** i.e.,a diffeomorphic embedding $\phi_3 : B \longmapsto \mathbb{R}^3$) (Marsden and Hughes 1983).

Cylinder $P = \mathbb{R} \times B$ (with the coordinates $(X^0 = T, X^I, I = 1, 2, 3)$) is considered as equiped with the 4D Riemannian metric G (Material Metric Tensor, shortly MMT) with the components G_{IJ} relative to the coordinates $X^0 + T,\ X^I, I = 1, 2, 3$. Space (P, G) is further refered to as the **material "space-time"**. Metric G defines the 4D volume form $dV = \sqrt{|G|}d^4X$, where $|G|$ is the determinant of the matrix (G_{IJ}).

An example of such a material metric G can be constructed as follows. Extend the "reference configuration" ϕ_3 to the diffeomorphic embedding $\phi_0 : P \longmapsto M,\ \phi_0(X, T) = (\phi_3(X), T)$. Let G_0 be the metric $\phi_0^*(H)$ (here and below we denote by ϕ^*Q the pullback of a covariant tensor Q by the differentiable mapping ϕ). Notice that in coordinates (X, T) the matrix of the metric G_0 is $\begin{pmatrix} 1 & 0 \\ 0 & h_{IJ} \end{pmatrix}$. Denote by $dV_0 = \sqrt{|G_0|}d^4X = \phi_0^*(dv^4)$ the 4D-volume element defined by the metric G_0.

Projection $\pi : P \longmapsto B$ along T-axes plays the same role in the construction below as in the relativistic elasticity theory (Carter and Quintana 1972). In particular, we require below the invariance of Lagrangian theory with respect to the authomorphismes of the bundle (P, π, B) (diffeomorphismes of material space-time P onto itself, projecting to B, so that material points do not lose their identity during the material evolution) preserving the direction of the flow of the "intrinsic" time (see below), but not with regard to the whole group of diffeomorphisms of P as in Gravity Theory.

2.2 Deformation History

The **history of the deformation** of the body B is represented by a **diffeomorphic embedding** $\phi : P \longmapsto M$ of the material space-time P into the laboratory space-time M (see Fig.1).

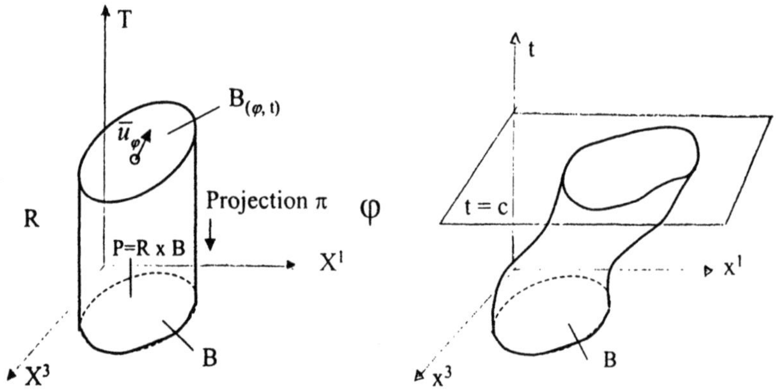

Figure 1. 4D Deformation History and the Field of Flow Vector

Deformations ϕ for which $t = \phi^0(X) = T$ is called **"synchronized"**.

Using the deformation ϕ, we introduce the slicing of the material space-time P by the level surfaces of the zeroth component of ϕ

$$B_\phi = \{B_{\phi,t} = \phi^{0\,-1}(t)\} = \{(T, X) \in P | \phi^0(T, X) = t\}. \tag{1}$$

For a synchronized deformation $B_{\phi,t} = B_{T=t}$.

There is a **flow vector field** u_ϕ in P, associated with the slicing $B_{\phi,t}$ of the space-time P (Misner et al. 1973; Carter and Quintana 1972). This vector field represent the flow of "intrinsic" (proper in Relativity Theory) time in the material.

Lifting the index in the 1-form $d\phi^0$ with the help of the metric G, we obtain

$$u_\phi = \frac{(d\phi^0)^\#}{||d\phi^0||}, \tag{2}$$

the norm of the 1-form $d\phi^0$ is defined as $||d\phi^0||^2 = (G^{AB}\phi^0_{,A}\phi^0_{,B})^{1/2}$ (summation agreement by repeating indices is used in the article). Thus, u_ϕ is the unit vector orthogonal to the slices $B_{\phi,t}$. In the local coordinates X^I,

$$u_\phi = \frac{G^{AB}\phi^0_{,B}}{\sqrt{G^{AB}\phi^0_{,A}\phi^0_{,B}}}\frac{\partial}{\partial X^A}. \tag{3}$$

For the synchronized deformations, $u_\phi = u_G$ does not depend on ϕ:

$$u_G = \frac{G^{I0}}{\sqrt{G^{00}}}\frac{\partial}{\partial X^I}. \tag{4}$$

If in addition, the metric G has the block-diagonal form in the coordinates $(X^0 = T, X^I)$ (shortly, BD - metric), we have $u_\phi = [G_{00}]^{-\frac{1}{2}}\frac{\partial}{\partial T}$.

Let $u_0 = \frac{\partial}{\partial T}$ be the flow vector associated with the metric G_0 and the corresponding 3D slicing B_0.

We require fulfillment of the following condition ensuring the irreversability of the flow of time:

$$< u_\phi, u_0 >_G > 0. \tag{5}$$

Deformation historiy ϕ for which the condition (5) is satisfied is called **admissible**.

In coordinates (X^I) this condition reduces to the following simple inequality

$$\phi^0_{,0} > 0 \tag{6}$$

and, therefore is a restriction on the deformation history only.

For all synchronized deformation ϕ, this condition is trivially satisfied.

Time component ϕ^0 of the deformation history may be excluded from the list of dynamical variables by an appropriate "gauging". Namely, we use the invariance of Lagrangian under the authomorphisms of the bundle (P, π, B) to make the deformation history synchronized.

An authomorphism $F : P \longmapsto P, X^I = F^I(Y^A)$ of the bundle (P, π, B) determines the change of variables $X^I = F^I(Y^A)$. In the new variables, the condition (6) takes the form $\frac{\partial \phi^0}{\partial Y^0} = \phi^0_{,I} F^I_{,0} = \phi^0_{,0} F^0_{,0} > 0$ (since F^I, $I = 1, 2, 3$ do not depend on Y^0). Thus, the class of admissible deformation histories ϕ is stable under the action of the subgroup $Aut^+(P)$ of all authomorphisms of P with $F^0_{,0} > 0$.

The group $Aut^+(P)$ of authomorphisms of the bundle contains two subgroups. One is the subgroup TC of the "time change" diffeomorphismes $(X^0 = T, X^1, X^2, X^3) \longmapsto (F(T, X^1, X^2, X^3), X^1, X^2, X^3)$ for arbitrary smooth function $F(X^I)$ with $F_{,0} > 0$. The other one (denoted $D(B)$) consists of the lifts to the slices $B_{\phi,t}$ of the bundle P of the orientation preserving diffeomorphismes of the base B (group of such transformations of B is denoted $Diff^+(B)$). To do this one is using (diffeomorphic) projectons $\pi_{\phi,t} = \pi|_{B_{\phi,t}} : B_{\phi,t} \longmapsto B$. If ϕ is synchronized, lifted diffeomorphismes do not depend on T.

Any authomorphism of the bundle $\phi \in Aut^+(P)$ generates the time independent diffeomorphism ϕ_B of the base B, that is element of $Diff^+(B)$. Lifting this element to the element of $D(B)$ we see that the group $Aut^+(P)$ is the semidirect product of the normal subgroup $TC(P)$ and the subgroup $D(B)$.

Let now ϕ be an admissible history of deformation. Define the element $F \in TC(P)$ as follows: $F : (X^0 = T, X^1, X^2, X^3) \longmapsto (\phi^0(T, X^1, X^2, X^3), X^1, X^2, X^3)$. Then, $\phi = \phi_1 \circ F$ where ϕ_1 is another admissible history of deformation with the same components $\phi^i, i = 1, 2, 3$ and the zero component $\phi^0_1(T, X^N) = T$. Transformation F is admissible since $F^0_{,0} = \phi^0_{,0} > 0$, thus $F \in TC(P)$. Apparantly, the deformation history ϕ_1 is synchronized.

Therefore we may restrict to consideration of the synchronized histories of deformation keeping in mind that the invariance group of the theory reduces to the group $D(B)$ of time-independent orientation preserving diffeomorphismes of slices.

2.3 Lapse and Shift of Material Metric

In what follows we assume that the 4D-deformation history ϕ is synchronized, thus slices $B_{\phi,t}$ has the form $T = const$.

Slicing $B_{\phi,t}$, generates the (3,1)-decomposition of the material metric G suggested (for the case of a Lorentz type metric) in (Arnovitt et al. 1962). Specifically, sandwich structure of the part of the manifold P bounded by the surfaces $B_{\phi,t}$ and $B_{\phi,t+\Delta t}$ allows one to introduce the time-dependent lapse function N and the shift vector field $\mathbf{N}$ tangent to the slices $B_{\phi,t}$ such that the metric is block-diagonalized in the coframe $(dT, dX^A + N^A DT)$:

$$dS^2 = G_{IJ}dX^I dX^J = G_{IJ}(dX^I + N^I dT)(dX^J + N^J dT) + NdT^2. \tag{7}$$

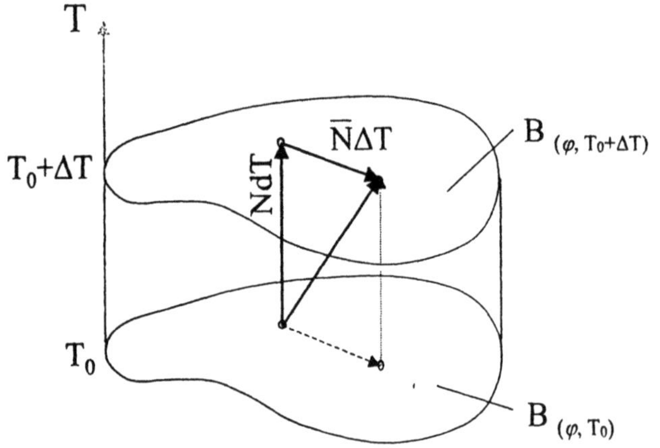

Figure 2. Lapse Function N and Shift Vector $\mathbf{N}$

Matrix representation of the material metric tensor G and inverse tensor G^{-1} have in these notations, the forms

$$\begin{pmatrix} G_{00} & G_{0J} \\ G_{I0} & G_{IJ} \end{pmatrix} = \begin{pmatrix} N_A N^A + N^2 & N_J \\ N_I & g_{IJ} \end{pmatrix}, \quad \begin{pmatrix} G^{00} & G^{0J} \\ G^{I0} & G^{IJ} \end{pmatrix} = \begin{pmatrix} \frac{1}{N^2} & -\frac{N^J}{N^2} \\ -\frac{N^I}{N^2} & g^{IJ} + \frac{N^I N^J}{N^2} \end{pmatrix}, \tag{8}$$

where g is the 3D-metric induced by G on the slices $B_{\phi,t}$ and g^{-1} is the corresponding inverse tensor.

It is easy to see that, in these notations

$$\sqrt{|G|} = N\sqrt{|g|}, \tag{9}$$

and for the flow vector u_G (for a sinchronized case) we have

$$\begin{aligned} u_G^* &= \frac{1}{N}\partial_T - \frac{N^A}{N}\partial_{X^A}, \\ u_{G\,*} &= NdT. \end{aligned} \tag{10}$$

The last formula gives the "proper time differential" $d\tau = NdT$ for the material metric. Thus, in our context, the lapse function $N(T, X)$ describes the inhomogeneity of the material aging at different points of the body as well as the different rate of aging processes in the body. On the Fig. 3 the local observer at different points of body at different moments of time T sees the different rate of the local time in comparision with the laboratory clocks.

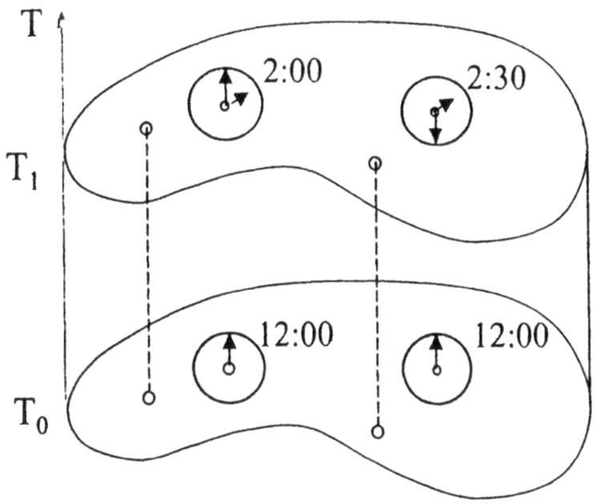

Figure 3. Illustration of Time Rate N (Lapse Function) by Variation of Time Interval $(\tau - \tau_0) = N(T_1 - T_0)$.

Presence of the shift vector field **N** in the metric reflects a propagation of the phase transition or chemical transformation boundary through the material as reflected, for instance in the mass "balance law" (see below).

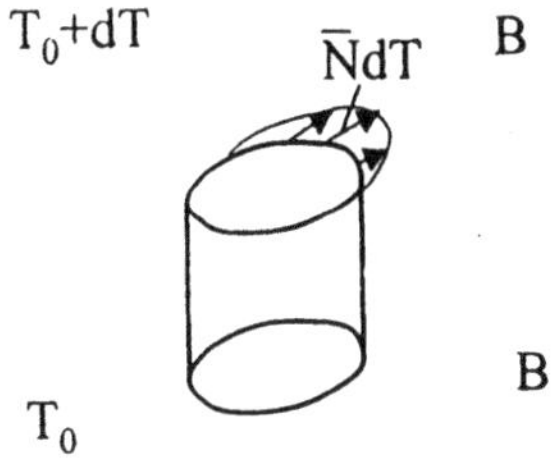

Figure 4. Propagation of Phase Boundary Represented by Shift Vector **N**

It shall be noted that the separation of the evolution of 3D material metric g, intrinsic translation processes in B_t described by **N** and the inhomogeneity of the local time $d\tau = NdT$, was the main reason for introduction (3+1) ADT-representation of 4D-metric in Gravity Theory (Arnovitt et al. 1962; Misner et al. 1973). In addition an adoption of this view leads to a very clear separation of the "fisical" degrees of freedom in the canonical formalism and to the explicit hyperbolic formulations of Einstein Equations (Anderson et al. 1999).

2.4 Mass form and Mass Conservation Law

In addition to the volume form dV, the **mass form** $dM = \rho_0 dV$ is defined in P. The reference mass density ρ_0, defined by this representation, satisfies the **mass conservation law** (Marsden and Hughes, 1983; Carter and Quintana 1972)

$$\mathcal{L}_{u_\phi} dM = 0. \tag{11}$$

Here $\mathcal{L}_{u_\phi}$ is the Lie (substantial) derivative of a differential form dM in the direction of the vector field u. Recall that the Lie derivative of a differential form ω along a vector filed u is defined as

$$\mathcal{L}_u \omega = \frac{d}{dt} \phi_t^* \omega|_{t=0},$$

where $\phi_t^* \omega$ is the pullback of the form ω by the flow $t \longrightarrow \phi_t = exp(tu)$ of the vector field u. Usage of Lie derivatives in Continuum Mechanics (in the form $\frac{\partial}{\partial t} + \mathbf{v} \cdot \nabla$) as well as in Relativistic Physics gives adequite and geometrically natural way to represent the evolution properties in media (Marsden and Hughes 1983).

Condition (11) is equivalent to the condition $div_G(\rho_0 u_\phi) = 0$, where divergence is taken with respect to the volume form dV. In local coordinates the Mass Conservation Law has the form

$$\left(\frac{G^{IB} \phi^0_{,B}}{||d\phi^0||} \rho_0 \sqrt{|G|} \right)_{,I} = 0. \tag{12}$$

Due to the properties of the metric G and the deformation ϕ, the material space-time manifold P is filled by the phase curves of the flow vector field u_ϕ and thus the value of the reference mass density $\rho_0(0, X)$ at $T = 0$ uniquely defines its values for all $T > 0$.

In the synchronized case $\phi^0 = T$, $||d\phi^0|| = \sqrt{G^{00}} = N^{-1}$, $G^{M0} = -\frac{N^M}{N^2}$ and (12) take the form of the following balance law

$$\left(\frac{\rho_0 \sqrt{|g|}}{N}\right)_{,0} = \sum_{A=1}^{3} \left(\frac{\rho_0 \sqrt{|g|}}{N} N^A \right)_{,A}. \tag{13}$$

As can be seen from (14) the shift vector field $\mathbf{N}$ describes the **matter (density) flow due to the some internal processes**such as the phase or chemical transformations.

If, in addition to being synchronized, the material metric G is also in the BD-form ($\mathbf{N} = 0$), the flow term in (13) disappears and the mass conservation law is equivalent to the following representation of the reference mass density in terms of its initial value $\rho_0(0, X) = \rho_0$:

$$\rho_0(T, X^I) = \rho_0(0, X) \sqrt{\frac{G_{00}}{|G|}} = \frac{\rho_0(0, X) N(T, X)}{\sqrt{|g(T, X)|}}, \tag{14}$$

where $G(0, X)$ is assumed to be Euclidean metric. If metric G does not changes with time T, we get the classical law $\frac{\partial \rho_0}{\partial T} = 0$.

The current (laboratory) mass density ρ is defined, conventionally, by the condition $\phi^*(\rho dv) = \rho_0 dV = \rho \circ \phi dV_0$, that gives

$$\rho \circ \phi = \frac{\rho_0 \sqrt{|G|}}{J(\phi)},$$

where $J(\phi)$ is the Jacobian of the deformation ϕ.

2.5 Elastic, Inelastic and Total Strain Tensors

Slicing $B_{\phi\ ,t}$ defines the covariant tensor $\gamma = G - u_\phi \otimes u_\phi = \begin{pmatrix} N_A N^A & N_J \\ N_I & g_{IJ} \end{pmatrix}$ (Misner et al. 1973; Carter and Quintana 1972). The tensor γ induces the time dependent 3D-metric g_t on the slices $B_{\phi\ ,t}$ (Fisher and Marsden 1972).

To obtain the expression for $g(t)$ in material coordinates X^I, we notice that the following vectors

$$\xi_I = -\frac{\phi^0_{,I}}{\phi^0_{,0}} \partial_{X^0} + \partial_{X^I},\ I = 1,2,3 \tag{15}$$

form the tangent space to the slices of the foliation B_ϕ. Thus, one can write

$$g(\xi_M, \xi_N) = G_{MN} - G_{0N}\frac{\phi^0_{,M}}{\phi^0_{,0}} - G_{M0}\frac{\phi^0_{,N}}{\phi^0_{,0}} + G_{00}\frac{\phi^0_{,M}\phi^0_{,N}}{\phi^{0\,2}_{,0}},\ M, N = 1,2,3. \tag{16}$$

For a synchronized deformation $\gamma = G - (G^{00})^{-1} dT \otimes dT$ and g_t is just the restriction of 4D-metric G to the slices $B_{T=c}$.

We do not apply any futher limitations to the metric g_t. For instance, it may have non-zero curvature i.e incompatible deformation associated with the residue stresses.

Denote by h_t the 3D metric on the leaves $B_{\phi,t}$ induced by the metric G_0 (or, by the tensor $\gamma_0 = G_0 - u_0 \otimes u_0$).

There is the projector ((1,1)-tensor)

$$\Pi = G^{-1}\gamma = I - u^*_\phi \otimes u_\phi \tag{17}$$

on the tangent spaces to the slices $B_{\phi,t}$ associated with the tensor γ. For synchronized deformation ϕ the projector Π has the following simple form

$$\Pi = I - \frac{G^{I0}}{G^{00}} \partial_{X^I} \otimes dT = \begin{pmatrix} 0 & 0 \\ N^I & I_3 \end{pmatrix}. \tag{18}$$

Let us consider the pullback $C_4(\phi) = \phi^* \hat{h}$ of the degenerate tensor $\hat{h}$ by the deformation mapping ϕ. Tensor $C_4(\phi)$ is degenerate in P, its kernal is generated by the vector $\phi_*^{-1}(\frac{\partial}{\partial t})$. In coordinates (X^I) we have

$$C_4(\phi)_{IJ} = \begin{pmatrix} h_{ij}\phi^i_{,0}\phi^j_{,0} = \|V\|^2_h & h_{ij}\phi^i_{,0}\phi^j_{,J} \\ h_{ij}\phi^i_{,I}\phi^j_{,0} & \sum_{i,j=1,\ldots,3} h_{ij}\phi^i_{,I}\phi^j_{,J} \end{pmatrix}. \tag{19}$$

The spacial part of this tensor is the conventional Cauchy strain tensor $C_3(\phi)$ of the Elasticity Theory. Components of this tensor with indices (0J) and (I0), $I, J = 1, 2, 3$ have the form *velocity x deformation covector* (Eselby 1970). (00)-component of $C_4(\phi)$ is the square of the material velocity.

The 4D *(1,1)-elastic strain tensor* in P is defined as follows

$$\hat{E}^{el\,\cdot}_{\cdot} = \frac{1}{2} G^{-1}(C_4(\phi) - \gamma). \tag{20}$$

This tensor containes the square of material velocity and the shift vector in its (00) and (I0,(0J) components. Having in mind the general, dynamical situation it is more appropriate to use the following tensor as the proper Elastic Strain Tensor

$$E^{el\,\cdot}(\phi) = \frac{1}{2} \Pi G^{-1}(C_4(\phi) - \gamma))\Pi = \begin{pmatrix} 0 & 0 \\ g^{IK}\phi^i_{,K}\phi^i_{,0} - I^I & g^{IK}C_3(\phi)_{KJ} - \delta^I_J \end{pmatrix}, \tag{21}$$

Here Π is the projector on the slices $B_{\phi,t}$ defined in (17). The last equality is valid in the synchronized case. Notice that the basic invariants for the tensor (21) are the same as of the tensor $C_3(\phi) - g$.

For the simplicity we use the same symbol E^{el} for the restriction of this tensor to the slices $B_{\phi,t}$.

Tensor $E(\phi)^{el}$ is a natural measure of the deviation of the metric of the actual state from the "ground state." For a synchronized deformation ϕ and a material metric G with the zero shift vector, $E^{el} = \frac{1}{2}(C(\phi) - g_t)$ represents the conventional measure of deformation (strain tensor).

Notice that strain tensor $\hat{E}^{el}$ also contains the information regarding the time scale evolution of the deformation history ϕ. More specifically, kernals of tensors $C_4(\phi)$ and γ are (in general) different 1D distributions (direction fields) on the material space-time P. These two direction fields coincide if and only if the vector fields u_ϕ and $\phi_*^{-1}(\frac{\partial}{\partial t})$ are parallel to one another.

For a synchronized deformation history ϕ this condition takes the following form

$$\phi^i_{,0} = N^K \phi^i_{,K}, \; i = 1, 2, 3. \tag{22}$$

Thus, in this case the time dependence of the deformation is determined uniquelly by the shift vector field N^K of the 4D material metric G, i. e. by the internal flow in the body.

We also introduce the *Inelastic Strain Tensor*

$$\hat{E}^{in} = \frac{1}{2} G^{-1}(\gamma - \gamma_0) = \begin{pmatrix} 0 & \frac{N^K \delta_{KB}}{N^2} \\ N^A & \delta^A_B - g^{AK}\delta_{KB} - \frac{N^A N^K \delta_{KB}}{N^2} \end{pmatrix}, \tag{23}$$

(last equality being true in synchronized case) and the *Total Strain Tensor* $\hat{E}^{tot}$ of the body at each given moment T to characterize the deviation of the deformed Euclidean metric $\phi^* h|_{B_{\phi,t}}$ from the initial (Euclidean) 3D-metric h (h being the restriction of G_0 to the slices $B_{\phi,t}$)

$$\hat{E}^{tot} = \frac{1}{2} G^{-1}(C_4(\phi) - \gamma_0). \tag{24}$$

Tensor $\hat{E}^{tot}$ can be represented as the sum of the Elastic Strain Tensor $\hat{E}^{el}$ and of Inelastic Strain Tensor $\hat{E}^{in}$:

$$\hat{E}^{tot} = \hat{E}^{el} + \hat{E}^{in}. \tag{25}$$

To obtain the corresponding decomposition for the 3D strain tensors we apply projector Π to the total and inelastic strain tensors. In particular we introduce

$$E^{in} = \Pi \hat{E}^{in} \Pi = \begin{pmatrix} 0 & 0 \\ N^A - g^{AK}\delta_{KB}N^B & \delta^A_B - g^{AK}\delta_{KB} \end{pmatrix}, \tag{26}$$

As a result, we get from (25) the corresponding decomposition of "3D total strain tensor"

$$\Pi E^{tot} \Pi = E^{el} + E^{in}. \tag{27}$$

Restriction of these tensors on the 3D slices $B_{\phi,t}$ leads to the more conventional (t-depėndent) version of this decomposition.

For a synchronized deformation history, restriction of E^{in} to the slices B_t takes the form

$$E^{in}|_{B_t} = \frac{1}{2}g^{-1}(g-h) = \frac{1}{2}(I - g^{-1}h), \tag{28}$$

that describes the decline of 3D material metric g from its initial value h.

Remark 1. One can also define the material elastic strain rate tensor as follows

$$\dot{E}^{el} = \mathcal{L}_{u_\phi} E^{el}, \tag{29}$$

as well as inelastic strain rate tensor

$$\dot{E}^{in} = \mathcal{L}_{u_\phi} E^{in}. \tag{30}$$

It is easy to see that in the case where $G = G_0$ and $\phi^0 = T$, elastic strain rate tensor defined in (29) has, the same spacial components as the conventional strain rate tensor (Marsden and Hughes 1973).

Denote by $\dot{G}$ the the Lie derivative $\dot{G} = \mathcal{L}_{u_\phi} G$ of the metric tensor G with respect to the flow vector $\mathbf{u}_\phi$. Then the calculation of the Lie derivative in (29) results in the following relation

$$\dot{E}^{el} = \mathcal{L}_{u_\phi} E^{el} = -G^{-1}\dot{G}E^{el} + \frac{1}{2}G^{-1}(\dot{C_4(\phi)} - \mathcal{K}), \tag{31}$$

where $\mathcal{K} = \mathcal{L}_{u_\phi}\gamma$ is the **extrinsic curvature tensor** of the slices $B_{\phi,t}$.

Remark 2. Here we are using material coordinates and tensors only. In order to obtain the corresponding "laboratory" quantities (seen by an external observer), one defines the laboratory (Euler) Elastic Strain Tensor

$$\epsilon^{el\ i}_j = \phi^i_{,A}\phi^{-1\ B}_j E^{el\ A}_B \tag{32}$$

and recalculate all the other quantities accordingly.

The diagram below presents the above decomposition.

The actual state under the load at any given moment T results from both elastic (with the variable elastic moduli) and inelastic (irreversible) deformations. The "ground state" of the body is characterized by the 3D-metric g_t. This state is the background to which the elastic deformation is added to reach the actual state (Simo and Ortiz 1985)).

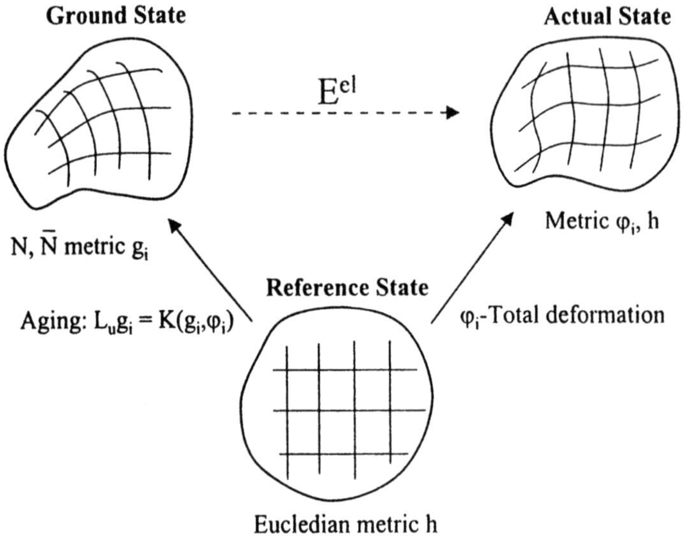

Figure 5. Decomposition of Deformation History

Transition from the reference state to the "ground state" that manifests in the evolution of the (initial) Euclidean metric h to the metric g_t cannot be described, in general, by any point transformation. Transition from the "ground state" to the actual state at the moment t also is not compatible in general. Yet the transition from the reference state to the actual state is represented by a diffeomorphism ϕ_t.

In this paper we consider the **material 4D-metric** G and the **deformation** ϕ (or elastic strain tensor $E^{el}(\phi)$) to be the dynamical variables of the field theory. The reference mass density ρ_0 is found (for the synchronized deformation ϕ and the BD-metric G) by the formula (14) if its initial value $\rho_0(T=0)$ is known. In this study we consider mainly the quasi-static version of the theory, i.e. inertia forces and kinetic energy are assumed to be negligable.

3 Lagrangian formulation of elasticity with a variable material metric.

In this section and in the following one we consider the processes of deformation and ageing of a solid with a synchronized deformation history. We use both general (G_{IJ}) and ADM $(N, \mathbf{N}, g)$ notations for the 4D material metric G.

Following the framework of Classical Field Theory (Landau and Lifshitz 1971) we take a Lagrangian density [1] $\mathcal{L}(G, \phi)$ refered to the **volume form** dV as a function of 4D-material metric G and the Elastic Strain Tensor: $\mathcal{L}(G, E^{el}) = L(G, E^{el})dV$. We represent the Lagrangian $L(G, E^{el})$ as the sum of the two parts: the "ground state" (shortly GS) Lagrangian $L_g(G)$ and the elastic part $L_e(G, E^{el})$ that is considered to be a perturbation of L_g due to the presence of an elastic deformation E^{el} ($L_e(G, E^{el})$ is assumed to be zero for $E^{el} = 0$).

$$L = L_g(G) + L_e(E^{el}, G). \tag{33}$$

The ground state (GS) Lagrangian $L_g(G)$ in (33) is introduced to account for the "cohesive energy" or strength of the solid state, strain energy of "residual strain" and the energy change associated with a variation of material properties in time, i.e., material ageing.

We assume that the GS Lagrangian L_g depends on 4D and 3D volume factors $\frac{|G|}{|G_0|}, \frac{|g_t|}{|h_t|}$ or, in more convenient but fuzzy notations, on the 3D volume factor $|g_t|$ (assuming $|h| = 1$) and the lapse function N (see (9)). Usage of these volume factors is associated with the solid state ability to resist the variations of its intrinsic topology in contrast to the fluid and gaseous states.

The tensor of extrinistic curvature $K = \mathcal{L}_{u_G}\gamma$ of the slices B_T in the material space-time P (Fisher and Marsden 1972; Misner et al. 1972) describes the evolution of the 3D material metric g_t.

The (1,1) mixed version of tensor $\mathcal{K}$ has the following form in the ADM notations

$$(K^I_J) = \begin{pmatrix} 0 & 0 \\ (*) & N^{-1}g^{IS}(\xi \cdot g)_{SJ} \end{pmatrix}, \tag{34}$$

where $(*)$ stays for the terms which does not affect the GS-Lagrangian L_g and $\xi \cdot g = \frac{\partial}{\partial T}g - \mathcal{L}_{\mathbf{N}}g$. Lie derivative $\mathcal{L}_{\mathbf{N}}g$ of the tensor g with respect to the vector field $\mathbf{N}$ is calculated on each 3D slice B_T for a fixed T.

In the case of a block-diagonal metric G (no shift), $\gamma = \begin{pmatrix} 0 & 0 \\ 0 & g_t \end{pmatrix}$; therefore, K is, essentially, the time derivative of the metric g_t:

$$K^I_J = \begin{pmatrix} 0 & 0 \\ 0 & N^{-1}g^{IA}\frac{\partial}{\partial T}g_{AJ} \end{pmatrix}. \tag{35}$$

Tensor K is interpreted as the rate of change of intrinsic length scales that reflects the ageing processes dependent on the elastic deformation.

[1] We are using the term scalar or tensor "density" in its mathematical meaning (Schouten, 1954, Chapter 1, Sec.2) as opposite to the traditional usage of this term in Continuum Mechanics, thus $L(\phi, G)$ is the scalar function of its arguments, while $\mathcal{L} = LdV$ is the corresponding scalar density ets.

In the spirit of variational approach we assume that L_g depend on the first two invariants of the tensor K: $Tr(K)$, $Tr(K^2)$.

$L_g(G)$ also depends on the shift vector field N through its norm $N_A N^A$ and, possibly, divergence $div_g(\mathrm{N})$ and the time derivative $\mathcal{L}_u\mathrm{N}$.

In addition we also include the term reflecting the residual strain energy (incomapatability) which is accounted for by the scalar curvature $R(g)$ of 3D material metric g.

Summarizing the above assumptions we construct the "ground state" Lagrangian L_g as the scalar function of listed above parameters.

One of the simplest Lagrangians L_g that includes all the mentioned quantities is:

$$L_g(G) = F_0 + a_1 Tr(K^2) + a_2 Tr(K)^2 + a_3 R(g_t). \tag{36}$$

Here $F_0 = F_0(|g_t|, N, ||\mathbf{N}||^2)$ is the initial ("cohesive") energy (per unit volume). The second and third term is the expression represents quadratic and linear invariants of tensor K. Coefficients a_1, a_2, a_3 are functions of two volume factors $N, |g|$.

In some cases, homogeneous media or 1D case, the scalar curvature $R(g_t)$ of the metric g_t is zero and the corresponding term in (36) disappears.

The scalar curvature $R(G)$ of the 4D metric G can be expressed, by the Gauss equation, as the combination of 3D invariants $-(tr(K^2) - (trK)^2) + R(g_t)$, up to a divergence term (Kobaiashi and Nomizu 1969), Sec.21.6; Fisher and Marsden 1972)). As a result, the above form of Lagrangian for an ageing media (36) is an extension of the Hilbert-Einstein Lagrangian $R(G)\sqrt{|G|}$ of the General Relativity (Misner et al. 1972).

The deformation ϕ is **essentially 3-dimensional** in the sense that it deforms only the spacial Euclidean metric h in M. As a result the 4D-tensor $C_4(\phi) = \phi^* h$ defines the degenerate metric in the material space-time P. It is instructive to compare $C_4(\phi)$ with the (degenerate) tensor $\gamma = G - u_\phi \otimes u_\phi$. The Elastic Strain Tensor E^{el} measures the deviation of $C_4(\phi)$ from γ on the slices $B_{\phi,t}$. Thus, the scheme presented here is essentially different from Relativistic Elasticity Theory (Carter and Quintana 1972; Kijowski and Magli 1992) as well as from 4D version of conventional Elasticity theory.

The coincidence of $\gamma|_{B_T}$ and $C(\phi)|_{B_T}$ is possible if there is no elastic deformation and the deformation ϕ describes just an evolution of the metric g_t: $g_t = \phi_t^* h$.

The form of Elastic Lagrangian steams from the Lagrangian of Classical Elasticity

$$L_e(E^{el}, G) = \frac{\rho_0}{2}||V||^2 - \rho_0 f(E^{el}, G) + \rho_0 F \circ \phi. \tag{37}$$

Here $\rho_0||V||^2 = \rho \sum_{ij} h_{ij}\phi^i_{,0}\phi^j_{,0}$ is the density of kinetic energy, f is the strain energy per unit of mass, F is the potential of the body forces. Strain energy f is assumed to be a function of two first invariants of the (1,1)-strain tensor E^{el}. Strain energy may depend on the metric G through the scalars $|g_t|$, N, vector field $\mathcal{N}$, scalar curvature $R(g)$ etc.

In a quasistatic theory we ignore the inertia effects and, therefore, the kinetic energy term in (37) disappears.

For the Linear Elasticity the strain energy density is conventionally presented as follows $f(E^{el}) = \frac{\mu}{2} Tr(E^{el\,2}) + \frac{\lambda}{2}(Tr(E^{el}))^2$, where μ, λ are the initial values of the Lame constants (Landau and Lifshitz 1970).

The Action functional is formulated as an integral of the Lagrangian density over a 4D domain $U = [0, T_0] \times V$, with $(V, \partial V)$ being an arbitrary subdomain of B with the boundary ∂V, combined with the 3D boundary integral that accounts for the work W of surface traction (Marsden and Hughes 1983).

$$A_U(G, \phi) = \int_U (L_g(G) + L_e(E^{el}, G))dU + \int_{\partial U} W(\phi, G)d^3\Sigma. \tag{38}$$

Here $d^3\Sigma$ is the area element on the 3D boundary ∂U of the cylinder U (Marsden and Hughes 1983). A zero condition for the variation of the action near the lateral sides of cylinder U leads to the expression of the first Piola-Kirchoff stress tensor P_j^I in terms of the strain energy (material form of the Hooke's law)

$$P_j^I = \frac{\partial L_e}{\partial \phi^j_{,I}}. \tag{39}$$

Notice that if the kinetic energy is included into L_e, Piola-Kirchoff Tensor has the density of linear momentum vector as its P_i^0 components.

It is useful to recall that the (laboratory) Cauchy stress tensor σ_{ij} is related to the first Piola-Kirchoff tensor by the following formula

$$\sigma_{ij} = J^{-1} h_{ik} \phi^k_{,I} P_j^I, \tag{40}$$

J being the Jacobian of the deformation ϕ.

Zero condition for the variation at the top ($T = T_0$) and the bottom ($T = 0$) of the cylinder lead to the relation between the linear momentum and the kinetic energy in the classical case (Legendre transformation). In the scheme presented here these variations also includes terms related to the aging processes.

The variation principle of the extremal action $\delta A = 0$ taken with respect to the dynamic variables ϕ and G results in a system of Euler-Lagrange equations that represent the coupled Elasticity and Ageing Equations

$$\frac{\partial}{\partial T}(\rho_0 \phi^m_{,0}) + \frac{\partial \mathcal{L}_e}{\partial \phi^m} - \sum_{I=1}^{I=3} \frac{\partial}{\partial X^I}\left(\frac{\partial \mathcal{L}_e}{\partial \phi^m_{,I}}\right) + \rho_0 \sqrt{|G|}(\nabla F)_m = 0, \qquad m = 1,2,3. \tag{41}$$

$$\frac{\delta \mathcal{L}_g}{\delta G^{IJ}} = -\frac{\sqrt{|G|}}{2} T_{IJ}, \qquad I = 0,1,2,3. \tag{42}$$

The Elasticity Equations (41) are obtained by taking the variation δA with respect to the components ϕ^i within the domain U. In the case of a BD metric G and the syncronized deformation ϕ, these equations coincide with the conventional dynamical equations of Elasticity Theory. However, their special features are associated with the different form of the Elastic Strain Tensor E^{el} and with the dependence of the elastic parameters on time through the invariants of the metric G. The evolution of these parameters is defined by the equations (42) (refered to as Ageing Equations).

The Ageing Equations.

Ageing Equations (42) results from the variation of action with respect to the metric tensor G. These equations describe the evolution of the material metric G for a given initial and boundary conditions. The right side of the equations (42) is the **"Canonical Energy-Momentum Tensor"** $\frac{1}{2}\sqrt{|G|}\,T_{IJ} = \frac{\delta\mathcal{L}}{\delta G^{IJ}}$ (Landau and Lifshitz 1971). This tensor is symmetric and is closely related to the **Eshelby Tensor** b_{IJ} (Eshelby 1951) (see Sec.VII below).

For the Lagrangian $L = L_g + L_e$ defined by (36-37) the Aging Equations (42) can be rewritten in the more convenient ADM notations. In order to do it we introduce the following notations. For both Lagrangians L_g, L_e we denote by $L_{(H)}$ the expressions obtained by taking derivatives of the coefficients F_0, a_i, λ, μ with respect to a variable H where H stands for $N, N^A, |g|$ or g^{MN}. Thus, for example

$$L_{g\,(N)} = \frac{\partial}{\partial N}F_0 + (\frac{\partial}{\partial N}a_1)Tr(K^2) + (\frac{\partial}{\partial N}a_2)(Tr(K))^2 + (\frac{\partial}{\partial N}a_3)R(g).$$

In addition to this we use short notation ∂_{X^A} for partial derivatives $\frac{\partial}{\partial X^A}$.

The Ageing Equations (42) takes the form of the system of PDE for the lapse function N, shift vector field $\mathbf{N}$ and the 3D material metric g:

$$\begin{aligned} &\frac{\delta\mathcal{L}_g}{\delta N} = 0;\\ &\frac{\delta\mathcal{L}_g}{\delta N^A} = 0,\ A = 1,2,3;\\ &\frac{\delta\mathcal{L}}{\delta g^{AB}} = 0,\ A,B = 1,2,3. \end{aligned} \tag{43}$$

The explicit form of the above equations can be readily obtained for the Lagrangian defined by (36-38):

$$L_{g\,(N)} + N^{-1}[F_0 + a_3R(g) - a_1Tr(K^2) - a_2(Tr(K))^2] = -N^{-1}L_e - L_{e\,(N)}. \tag{44}$$

$$\begin{aligned} &2\frac{\partial F_0}{\partial(\|\mathbf{N}\|^2)}N_A + a_1N^{-2}[(\partial_{X^A}g^{KS})(\xi\cdot g)_{SK} + (\partial_{X^A}g_{KS})(\xi\cdot g^{-1})^{SK}]\\ &\quad + \frac{4}{N\sqrt{|g|}}\frac{\partial}{\partial X^S}(a_1N\sqrt{|g|}K^S_A) + 4N^{-1}\frac{\partial}{\partial X^A}(a_2Tr(K)) = 0. \end{aligned} \tag{45}$$

$$\begin{aligned} &[\Delta_g(Na_3)g_{AB} + Hess(Na_3)_{AB} - (Na_3)(Ric(g)_{AB} - \frac{1}{2}R(g)g_{AB}]\\ &+ 2a_2N^{-2}Tr(K)(\xi\cdot g)_{AB} - \frac{1}{N\sqrt{|g|}}g_{AM}g_{BN}\xi\cdot(2a_2N^{-1}\sqrt{|g|}Tr(K)g^{MN})\\ &+ \frac{1}{N\sqrt{|g|}}[(\xi\cdot(a_1N^{-1}\sqrt{|g|}\xi\cdot g))_{AB} - (\xi\cdot(a_1N^{-1}\sqrt{|g|}\xi\cdot g^{-1}))^{MN}g_{MA}g_{NB}]\\ &\quad - \frac{1}{2}L_g g_{AB} + L_{g\,(g^{AB})} = (\frac{1}{2}L_e - \rho_0|g|f_{(|g|)})g_{AB} + 2S_{(AB)}. \end{aligned} \tag{46}$$

In the first line of equations (46) (left side) Δ_g is the 3D Laplace operator corresponding to the metric g, $Hess(f)$ stands for the Hessian of the function f (double covariant derivative tensor $f_{;M;N}$).

The expression $\xi = \partial_T - \mathcal{L}_{\mathbf{N}}$ In the second and third lines of (48) is the first order differential operator (see (34)) acting on tensors.

The last term in the right side of (46) is the symmetrized Second Piola-Kirchoff Stress Tensor S (Here and thereof $S_{(AB)} = \frac{1}{2}(S_{AB} + S_{BA})$ means symmetrization by indices A, B). The term $S_{(AB)}$ together with the diagonal tensor $(\frac{1}{2}L_e - \rho_0|g|f_{(|g|)})g_{AB}$ form the spacial part of the Canonical Energy Momentum Tensor. It is shown in Section VII that $S_{(AB)} = b_{(AB)} + \mathcal{L}_e g_{AB}$ for the Eshelby Tensor b. Since $S_{(AB)}$ is the driving force of the evolution of material metric g, it is apparant that the Eshelby Tensor b is the major part of this force (Maugin, Epstein, 1990).

Condition of synchronization $\phi^0 = T$ that we assume in this Section is similar to the gauging conditions in Electrodynamics where one request that the vector-potential A satisfy one of the conditions: $div(A) = 0$ or $A_0 = 0$. This removes the gauge invariance freedom of the model at the price of breaking the Lorentz invariance. Here we lose the invariance of the Lagrangian with respect to the full group of (time direction preserving) authomorphismes of the bundle (P, π, B). The resulting group of symmetries is the group $D(P)$ of orientation preserving diffeomorphismes of the base B. As a result, 4D Canonical Energy-Momentum Tensor splits into the components: the scalar N-component, 3D vector N-component and 3D tensor g-component conjugate to the lapse function, shift vector field and the 3D material metric respectively.

Equations (41-42) together with the equation (13) for the reference density form a closed system of equations for dynamic variables (G_{IJ}, ϕ^i, ρ_0). These equations, complemented with the initial and boundary conditions, provide a closed non-linear boundary value problem for the deformation of solid and evolution of the material properties.

In general, the system (41-42) seems rather complex, especially if L_e depends on $Ric(g_t)$ and/or K. Nevertherless, leaving a detailed analysis of this system for future studies, in the next section we briefly consider some special cases where system (42) can be effectively simplified.

4 Special cases of Ageing Equations

4.1 A. Block-diagonal metric G

In a case of a BD-metric, N = 0 (no shift). Therefore, GS-Lagrangian has the form $L_g = F_0(N,|g|) + a_1 Tr(K^2) + a_2(Tr(K))^2 + a_3 R(g)$ that includes time derivatives of 3D metric g (terms with K) and the space derivatives of g in the curvature term $R(g_t)$.

No derivatives of the lapse function N appear anywhere in $\mathcal{L}$. In particular, equation obtained by variation of N **is not a dynamical equation but rather a condition**, similar to the "energy condition" in the Gravity Theory (Fisher and Marsden 1972).

In the case, when the elastic coefficients do not depend on N, equation (44) takes the form

$$\frac{\partial \mathcal{L}_g}{\partial N} = -2\rho_0 f \sqrt{|g|}, \tag{47}$$

where $-\rho_0 f\sqrt{|g|}N$ is the density of strain energy (per unit of unperturbed volume)

For a given material $\mathcal{L}_g$ is a certain function of its arguments. Thus, if the function L_g and its derivative with respect to N are bounded for the admissible range of values of g, there is a limit for the elastic energy density at the material point. From physical point of view it corresponds to the energy criterion of strength. A more detailed discussion of this criterion is presented elsewhere.

It is interesting to notice that as $f \to 0$, the system tends to the "natural" limit state which, as (47) shows, corresponds to the critical points of $F_0(N,|g|)$ by N.

The spacial part of aging equations (46) can be written in the form

$$N^{-2}\left(3a_1\delta^M_A\delta^N_B + 2a_2 g^{MN} g_{AB}\right)\ddot{g}_{MN} + q_{IJ}(g,\dot{g},N) + \tilde{\mathcal{E}}_{IJ}(g_t) = S_{(IJ)}. \tag{48}$$

Here $\tilde{\mathcal{E}}_{IJ}(g) = [\Delta_g(Na_3)g_{AB} + Hess(Na_3)_{AB} - (Na_3)(Ric(g)_{AB} - \frac{1}{2}R(g)g_{AB}]$ is the analog of the Einstein tensor for the 3D-metric g_t. The difference with the conventional Einstein tensor is due to the presence of the factor Na_3 in the Lagrangian $a_3NR(g)$ (Fisher and Marsden 1972).

The term q inn the left side depends on the metric coefficients, function N and their first derivatives with respect to time.

The right side of this equation contains no derivatives of components of metric tensor g_{IJ}. The third term in the left side contains only derivatives of g_{IJ}, $I,J = 1,2,3$ with respect to X^I, $I = 1,2,3$. The first two terms in the left side on the contrary, contain only derivatives in time but no space derivatives. This equation is of the second order with respect to time. In the case where the tensor $\xi^{MN}_{AB} = 3a_1\delta^M_A\delta^N_B + 2a_2 g^{MN} g_{AB}$ is invertible this equation can be easily transformed to the normal hyperbolic form.

In the case, where $Ric(g_t) \approx 0$, a good approximation of the general system (41-42) can be proposed. If the total deformation ϕ is approximated by the "*ground deformation*" $\bar{\phi}$ (i.e. deformation $\bar{\phi}(X,T)$ such that $\bar{\phi}^*h = g_T$, recall that this is synchronous case!) in the evaluation of the EMT T_{IJ} on the right side of (42), the latter becomes decoupled from the equilibrium equation (41). This allows one to study the ageing equations separately from the Elasticity Equations and, after obtaining solution for G, substitute them into the elastic equilibrium equation (41) and solve it as the elasticity equation **with variable elastic moduli**.

Below we breafly discuss 3 special cases.

1) Homogeneous media

In the case of a homogeneous material (Maugin 1993); Wang and Truesdell 1973)) metric G depends on T only, and Einstein tensor $\tilde{\mathcal{E}}_{IJ}(g)$ is identically zero. As a result, (42) becomes a system of quasilinear **ordinary** differential equations of the second order for the lapse function N and the material 3D metric g_{IJ}. The Cauchy problem for this system is correct one provided the matrix of the coefficients ξ_{AB}^{MN} of second derivatives in time is nondegenerate.

2) Static case

Let us consider the case of a time-independent metric G. Let also assume that $a_3 = 1,\ N = 1,\ \mathbf{N} = 0$.

Equation (46) becomes the equation describing the self-equilibrated stress resulted from the curvature of the metric g and related to the incompatibility of embedding of the solid into the laboratory space:

$$-\frac{1}{2}(F_0 + R(g))g_{AB} + \frac{\partial F_0}{\partial g^{AB}} - \mathcal{E}(g) = -\frac{1}{2}\rho_0 f g_{AB} - \rho_0 |g| f_{(|g|)} g_{AB} + 2S_{(AB)}. \quad (49)$$

In this equation first terms on the left and right sides are related to the volume redistribution of energy while second terms on both sides - to the deviation of the total energy from its stationary value.

3) Homogeneous rod (1D media case)

In the case of a 1D media (an elastic rod) the curvature of metric g is identically zero. Then equations (42) are reduced to the nonlinear dynamical system for the lapse function N and the metric $g = g_{11}$. Linearized vesion of this problem is discussed in (Chudnovsky and Preston 1995).

We conclude this section with two 1D-media examples that show the type of material behaviour that can be studied using this model.

Example 3. Consider 1D-media case, with the lapse function $N = 1$ and the shift vector field $\mathbf{N}$ being constant with the only non-zero component N^1. Denote by $g = g_{11}(T, X)$ the only component of material metric. GS-Lagrangian is reduced to $L_g = F_0(g) + Tr(K)^2$. It is easy to see that in this case $Tr(K) = N^{-1}g^{-1}(u \cdot g)^2$, where $u = \partial_T - N^1\partial_X$. Thus, $\mathcal{L}_g$ takes the form $[F_0(g) + a_1(g)g^{-2}((\partial_T - N^1\partial_X)g)^2]\sqrt{g}$. The only nontrivial material ageing equation (48) takes the form

$$(\partial_T - N^1\partial_X)^2 g + (u \cdot a_1)(\partial_T - N^1\partial_X)g + \frac{a_1((\partial_T - N^1\partial_X)g)^2}{g^3} - V(g) = 2S_{11}, \quad (50)$$

where $V(g) = \frac{1}{2\sqrt{g}}\frac{\partial\sqrt{g}F_0(g)}{\partial g}$ and S_{11} - the second Piola-Kirchoff Stress Tensor.

Assuming that S_{11} depend on g but not on its derivatives we get equation of a non-linear oscillator with the time derivative $\partial_T - N^1\partial_X$. If for a given $F_0(g)$ this equation admits separatrix solution, then, for the metric $g(X, T)$ (and density $\rho_0(X < T)$) we get the "traveling wave solutions" (Rajaraman 1982) in the form of "kink", propagating with the speed N^1 along the "rod":

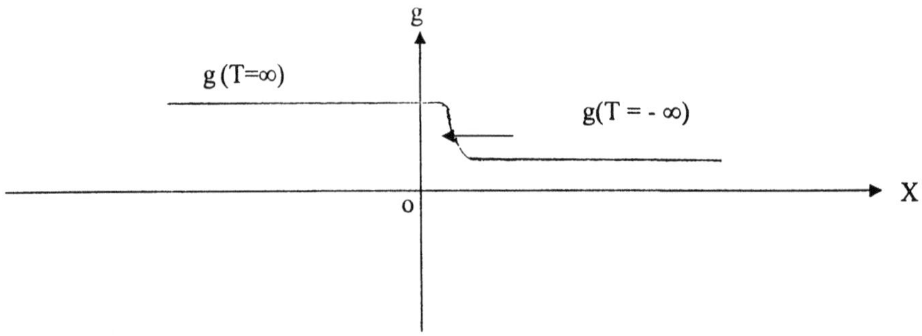

Figure 6. Necking of a 1D Rod

Delayed Necking, observed in various engineering thermoplastics, is a pictorial illustration of traveling wave solution. Necking in general is a localized large deformation (drawing) of a polymer with a distinct boundary between the drawn and undrawn material domains (Zhou et. al. 1994, W. Zhou et.al. 1999; J. Liu et al. 1999). A length scale l_0 selected in the reference (undrawn) state becomes $\lambda_n l_0$ in the drawn (oriented) state, where λ_n stands for the natural draw ratio, a material parameter that is of order of 2 for amorphous and varies between 2 and 10 for semi-crystalline thermoplastics. The length scale is usually monitored via equally spaced marking of the material particles. The delayed necking takes place in a rod under creep conditions, i.e., under constant applied load when the initial Piola-Kirchoff stress $S_{11}(0)$ is less then the yield stress. At first a uniform creep takes place, i.e., a uniform drawing with a draw ratio $\lambda = l/l_0$, where l stands for an actual (current) length scale. The stress S_{11} increases with λ (can be considered proportional to λ_n, in the first approximation $S_{11} = \lambda S_{11}(0)$, since the volume change in thermoplastics under uniaxial tension is small). Then, after certain time interval when the increasing stress S_{11} reaches the yield stress value, a neck (strain localization) is formed and propagates along the rod with a constant speed N^1 (necking process). There is no strain variation within the drawn and undrawn domains of the rod during the necking. The observed elongation results exclusively from a transformation of the original material adjacent to the neck boundary into the drawn (oriented) state and propagation of the boundary along the rod, as depicted in Fig. 6.

Example 4 (1D-media, free ageing). Here we take N = 0 and assume that all quantities depend on time only (homogeneous case). Consider the Metric Lagrangian density of the form $\mathcal{L}_g = N\sqrt{g}(F(N,g) + Q(N,g)Tr(K) + aTr(K^2))$ with $a - const$. It is easy to rewrite it in the form $f(N,g) + q(N,g)\dot{g} + \alpha(N,g)\dot{g}^2$, with $\alpha(N,g) = aN^{-1}g^{-3/2}$.

We take $f(x,y) = -g + c(N-1)^k$, $q(N,g) = q_0(N-1)^m$, $m < 1$. This choice reflect the view that the aging evolution is driven by the proper time rate N.

System of ageing equations for N, g with the zero right side (free ageing - no strain) takes the form of 2D dynamical system in the half plane $N > 1$

$$\dot{N} = \frac{f_{,g} - \alpha_{,g}\eta^2 - 2\alpha\eta\eta_{,g}}{q_{,N} + 2\alpha_{,N}\eta + 2\alpha\eta_{,N}},$$
$$\dot{g} = \eta(N,g) = \frac{2f_{,N}}{q_{,N} \pm \sqrt{q_{,N}^2 - 4\alpha_{,N}f_{,N}}} \tag{51}$$

Here function η is determined in the second equation. We also have used notation $f_{,x}$ for the derivative of function $f(x, y)$ by variable x.

Phase picture of this system is presented on the Fig.7.

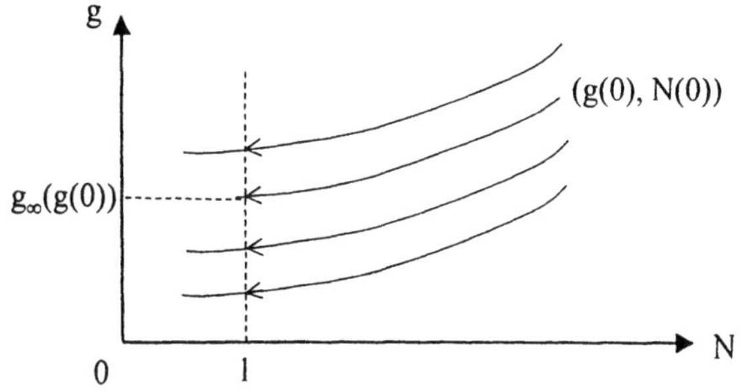

Figure 7. Free Ageing of a 1D Rod

Thus, as $T \to +\infty$ material metric approaches the form $\begin{pmatrix} 1 & 0 \\ 0 & g_\infty(g(0)) \end{pmatrix}$ with the asymptotic value of g depending on the initial state. These types of processes are well studied in ageing of organic glasses (amorphous polymers). During an unconstrained ageing the value of $g = g_{11}$ decreases, i.e., the density of glass ρ_0 increases approaching some ultimate values of the equilibrium glassy state.

Apparantly, the above cases are readily available for detailed analysis and for experimental examination/verification of the model.

5 Space and Material Balance Laws

As it is usual for a Lagrangian Field Theory, action of any one-parameter group of transformations of the space $P \times M$, commuting with the projector to P, leads to the corresponding balance law (See. Marsden and Hughes 1983)). In particular, translations in the "laboratory" space-time M lead to the equation of motion (41), rotations in M lead to the angular momentum balance law (conservation law in an isotropic case). Respectivelly, translations in the "material space-time" P lead to the **energy balance law** (translations along the time T axis) and to the **material momentum balance law** ("pseudomomentum" balance, (Maugin 1993; Golebiewska-Herrmann 1981; Pak and Herrmann 1986), rotations in the material space B lead to the "material angular momentum" balance law (Maugin, 1993; Simo and Marsden 1984).

In the table below we present basic balance laws together with the transformations generating them. It is instructive to compare the space and material balance laws as it has been considered previously by several authors (Golebiewska-Herrmann 1981; Pak and Herrmann 1986).

Table 1. Space and Material Balance Laws

Symmetry	Laboratory space-time (Material independent)	Material space-time (Space independent)
Homogeneity of 3D-space	Linear momentum balance law (equilibrium equations) $\mathrm{div}(\sigma)=f$	Material momentum (pseudomomentum) balance law $\mathrm{div}(b)=f_{mat}$
Time homogeneity	Energy balance law: $\partial_t \mathcal{E}^{tot} = \mathrm{div}(P^{tot})$	Energy balance law: $\partial_t (\mathcal{E}^{el}+\mathcal{E}^{m}+\mathcal{E}^{dis})=\mathrm{div}(P^{el}+P^{dis})$
Isotropy of 3D-space	Angular momentum balance law ≡ h-symmetry of Cauchy stress-tensor σ $I:\sigma = \sigma:I$	Material angular momentum balance law ≡ C-symmetry of Eshelby stress tensor b $b:C=C:b$

Space and Material balance (conservation) laws are related via the deformation gradient $d\phi$. In the synchronized case denoting by $\eta_I = 0$ the material balance laws and by $\nu_i = 0$ their "laboratory" counterparts, we get the relation in the form

$$\begin{pmatrix} \eta_0 \\ \dots \\ eta_4 \end{pmatrix} = \begin{pmatrix} 1 & \phi^1_{,0} & \dots & \phi^3_{,0} \\ 1 & \phi^1_{,1} & \dots & \phi^3_{,1} \\ \dots & \dots & \dots & \dots \\ 0 & \dots & \dots & \phi^3_{,3} \end{pmatrix} \cdot \begin{pmatrix} \nu_0 \\ \dots \\ nu_4 \end{pmatrix} \tag{52}$$

Similar to the relativistic elasticity (Kijowski and Magli 1992), the system of material balance laws $\eta_I = 0,\ I = 1, 2, 3$ is equivalent to the elasticity equations $\nu_i = 0,\ i = 1, 2, 3$, while the energy balance law $\eta_0 = 0$ (which here is the **material** balance law!) follows from any of these two systems: $\eta_0 = \sum_{i=1}^{i=3} \phi^i_{,0} \nu_i$. This reflects the fact that the deformation we consider here are not truely 4-dimensional.

Energy-Momentum Balance Law plays special role in our considerations and its detailed discussion is presenting in the next section

6 Energy-Momentum Balance Law and the Eshelby Tensor.

In this section we consider the Energy-Momentum balance law resulting from the Least Action Principle and the space-time symmetries.

Consider local rigid translations in the material space-time $X^J \longmapsto X^J + \delta X^J$. They generate a variation of components ϕ^i of the deformations, components G_{IJ} of material metric and their derivatives. Calculating the variation of the Lagrangian we get the material balance equations. We follow the arguments of J.Eshelby (Eshelby, 1970; Golebiewska-Herrmann, 1981).

The Lagrangian density has the form $\mathcal{L} = \sqrt{|G|}L = \sqrt{|G|}(L_g(G) + L_e(G, E^{el}))$ where L is decomposed into the ground state Lagrangian $L_g(G)$ and its perturbation L_e due to elastic deformations (see (33)). The Elastic Lagrangian depends on the space gradient $\phi^i_{,J}$, $i, J = 1, 2, 3$ of the deformation ϕ and on the metric G.

Taking the variation of the Lagrangian density with respect to the material coordinates X^J, one obtains

$$\frac{\delta \mathcal{L}}{\delta X^J} = \frac{\partial \mathcal{L}}{\partial \phi^i}\frac{\partial \phi^i}{\partial X^J} + \frac{\partial \mathcal{L}}{\partial \phi^i_{,I}}\frac{\partial \phi^i_{,I}}{\partial X^J} + \frac{\partial \mathcal{L}}{\partial G^{AB}}\frac{\partial G^{AB}}{\partial X^J} + \frac{\partial \mathcal{L}}{\partial G^{AB}_{,I}}\frac{\partial G^{AB}_{,I}}{\partial X^J} + \frac{\partial \mathcal{L}}{\partial G^{AB}_{,IK}}\frac{\partial G^{AB}_{,IK}}{\partial X^J}$$

$$= \sum_{i=0}^{i=3} \frac{\delta \mathcal{L}}{\delta \phi^i}\phi^i_{,J} + \frac{\partial}{\partial X^I}\left(\sum_{i=1}^{i=3} \frac{\partial \mathcal{L}}{\partial \phi^i_{,I}}\phi^i_{,J}\right) + \frac{\delta \mathcal{L}}{\delta G^{AB}} G^{AB}_{,J} +$$

$$\frac{\partial}{\partial X^I}\left(\mathcal{E}(G)^I_J = \frac{\partial \mathcal{L}}{\partial G^{AB}_{,I}} G^{AB}_{,J} + \frac{\partial \mathcal{L}}{\partial G^{AB}_{,IK}} G^{AB}_{,KJ} - \frac{\partial}{\partial X^K}\left(\frac{\partial \mathcal{L}}{\partial G^{AB}_{,IK}}\right) G^{AB}_{,J}\right). \quad (53)$$

where $\frac{\delta \mathcal{L}}{\delta \phi^i} = \frac{\partial \mathcal{L}}{\partial \phi^i} - \frac{\partial}{\partial X^I}\left(\frac{\partial \mathcal{L}}{\partial \phi^i_I}\right), i = 1, 2, 3$ and the variational derivatives $\frac{\delta \mathcal{L}}{\delta G}$ and $\frac{\delta \mathcal{L}}{\delta \phi^0}$ have similar forms.

Then, employing the Euler-Lagrange equations (41-43), we obtain for the Total Energy-Momentum Tensor (density)

$$\mathcal{E}(tot) = -\mathcal{L}\delta^I_J + \sum_{i=1}^{i=3} \frac{\partial \mathcal{L}}{\partial \phi^i_{,I}}\phi^i_{,J} + \mathcal{E}(G)^I_J, \quad (54)$$

the following conservation law

$$div_{G_0}(\mathcal{E}(tot)) = \frac{\partial}{\partial X^I}\mathcal{E}(tot)^I_J = 0, \; J = 0, 1, 2, 3, \quad (55)$$

(divergence is taken with respect to the 4D "reference" metric G_0). Notice that the second term in the expression of the total energy density (53) can be written as

$$S^I_J = \sum_{i=1}^{i=3} \frac{\partial \mathcal{L}}{\partial \phi^i_{,I}}\phi^i_{,J} = \sum_{i=1}^{i=3} P^I_i N \sqrt{|g|}\phi^i_{,J}, \quad (56)$$

which represents 4D-version of the (density of) Second Piola-Kirchoff Stress Tensor.

Althogether, material space-time homogeneity, i.e., the invariance of the Lagrangian with respect to 4D translations δX^I leads to the conservation of $\mathcal{E}(tot)$.

In his celebrated works J.Eshelby (1951,1970), introduced the 3D and then 4D dynamical energy-momentum tensor (**Eshelby Tensor**) b (denoted P^*_{lj} in (Eshelby 1970).

$$b^I_J = -\rho_0 f \delta^I_J + \sum_{i=1}^{i=3} \rho_0 \frac{\partial f}{\partial \phi^i_{,I}} \phi^i_{,J}, \tag{57}$$

f being the elastic energy per unit volume. We denote by $\mathcal{B}$ the corresponding tensor density $\mathcal{B} = b\sqrt{|G|}$.

The tensor b unifies in itself the 3D-Eshelby tensor (Eshelby 1970,1975; Maugin 1993; Maugin and Trimarco 1992), the 1-form of *quasi-momentum (pseudomomentum)* $\mathcal{P} = b^0_J$, $J = 1, 2, 3$ (Eshelby 1975; Pak and Herrmann 1981; Maugin 1993), strain energy density $b^0_0 = -L_e = f$ (plus kinetic energy, if the last one is present) and the energy flow vector $s = b^I_0 = -P^I_i \phi^i_{,0}$, $I = 1, 2, 3$ ($P^I_i = -\frac{\partial L_e}{\partial \phi^i_{,I}}$ is the first Piola-Kirchoff stress tensor).

In the quasi-static case $b^0_J = 0$ for $J = 1, 2, 3$.

In the case of a BD metric ($\mathbf{N} = 0$) G we have $\mathcal{P}_B = b_{0B} = 0$, $B = 1, 2, 3$.

Tensor b_{IJ} is, in general, not symmetric (although its 3x3 space part is symmetric with respect to the Cauchy metric $C_3(\phi)$, (Maugin 1993)).

Decompose $\mathcal{L} = \mathcal{L}_g + \mathcal{L}_e$ and, correspondingly, $\mathcal{E}(G) = \mathcal{E}_e(G) + \mathcal{E}_g(G)$ into elastic and metrical parts. Metric part of Lagrangian L_g does not depend on the deformation ϕ and, therefore, is absent from the second term in the right side of (53). Using Eshelby Tensor density $\mathcal{B}$ we can rewrite (54) as follows

$$div_{G_0} \mathcal{B} = \mathcal{B}^I_{J,I} = div_{G_0}(\mathcal{L}_g \delta + \mathcal{E}_e(G) + \mathcal{E}_g(G)). \tag{58}$$

The first and third terms in the right side of (57) are related to the ground state of the Lagrangian density i.e. to the inhomogeneity of "cohesive energy" and the "material flows". The second term on the right is related to a variation of elastic moduli if these moduli depend on the derivatives of the metric G through the scalars $div(\mathbf{N})$, $u \cdot N$, tensors K and $Ric(g_t)$ etc.

The equality (57) can be easily rewritten in terms of covariant derivatives with respect to the metric G (Landau and Lifshitz 1971).

For $J = 0$ we arrive at the energy conservation law (using T instead of X^0)

$$\frac{\partial}{\partial T} \mathcal{E}^{tot\ 0}_0 = -\sum_{I=1}^{I=3} \frac{\partial}{\partial X^I} \left(\sum_{i=1}^{i=3} P^I_i \sqrt{|G|} \phi^i_{,0} + \mathcal{E}(G)^I_0 . \right) \tag{59}$$

Apparently it has the form $\frac{\partial(TotalEnergyDensity)}{\partial T} = TotalFlowDensity$, with the total (inner) energy density given by

$$\mathcal{E}^{tot\ 0}_0 = -\mathcal{L}_e - \mathcal{L}_g + (\mathcal{E}(G)^0_{g\ 0}) + (\mathcal{E}(G)^0_{e\ 0}. \tag{60}$$

The total energy consists of the following parts: elastic and cohesive energy (the first term and the term $F_0(|g|, N)$ of $-L_g$), "kinetic metric energy" that is defined by the next two terms of L_g and reflects irreversible deformation, aging (kinetic terms in $\mathcal{L}_g$ and the second term), as well as the energy associated with a variation of the elastic moduli (last term).

The sum on the right side of (58) consists of the flow of the Piola-Kirchoff stress tensor density $\sum_{I=1}^{I=3}(\frac{\partial\mathcal{L}_e}{\partial\phi^i_{,I}}\phi^i_{,0})_{,I}$ and the flows related to the change of the material metric and elastic moduli.

If the metric G does not depend on time (i.e. $K = 0$) and if $Ric(g_t) = 0$, one obtains the conventional energy conservation law of Elasticity Theory (Marsden and Hughes 1983, Chapter 5, Sec.5): $\frac{\partial e}{\partial T} = -\sum_{I=1}^{I=3}\frac{\partial}{\partial X^I}(P^I_i\sqrt{|G|}\phi^i_{,0})$. In this particular case intrinsic time coincide with the external time, thus the conventional conservation of energy can be attributed to the absence of ageing phenomena and therefore to the homogeneity of both times.

Example 5 (Block diagonal metric G, synchronous deformation and homogeneous media). In this case we have $Ric(g_t) = 0$, the extrinsic curvature has the form $K^I_J = \begin{pmatrix} 0 & 0 \\ 0 & N^{-1}g^{IK}_{,0}g_{KJ} \end{pmatrix}$.

Thus $\frac{\partial\mathcal{L}}{\partial G^{AB}_{,0}}G^{AB}_{,0} = \frac{\partial\mathcal{L}}{K^A_B}K^A_B$. In addition to this, no flow terms except the usual Piola-Kirchoff flow appear on the right side in (58) which takes the form

$$\frac{\partial}{\partial T}\left(\mathcal{E}^{el} - \mathcal{L}_g + \frac{\partial\mathcal{L}_e}{K^A_B}K^A_B + \frac{\partial\mathcal{L}_g}{K^A_B}K^A_B\right) = -\sum_{I=1}^{I=3}(P^I_i\phi^i_{,0})_{,I}. \tag{61}$$

If elastic Lagrangian L_e does not depend on K and if GS-Lagrangian has the form: $L_g = F_0(N,|g|) + a_0TrK + a_1Tr(K^2) + a_2(TrK)^2$, we obtain the energy conservation law

$$\frac{\partial}{\partial T}\left(\mathcal{E}^{el} - [F_0(N,|g|) - \alpha Tr(K^2) - \beta(TrK)^2]N\sqrt{|g|}\right) = -\sum_{I=1}^{I=3}(P^I_i\phi^i_{,0})_{,I}. \tag{62}$$

This equation describes how the energy supplied by the boundary load leads not just to the increase of the strain energy, but also to the diminishing of its "cohesive energy" of the material $(F_0(N,|g|))$ and to the acceleration of the degradation.

Example 6. If SG-Lagrangian L_g depends on the shift vector $\bar{N}$ throught its divergence $div_g(\mathrm{N})$, this dependence does not give any input into the metric part $\mathcal{E}(G)^0_0$ of the total energy but gives the input $\frac{\partial\mathcal{L}}{\partial div_g(\mathrm{N})}N^K_{,T}$ into the flow part $\mathcal{E}(G)^K_0$ of the energy balance law (61). Apparantly it is related to the acceleration of the material flows in the body.

Example 7. If Lagrangian $\mathcal{L}$ depends on the lapse function N throught its rate of change $u_G \cdot N$, this dependence gives the input $\frac{\partial\mathcal{L}}{\partial u_G\cdot N}N^{-1}\partial_T N$ into the metric part $\mathcal{E}(G)^0_0$ of the total energy and the input $\frac{\partial\mathcal{L}}{\partial u_G\cdot N}(N^{-1}\partial_T N)N^K$ into the flow part $\mathcal{E}(G)^K_0$ of the energy balance law (61). Apparantly it is related to the transport of the changes in the rate of aging throught the media.

7 Relationship between the Canonical Energy-Momentum, Piola-Kirchoff and Eshelby Tensors.

In this section we discuss the relation between the Eshelby Tensor b that appears in the Energy-Momentum Balance Law (60) and the canonical Energy-Momentum Tensor T, defined by the relation $\mathcal{T} = T\sqrt{|G|} = \frac{\delta \mathcal{L}_e}{\delta G}$ of Lagrangian field theory with the 4D-metric G.

Tensor T is symmetric by definition and, in the case of a BD metric G, has the form $T = \begin{pmatrix} T_{00} & 0 \\ 0 & T_{IJ}, \end{pmatrix}, I, J = 1, 2, 3.$

On the other hand tensor b is not symmetric and its $I0$ terms are non-zero even for a block-diagonal metric. Below we compare spacial components of these tensors.

According to (37), Elastic Lagrangian density $\mathcal{L}_e$ for the quasistatic case has the form $\mathcal{L}_e = L_e\sqrt{|G|} = -f(E^{el}, G)\sqrt{|G|}$, where f is the strain energy per unit material volume.

As the first step we calculate the second Piola-Kirchoff tensor density $\mathcal{S} = S\sqrt{|G|}$. Tensor S is defined as $S^I_J = P^I_i \phi^i_{,J}$. It differs from the tensors defined in (Maugin 1993; Marsden and Hughes 1983) by the factor $C(\phi)$. It is easy to check that the tensor S is symmetrical in its (3x3) spacial part in the case of synchronized deformations and BD-metric G.

$$\mathcal{S}_{IJ} = P^M_m \sqrt{|G|}\phi^m_{,J} G_{MI} = \frac{\partial \mathcal{L}_e}{\partial \phi^m_{,M}} \phi^m_{,J} G_{MI} = \frac{\partial \mathcal{L}_e}{\partial E^A_B} \sum_{m=0}^{m=3} \frac{\partial E^A_B}{\partial \phi^m_{,M}} \phi^m_{,J} G_{MI}. \tag{63}$$

In the Appendix we derive the following expression for $\mathcal{S}_{IJ}$ (81):

$$\mathcal{S}_{IJ} = \frac{\partial \mathcal{L}_e}{\partial E^I_B} \cdot C_3(\phi)_{BJ}, \tag{64}$$

We present the total variation of the Elastic Lagrangian due to material ageing, i.e. variation with respect to the material metric as follows

$$\frac{\delta \mathcal{L}_e}{\delta G^{IJ}} = \left(\frac{\delta \mathcal{L}_e}{\delta G^{IJ}}\right)_{exp} + \frac{\partial \mathcal{L}_e}{\partial E^{el\ A}_B} \frac{\partial E^{el\ A}_B}{\partial G^{IJ}}, \tag{65}$$

Here we have utilized the fact that Elastic Strain Tensor E^{el} depends on the metric components G_{IJ} but not on its derivatives. Now, we calculate

$$\frac{\partial E^A_B}{\partial G^{IJ}} = \frac{1}{2}\left[\delta^A_I h_{ij}\phi^i_{,J}\phi^j_{,B} + \delta^A_J h_{ij}\phi^i_{,I}\phi^j_{,B}\right] = \frac{1}{2}(\delta^A_I C_3(\phi)_{JB} + (\delta^A_J C_3(\phi)_{IB}]. \tag{66}$$

Then, using simple transformations, we get the following expression for the Canonical Energy-Momentum Tensor density

$$\mathcal{T}_{IJ} = \frac{\delta \mathcal{L}_e}{\delta G^{IJ}} = \frac{\delta \mathcal{L}_e}{\delta G^{IJ}}_{exp} + \frac{1}{2}\left(\frac{\partial \mathcal{L}_e}{\partial E^I_B} C_3(\phi)_{JB} + \frac{\partial \mathcal{L}_e}{\partial E^J_B} C_3(\phi)_{IB}\right). \tag{67}$$

Next we calculate the Eshelby tensor density $\mathcal{B}$:

$$\mathcal{B}^I_J = -\mathcal{L}_e \delta^I_J + P^I_i \sqrt{|G|}\phi^i_{,J} = -\mathcal{L}_e \delta^I_J + \mathcal{S}^I_J. \tag{68}$$

The covariant tensor density $\mathcal{B}$ apparantly is

$$\mathcal{B}_{IJ} = G_{IS}\mathcal{B}^S_J = -\mathcal{L}_e G_{IJ} + \mathcal{S}_{IJ}, \tag{69}$$

and the symmetrical part of tensors $\mathcal{B}$ is

$$\mathcal{B}_{(IJ)} = -\mathcal{L}_e G_{IJ} + \mathcal{S}_{(IJ)}, \tag{70}$$

where we denote by $\mathcal{A}_{(IJ)} = \frac{1}{2}(\mathcal{A}_{IJ} + \mathcal{A}_{JI})$ the symmetric part of the covariant tensor density $\mathcal{A}$.

Comparing (63) and (66) one arrives at

$$\mathcal{T}_{IJ} = \frac{\delta \mathcal{L}_e}{\delta G^{IJ}}_{exp} + \mathcal{S}_{(IJ)}, \tag{71}$$

Combining this result with (69) we obtain the following relationship between the canonical Energy-Momentum Tensor T and the Eshelby stress tensor b and one between the corresdponding tensor densities :

For tensors

$$T_{IJ} = b_{(IJ)} + \frac{1}{2} L_e G_{IJ}' + \left(\frac{\delta L_e}{\delta G^{IJ}}\right)_{exp}. \tag{72}$$

For tensor densities

$$\mathcal{T}_{IJ} = \mathcal{B}_{(IJ)} + \mathcal{L}_e G_{IJ} + \left(\frac{\delta \mathcal{L}_e}{\delta G^{IJ}}\right)_{exp}. \tag{73}$$

Here we have used the fact that $\frac{\delta \mathcal{L}_e}{\delta G^{IJ}}_{exp} = \sqrt{|G|}\frac{\delta L_e}{\delta G^{IJ}}_{exp} - \frac{1}{2}\sqrt{|G|}L_e G_{IJ}$.

Taking divergence of the ageing equation (42), we obtain the relationship between the divergences of tensors T and $\frac{\delta L_g}{\delta G}$. The energy-momentum balance law (58) is obtained from this relation "on shell" - provided the Elasticity equations (41) are satisfied. The main difference with General Relativity (apart from the fact that G is Riemannian) is that the divergence of our tensor $\frac{\delta \mathcal{L}_g}{\delta G}$ is not zero and serves as the "source" of metric variation due to material ageing.

The symmetrical tensor T_{ij} is analogous to the symmetrised Eshelby tensor P^*_{IJ} (Eshelby 1975) and the tensor Σ introduced by P. Chadwick (Chadwick 1975). Correspondingly, its physical interpretation is similar to these tensors. For example considering of micromechanical modeling of damage, one finds that the integral of the covector-valued two-form defined by the tensor T on the surfaces B_t over the boundary of a domain D determine the "force on the defects" within the domain D. However, there is a significant difference between the tensor T one one side and tensors P, Σ on the other. CEM Tensor T, employed above, serves as the material force actively participating on the phenomenological level in modeling of inelastic deformation and ageing processes. In that sense the presented scheme is close to one developed in (Epstein and Maugin 1990]) where material forces appears due to the non-flat "material connection".

As an example, consider a case of a synchronized deformation where the elastic moduli in (37) are constant and all the explicit dependence of $\mathcal{L}_e$ on G is due to the volume element $\sqrt{|G|}$ only.

In this case the relation (71) between the Eshelby Tensor and the Canonical Energy-Momentum Tensor (CEM Tensor) takes the form $(I, J = 1, 2, 3)$:

$$T_{IJ} = b_{IJ} + \frac{1}{2}\mathcal{L}_e G_{IJ};\ (I, J = 1, 2, 3). \tag{74}$$

Remark 8. We would like to notice an analogy of (73) and the formula $2\frac{\delta L_e}{\delta G} = \Sigma_{IJ}$, obtained in (Simo and Marsden 1984) for the relation between the derivative (variation) of the strain energy with respect to the material metric (which is not a dynamical variable in (Simo and Marsden 1984)) and the **rotated stress tensor** Σ, although the relation between the canonical energy-momentum tensor and the second Piola-Kirchoff tensor (spacial part of the Eshelby tensor) has quite a different meaning here.

8 Conclusion

The intrinsic material metric tensor is considered to be an additional parameter of state, i.e., an internal variable that characterizes material degradation and ageing. The material metric tensor is a conjugate (with respect to a particular Lagrangian) to the Canonical Energy-Momentum Tensor (CEM Tensor).

Thus, the equation of metric evolution, i.e., the ageing equation, is introduced as the Euler-Lagrange equation of a corresponding variational problem and expressed in terms of CEM Tensor. This approach is an alternative to numerous phenomenological models. The phenomenological damage models that they usually have more adjustable parameters then a practical testing allows to determine. Thus it is difficult to examine the validity of the models since they can almost always be adjusted to reach an agreement with the experiment. In contrast with this, a variotional approach prescribes a functional form of the ageing equations, limits the number of constants (adjustable parameters) employed in the Lagrangian, provides a simple physical interpretation of the constants and admits an essential experimental examination of the validity of the basic assumptions of the model. Consideration of particular examples that can be analyzed theoretically and unambiguously tested in the experiments is an natural continuation of the present work.

A Appendix

Here we calculate the second Piola-Kirchoff tensor density $\mathcal{S} = S\sqrt{|G|}$ for the Elastic Lagrangian (6.1). We consider only quasistatic situation (thus, $\phi^i_{,)}$ does not appear in the elastic Lagrangian) and restrict calculations to $I, J = 1, 2, 3$ only. Thus, in calculations below we have $G_{IJ} = g_{IJ}$ and G^{AC} in expression for $E^{el\ A}_B$ are replaced with g^{AC} (see (2.17')). We have

$$\mathcal{S}_{IJ} = P^M_m \sqrt{|G|}\phi^m_{,J} G_{SI} = \frac{\partial \mathcal{L}_e}{\partial \phi^m_{,M}} \phi^m_{,J} G_{MI} = \frac{\partial \mathcal{L}_e}{\partial E^A_B} \sum_{m=0}^{m=3} \frac{\partial E^A_B}{\partial \phi^m_{,M}} \phi^m_{,J} G_{MI}. \quad (75)$$

Using expression (2.17') for E^{el}, (2.3) for u_G and noting that only space components of E^{el} participate in the Elastic Lagrangian, we have

$$E^{el\ A}_B = \frac{1}{2}[g^{AM} C(\phi)_{MB} - \delta^A_B] = \frac{1}{2}[g^{AM} h_{ij} \phi^i_{,M} \phi^j_{,B} - \delta^A_B] \quad (76)$$

Therefore

$$\frac{\partial E^A_B}{\partial \phi^m_{,L}} = \frac{1}{2} g^{AK} [h_{ij} (\delta^i_m \delta^L_K \phi^j_{,B} + \delta^j_m \delta^L_B \phi^i_{,K})] = \frac{1}{2} [G^{AI} h_{mj} \phi^j_{,B} + g^{AK} h_{im} \phi^i_{,K} \delta^L_B] \quad (77)$$

Using this in the formula (A.1) we get

$$\begin{aligned} \mathcal{S}_{IJ} &= \frac{\partial \mathcal{L}_e}{\partial E^A_B} \cdot \left\{ \frac{1}{2} [g^{AS} h_{mj} \phi^j_{,B} \phi^m_{,J} + g^{AK} h_{im} \phi^i_{,K} \phi^m_{,J} \delta^S_B] \right\} g_{IS} \\ &= \frac{\partial \mathcal{L}_e}{\partial E^A_B} \cdot \left\{ \frac{1}{2} [\delta^A_I h_{mj} \phi^j_{,B} \phi^m_{,J} + g^{AK} g_{SI} h_{im} \phi^i_{,K} \phi^m_{,J} \delta^S_B] \right\}. \end{aligned} \quad (78)$$

Now we notice that

$$\frac{\partial \mathcal{L}_e}{\partial E^A_B} \cdot g^{AK} g_{BI} = g_{AC} \frac{\partial \mathcal{L}_e}{\partial E_{CB}} \cdot g^{AK} g_{BI} = \frac{\partial \mathcal{L}_e}{\partial E_{CB}} \cdot \delta^K_C g_{BI} = \delta^K_C \frac{\partial \mathcal{L}_e}{\partial E^I_C} = \frac{\partial \mathcal{L}_e}{\partial E^I_K}, \quad (79)$$

since $\frac{\partial \mathcal{L}_e}{\partial E^A_B} = g_{AC} \frac{\partial \mathcal{L}_e}{\partial E_{CB}}$ and tensor $E_{CB} = E^A_B g_{CA}$ is symmetric. Therefore the two terms in (A.4) coincide and give input

$$\frac{\partial \mathcal{L}_e}{\partial E^I_B} \cdot h_{mj} \phi^j_{,B} \phi^m_{,J} \quad (80)$$

into $\mathcal{S}_{IJ}$ and, finally, we get

$$\mathcal{S}_{IJ} = \frac{\partial \mathcal{L}_e}{\partial E^I_B} \cdot h_{mj} \phi^j_{,B} \phi^m_{,J} = \frac{\partial \mathcal{L}_e}{\partial E^I_B} C_3(\phi)_{BJ}. \quad (81)$$

References

Anderson, A., Choquet-Bruhat, Y., and York, J. (1999). Einstein equations and equivalent hyperbolic dynamical systems. preprint gr-qc/990/099 V2.

Arnovitt, R., Deser, S., and Misner, C. (1962). The dynamics of general relativity. In L.Witten., ed., *Gravitation: An Introduction to Current Research*. Wiley, N.Y.

Ashby, M. F., C., G., and Taplin, D. M. (1979). *Acta Metallurgica* 27, 699.

Carter, B., and Quintana. (1972). Foundations of general relativistic high-pressure elasticity theory. *Proc. R. Soc. Lond.* A.331, 57–83.

Chadwick, P. (1975). Applications of an energy-momentum tensor in non-linear elastostatics. *Journal of Elasticity* 5(N.3-4):249–258.

Chudnovsky, A., and Preston, S. (1995). Geometrical modeling of material ageing. *Extracta Matematicae* 1–15.

Chudnovsky, A. (1996). Intrinsic time and ageing. In *Handbook of Modern Sensors, 2 ed., ed. J.Fraden*, Sec. 3.10.4. Woodbury, N.Y.,: American Institute of Physics Press.

Epstein, M., and Maugin, G. (1990). The energy-momentum tensor and material uniformity in finite elasticity. *Acta Mechanica* 83:127–133.

Epstein, M., and Maugin, G. (2000). Material evolution in plasticity and growth. in preparation.

Epstein, M. (1999). On material evolution laws. In *Geometry, Continua and Microstructure,Collection Travaux en Course*, volume Vol.60, pp.1–10. Paris: Hermann.

Eshelby, J. D. (1951). The force of an elastic singularity. *Phil. Trans. Roy. Soc. London* A244, 87.

Eshelby, J. D. (1970). Energy relations and the energy-momentum tensor in continuum mechanics, in *inelastic behaviour of solids*. 77–113. New York,: McGraw-Hill,.

Eshelby, J. D. (1975). The elastic energy-momentum tensor. *Journal of Elasticity* 5, Nos. 3-4:321–335.

Fisher, A., and Marsden, J. (1972). The einstein equations of evolution - a geometrical approach. *Journal of Mathematical Physics,* 13, No.4,:546–568.

Francfort, G., and Golebiewska-Herrmann, A. (1982). Conservation laws and material momentum in thermoelasticity. *Transactions of ASME* 49:710–714.

Herrmann, A. G. (1981). On physical and material conservation laws. *Proc.IUTAM Symp. on Finite Elasticity* 201–209.

Kachanov, L. *Introduction to Continuum Damage Mechanics*. Boston: Martinus Nijhoff Publications.

Kachanov, L. (1958). On the time to failure under creep conditions. *Izv. AN. SSSR* No.8.

Kijowski, J., and Magli, G. (1992). Relativistic elastomechanics as a lagrangian field theory. *Journal of Geometry and Physics* 9 no.3:207–233.

Kobayashi, S., and Nomizu, K. (1969). *Foundations of Differential Geometry*. New York: Interscinece.

Landau, L., and Lifshitz, E. (1970). *Elasticity Theory, 2nd ed.* Elmsford, N.Y.: Pergamon Press.

Landau, L., and Lifshitz, E. (1971). *The Classical Theory of Fields*. Reading, MA.: Addison-Wesley.

Liu, J. e. a. (1999). True stress-strain-temperature diagrams for polypropilenes. In *Proceedings SPE/ANTEC'99,III,N.Y, May 2-6*, 3338–3442. N.Y.: Plenum Press.

Marsden, J., and Hughes, T. (1983). *Mathematical Foundations of Elasticity*. Mineola, N.Y.: Dover Publications.

Marsden, J. (1983). Stress and riemannian metrics in nonlinear elasticity. Publications MSRI.

Maugin, G., and Trimarco, C. (1992). Pseudomomentum and material forces in nonlinear elasticity: variational formulations and application to to brittle fracture. *Acta Mechanica 94* 1–28.

Maugin, G., Epstein, M., and Trimarco, C. (1992). Pseudomomentum and material forces in inhomogeneous materials. *International Journal of Solids and Structures* 29, No.14/15:1899–1900.

Maugin, G. (1993). *Material Inhomogeneities in Elasticity*. London: Chapman Hall.

Misner, C., Thorne, K., and Wheeler, J. (1973). *Gravitation*. New York: Freeman.

Pak, Y., and Herrmann, A. G. (1986). Conservation laws and the material momentum tensor for the elastic dielectric. *International Journal of Engineering Sciences* 24, N 8:1365–1374.

Rajaraman, R. (1982). *Solitons and Instantons*. North Holland.

Rosen, N. (1980). Bimetric general relativity theory. In ed. by P.Bergmann, V. D. S., ed., *Cosmology and Graviation*, 383–403. N.Y.: Plenum Press.

Simo, J., and Marsden, J. (1984). On the rotated stress tensor and the material version of the doyle-ericksen formula. *Arch. Rational Mech. Analysis 86* 213–231.

Simo, J., and Ortiz, S. (1985). A unified approach to finite deformation elastoplastic analysis based on the use of hyperelastic constitutive equations. *Comp. Methods in Applied Math. and Engineering* 49:221–245.

Wang, C.-C., and Truesdell, C. (1973). *Introduction to Rational Elasticity*. Leyden: Noordhoff Int. Publishing.

Zhou, W. e. (1994). Cold-drawing (necking) behaviour of polycarbonate as a double glass transition. *Polymer Engineering Science* 35:304–309.

Zhou, W. e. a. (1999). The time dependency of the necking process in polyethilene. In *Proceedings SPE/ANTEC'99,III,N.Y, May 2-6*, 3399–3402. N.Y.: Plenum Press.